THE EDITOR

DAVID E. DAVIS is Head of the Department of Zoology at North Carolina State University, a position he has held since 1967. Prior to 1967, Dr. Davis was Professor of Zoology at Pennsylvania State University.

Professor Davis received his B.A. degree from Swarthmore College and his Ph.D. in Biology from Harvard University. He is a Past President of the Wildlife Disease Association and the American Institute of Biological Sciences. Dr. Davis has published over 230 papers and is the author or co-author of three books.

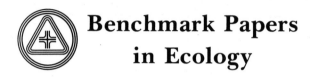

Benchmark Papers in Ecology

Series Editor: Frank B. Golley
University of Georgia

Published Volumes and Volumes in Preparation

CYCLES OF ESSENTIAL ELEMENTS
  Lawrence R. Pomeroy
BEHAVIOR AS AN ECOLOGICAL FACTOR
  David E. Davis
NICHE
  R. H. Whittaker and S. A. Levin
ECOLOGICAL ENERGETICS
  Richard G. Wiegert
SYSTEMS ECOLOGY
  Bernard C. Patten
DIVERSITY
  Ruth Patrick
PATTERNS OF PRIMARY PRODUCTION IN THE BIOSPHERE
  Helmuth F. H. Leith

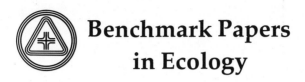

Benchmark Papers
in Ecology

A *BENCHMARK* ® Books Series

BEHAVIOR AS AN
ECOLOGICAL FACTOR

Edited by
DAVID E. DAVIS
North Carolina State University at Raleigh

Dowden, Hutchinson
& Ross, Inc.
Stroudsburg, Pennsylvania

Copyright © 1974 by **Dowden, Hutchinson & Ross, Inc.**
Benchmark Papers in Ecology, Volume 2
Library of Congress Catalog Card Number: 74–3006
ISBN: 0–87933–132–1

Manufactured in the United States of America.

Exclusive distributor outside the United States and
Canada: John Wiley & Sons, Inc.

74 75 76   5 4 3 2 1

Library of Congress Cataloging in Publication Data

Davis, David Edward, comp.
    Behavior as an ecological factor.

    (Benchmark papers in ecology, v. 2)
    1. Animals, Habits and behavior of—Addresses,
essays, lectures. 2. Zoology—Ecology—Addresses,
essays, lectures. I. Title. [DNLM: 1. Behavior,
Animal—Collected works. 2. Environment—Collected
works. W1 BE515E v. 2 1974 / [QL751 D261b 1974]]
QL751.6.D38   591.5   74–3006
ISBN 0–87933–132–1

Acknowledgments
and Permissions

ACKNOWLEDGMENTS
AMERICAN MUSEUM OF NATURAL HISTORY—*Bulletin of the American Museum of Natural History*
  The Courtship of Gould's Manakin (*Manacus vitellinus vitellinus*) on Barro Colorado Island, Canal Zone

AMERICAN ORNITHOLOGISTS' UNION—*Auk*
  Removal and Repopulation of Breeding Birds in a Spruce–Fir Forest Community
  Social Nesting Habits of the Smooth-Billed Ani
  The Social Order in Flocks of the Common Chicken and the Pigeon

AMERICAN SOCIETY OF MAMMALOGISTS—*Journal of Mammalogy*
  The Adreno–Pituitary System and Population Cycles in Mammals

LINNAEAN SOCIETY OF NEW YORK—*Transactions of the Linnaean Society of New York*
  Studies in the Life History of the Song Sparrow I

NATIONAL ACADEMY OF SCIENCES—*Proceedings of the National Academy of Sciences USA*
  Reproductive Cycles in an Equatorial Sparrow

NEW YORK ZOOLOGICAL SOCIETY—*Zoologica*
  Distribution and Status of the Mountain Gorrilla (*Gorilla gorilla beringei*)—1959

THE POULTRY SCIENCE ASSOCIATION—*Poultry Science*
  Number of Offspring Sired by Cockerels Related to Social Dominance in Chickens

SOCIETY FOR THE STUDY OF EVOLUTION—*Evolution*
  Natural Selection and Family Size in the Starling

PERMISSIONS
The following papers have been reprinted with the permission of the authors and copyright holders.

ACADEMIA SCIENTIARUM FENNICA (SUOMALAINEN TIEDEAKATEMIA)—*Annales Academiae Scientiarum Fennicae, Series A, IV: Biologica*
  The Life of the Swift, *Micropus apus* (L), in Relation to the Weather

AMERICAN MIDLAND NATURALIST—*American Midland Naturalist*
  A Study of Prairie Deer-Mouse Populations in Southern Michigan

BLACKWELL SCIENTIFIC PUBLICATIONS LTD. FOR THE BRITISH ECOLOGICAL SOCIETY/ *Journal of Applied Ecology*
  The Relationships Between Wood-Pigeons and Their Clover Food Supply and the Mechanism of Population Control

BRITISH ORNITHOLOGISTS' UNION—*Ibis*
  The Behaviour of the Robin: Population Changes over Four Years
  Territory in the Pied Flycatcher *Muscicapa hypoleuca*

CAMBRIDGE UNIVERSITY PRESS—*Bird Flocks and the Breeding Cycle*

COMMONWEALTH SCIENTIFIC AND INDUSTRIAL RESEARCH ORGANIZATION, AUSTRALIA
—*C.S.I.R.O. Wildlife Research*
A Study of the Biology of the Wild Rabbit, *Oryctolagus cuniculus* (L.), in Confined Populations: I. The Effects of Density on Home Range and the Formation of Breeding Groups

DOUBLEDAY & COMPANY, INC.—*Two Little Savages*

DUKE UNIVERSITY PRESS FOR THE ECOLOGICAL SOCIETY OF AMERICA—*Ecological Monographs*
Some Contributions of a Fifteen-Year Local Study of the Northern Bobwhite to a Knowledge of Population Phenomena
The Vicuña and the Puna

THE STONY BROOK FOUNDATION FOR THE QUARTERLY REVIEW OF BIOLOGY—*The Quarterly Review of Biology*
The Phylogeny of Social Nesting Habits in the Crotophaginae

Series Editor's Preface

Ecology—the study of interactions and relationships between living systems and the environment—is an extremely active and dynamic field of science. The great variety of possible interactions in even the most simple ecological system makes the study of ecology compelling but difficult to discuss in simple terms. Further, living systems include individual organisms, populations, communities, and ultimately the entire biosphere; there are thus numerous subspecialties in ecology. Some ecologists are interested in wildlife and natural history, others are intrigued by the complexity and apparently intractable problems of ecological systems, and still others apply ecological principles to the problems of man and the environment. This means that a Benchmark Series in Ecology could be subdivided into innumerable subvolumes that represented these diverse interests. However, rather than take this approach, I have tried to focus on general patterns or concepts that are applicable to two particularly important levels of ecological understanding: the population and the community. I have taken the dichotomy between these two as my major organizing concept in the series

In a field that is rapidly changing and evolving, it is often difficult to chart the transition of single ideas into cohesive theories and principles. In addition, it is not easy to make judgments as to the benchmarks of the subject when the theoretical features of a field are relatively young. These twin problems—the relationship between interweaving ideas and the elucidation of theory, and the youth of the subject itself—make the development of a Benchmark series in the field of ecology difficult. Each of the volume editors has recognized this inherent problem and each has acted to solve it in his or her unique way. Their collective efforts will, we anticipate, provide a survey of the most important concepts in the field. Thus we expect that the Benchmark Series in Ecology will be useful not only to the student who needs an authoritative selection of original literature but also the professional who wants to quickly and efficiently expand his or her background in an area of ecology outside his special competence.

David E. Davis has organized a series of benchmark papers in behavior to illustrate how behavioral interactions of the animal may control its response and interactions with its environment. "Environment" is usually defined as those factors outside the biological system under consideration which impinge upon it and alter its behavior. As Davis will show in his selections and comments, behavior becomes the link between the organism and environment; this a crucial aspect in understanding the ecology of the species. David E. Davis is especially able to guide us in these areas between two separate disciplines because he has had long experience in both behavioral and ecological studies of rodents, including the urban rat, and of birds and has written several texts. He is presently Professor and Head of the Department of Zoology, North Carolina State University at Raleigh.

Frank B. Golley

Preface

The theme of this Benchmark volume is the evolution of our understanding of the behavioral relation of an animal to its surroundings. Entwined in this development has been the understanding of how populations are regulated. The early view (about 1910) was that physical factors and disease killed some individuals and prevented others from breeding. In this way the population decreased. Population increase occurred when physical factors were favorable and disease was curbed. Behavior was a means for simple maintenance in such aspects as feeding, sleeping, hiding, drinking, and procreating. The readings will show that behavior as an ecological factor is much more complex.

While in pursuit of the primary objective—understanding relations to the surroundings—various investigators explored the physiological mechanisms that underlie behavior and in the process added new insights to physiology, which also had been considered to be primarily a maintenance operation. This set of readings emphasizes the involvement of physiology but does not include any truly physiological papers.

The basic principle considered in the selection of these papers was the extent to which facts were consolidated into principles. (The procedures used are described in the Introduction.) Many papers, especially the early ones, provided facts in abundance but failed to stimulate research or to consolidate the existing information. Perhaps the nature of growth of science requires a long period of collection of facts before the principles can be discerned. Then these principles, although often incomplete, serve to stimulate the collection of facts for the specific purpose of testing the principles. Now we have a set of principles about behavior as an ecological factor that can be used both to enlarge our understanding of ecology and also to apply to the welfare of the world's ecosystem.

It is impossible to list everyone who made a useful suggestion, but two merit special mention: Carl Koford had the happy thought of including Seton; and Jack Christian appraised the influence of various population papers.

David E. Davis

Contents

I. BEHAVIORAL ADJUSTMENTS TO HABITAT

II. REPRODUCTIVE BEHAVIOR

III. SOCIAL BEHAVIOR

IV. BEHAVIOR OF POPULATIONS

V. APPLIED BEHAVIOR

Contents by Author

Introduction

This Benchmark volume, *Behavior as an Ecological Factor,* emphasizes contributions that bring behavior and ecology together. *Behavior* can be defined as the action taken by an organism in response to a stimulus in the external or internal environment of the organism; the action is executed by physiological means (neurological, hormonal, etc.). Behavior is thus, in its broadest sense, one means of adjusting the organism to its environment. *Ecology* is the study of the relationship between an organism and its environment. Therefore, in this Benchmark volume we have considered behavior as a factor of ecology.

As late as 1910, behavior, physiology, and ecology were considered discrete matters for study, except in the minds of a few observers. By 1930, however, behavior and ecology were considered to be interrelated, and by 1950, physiological aspects of the two fields were regularly included in their study. The junction, and sometimes the union, of these three facets of biology has provided an increased level of understanding that permits us new insights into such fundamental problems of biology as evolution and inheritance.

The procedure for selection of these readings was highly personal. The publishers requested "influential" papers, which obviously means that I would choose papers that had influenced me and my colleagues. For six months I read or perused about 500 articles and books, in addition to asking many persons (too numerous to list) for suggestions. Naturally my own papers seemed influential, and thus are included, perhaps more than is justified by their influence on others.

Several problems in selection arose. Many books, especially those published before 1920, were influential. Although it is obvious that these could not be reprinted in their entirety, in only a few cases could even part be chosen and still make sense. Hence, influential books are listed at the end of each section. Another constraint on selection is the theme "Behavior as an ecological factor." Several very influential papers in behavior (e.g., Lorenz, 1935) were omitted because little ecology was included in them. Another problem was the initial date. One could quote Aristotle, Darwin, and many others, since perceptive observers have always been aware of behavior, but direct influence really began about 1910. Similarly, the terminal date needed consideration. Since many years are usually necessary to evaluate influence, few papers after 1960 are included.

Most of the papers concern birds, a situation that reflects several circumstances. My own interest in ornithology started about 1920 and thus biased the selection. More importantly, because birds are diurnal and spectacular, they have attracted more attention than did mammals, reptiles, and fish. The relation of behavior to ecology has been dominated, therefore, by examples from birds. Some early research on fish (Breder, 1936) was influential but soon was submerged by the flood of papers on birds. Illustrative of this trend is the fact that when the herpetologist, G. K. Noble, about 1938, became interested in behavior he shifted from reptiles to birds (gulls and herons).

Some persons will wonder why no selections are included about insects. In part the answer lies in my lack of training in entomology, but far more important is the fact that the extreme stereotypy of behavior of insects leaves little opportunity for behavioral interpretation. It is true that recent research is demonstrating more flexibility and opportunity for learning than had been thought possible; nevertheless the study of insect behavior generally consists of a description of intricate and elaborate actions, followed by comments on their significance for survival.

Thus, the criteria for choice of Benchmark papers were largely subjective. Unfortunately no quantitative measure of "influence" has been developed, although perhaps a form of "citation index" should be tried. Suggestions were culled from current and former graduate students and from colleagues. Those papers that, when read recently, seemed influential were included. Some papers, although "popular," were omitted.

I have also excluded certain topics which in my judgment demonstrated little relation between ecology and behavior. For example, predation is exceedingly important in both ecology and behavior, but the two are very simply linked at the behavioral level and the real significance of predation lies in its effect, if any, on population regulation. Similarly, migration is steeped in behavior and ecology, but migratory behavior is mainly understood as a physiological problem and the ecology is at best a survival history. Finally, for obvious reasons, sensory and other aspects of physiology have been omitted simply to reduce the scope of the book. Some people will consider these reasons for exclusion frivolous or unwise, and they may be correct!

References

Breder, C. M., Jr. 1936. The reproductive habits of the North American sunfish (Centrarchidae). *Zoologica 21:* 1–48.

Lorenz, K. 1935. Der Kumpan in der Umwelt des Vogels. *J. Ornith. 83:* 137–213, 289–413.

Noble, G. K., M. Wurm, and G. Schmidt. 1938. Social behavior of the black-crowned night heron. *Auk 55:* 7–40.

Editor's Comments on Paper 1

1 Seton: *Selection from* Two Little Savages

The influence of a scientific paper obviously depends upon the early nurture of an interest in behavior and ecology. Indeed, a pervasive conclusion in recent years from research on behavior and ecology is that the behavior of an adult depends very greatly on the environment during early life. A generation of young people was influenced by the writings of Ernest Thompson Seton, especially *Two Little Savages*, published in 1921 and written for 10- to 12-year-olds. Nearly every biologist can remember one or more episodes of the type described in this selection.

Seton (born in 1860) grew up in southern Ontario, where he presumably endured or enjoyed many of the episodes described in *Two Little Savages*. His first notable publication was the set of volumes *Life Histories of Northern Animals*, which appeared in several editions after 1909. In the introduction to this work, Seton lists the topics that should be included for each species. In addition to descriptive and taxonomic topics, he lists home range, food, property, relation to light, sociability, senses, sanitation, training the young, and many others. These topics were highlighted in all of Seton's many books on birds and mammals, which helped to train a generation of ecologists.

1

Spotted
Pipsissewa

V

The Collarless Stranger

Oh, sympathy! the noblest gift of God to man. The greatest bond there is twixt man and man.
The strongest link in any friendship chain.
The single lasting hold in kinship's claim.
The only incorrosive strand in marriage bonds.
The blazing torch where genius lights her lamp.
The ten times noble base of noblest love.
More deep than love—more strong than hate—the biggest thing in all the universe—the law of laws.
Grant but this greatest gift of God to man—this single link concatenating grant, and all the rest are worthless or comprised.

EACH year the ancient springtime madness came more strongly on Yan. Each year he was less inclined to resist it, and one glorious day of late April in its twelfth return he had wandered northward along to a little wood a couple of miles from the town. It was full of unnamed flowers and voices and mysteries. Every tree and thicket had a voice— a long ditch full of water had many that called to him. "*Peep-peep-peep*," they seemed to say in invitation for him to come and see. He crawled again and again to the ditch and watched and waited. The loud whistle would sound only a few rods away, "*Peep-peep-peep*," but ceased at each spot when he

38

4

The Collarless Stranger

came near—sometimes before him, sometimes behind, but never where he was. He searched through a small pool with his hands, sifted out sticks and leaves, but found nothing else. A farmer going by told him it was only a "spring Peeper," whatever that was, "some kind of a critter in the water."

Under a log not far away Yan found a little Lizard that tumbled out of sight into a hole. It was the only living thing there, so he decided that the "Peeper" must be a "Whistling Lizard." But he was determined to see them when they were calling. How was it that the ponds all around should be full of them calling to him and playing hide and seek and yet defying his most careful search? The voices ceased as soon as he came near, to be gradually renewed in the pools he had left. His presence was a husher. He lay for a long time watching a pool, but none of the voices began again in range of his eye. At length, after realizing that they were avoiding him, he crawled to a very noisy pond without showing himself, and nearer and yet nearer until he was within three feet of a loud peeper in the floating grass. He located the spot within a few inches and yet could see nothing. He was utterly baffled, and lay there puzzling over it, when suddenly all the near Peepers stopped, and Yan was startled by a footfall; and looking around, he saw a man within a few feet, watching him.

Yan reddened—a stranger was always an enemy; he had a natural aversion to all such, and stared awkwardly as though caught in crime.

39

The man, a curious looking middle-aged person, was in shabby clothes and wore no collar. He had a tin box strapped on his bent shoulders, and in his hands was a long-handled net. His features, smothered in a grizzly beard, were very prominent and rugged. They gave evidence of intellectual force, with some severity, but his gray-blue eyes had a kindly look.

He had on a common, unbecoming, hard felt hat, and when he raised it to admit the pleasant breeze Yan saw that the wearer had hair like his own— a coarse, paleolithic mane, piled on his rugged brow, like a mass of seaweed lodged on some storm-beaten rock.

"F'what are ye fynding, my lad?" said he in tones whose gentleness was in no way obscured by a strong Scottish tang.

Still resenting somewhat the stranger's presence, Yan said:

"I'm not finding anything; I am only trying to see what that Whistling Lizard is like."

The stranger's eyes twinkled. "Forty years ago Ah was laying by a pool just as Ah seen ye this morning, looking and trying hard to read the riddle of the spring Peeper. Ah lay there all day, aye, and mony anither day, yes, it was nigh onto three years before Ah found it oot. Ah'll be glad to save ye seeking as long as Ah did, if that's yer mind. Ah'll show ye the Peeper"

Then he raked carefully among the leaves near

10

6

the ditch, and soon captured a tiny Frog, less than an inch long.

"Ther's your Whistling Lizard: he no a Lizard at all, but a Froggie. Book men call him *Hyla pickeringii*, an' a gude Scotchman he'd make, for ye see the St. Andrew's cross on his wee back. Ye see the whistling ones in the water put on'y their beaks oot an' is hard to see. Then they sinks to the bottom when ye come near. But you tak this'n home and treat him well and ye'll see him blow out his throat as big as himsel' an' whistle like a steam engine."

Yan thawed out now. He told about the Lizard he had seen.

"That wasna a Lizard: Ah niver see thim aboot here. It must a been a two-striped *Spelerpes*. A *Spelerpes* is nigh kin to a Frog—a kind of dry-land tadpole, while a Lizard is only a Snake with legs."

This was light from heaven. All Yan's distrust was gone. He warmed to the stranger. He plied him with questions; he told of his getting the Bird Book. Oh, how the stranger did snort at "that driveling trash." Yan talked of his perplexities. He got a full hearing and intelligent answers. His mystery of the black ground-bird with a brown mate was resolved into the Common Towhee. The unknown wonderful voice in the spring morning, sending out its "*cluck, cluck, cluck, clucker,*" in the distant woods; the large gray Woodpecker that bored in some high stub and flew in a blaze of gold, **and the wonderful** spotted bird with red head and yellow

41

wings and tail in the taxidermist's window, were all resolved into one and the same—the Flicker or Golden-winged Woodpecker. The Hang-nest and the Oriole became one. The unknown poisonous-looking blue Hornet, that sat on the mud with palpitating body, and the strange, invisible thing that made the mud-nests inside old outbuildings and crammed them with crippled Spiders, were both identified as the Mud-wasp or *Pelopæus*.

A black Butterfly flew over, and Yan learned that it was a Camberwell Beauty, or, scientifically, a *Vanessa antiopa*, and that this one must have hibernated to be seen so early in the spring, and yet more, that this beautiful creature was the glorified spirit of the common brown and black spiney Caterpillar.

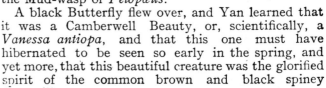

Vanessa antiopa

The Wild Pigeons were flying high above them in great flocks as they sat there, and Yan learned of their great nesting places in the far South, and of their wonderful but exact migrations without regard to anything but food; their northward migration to gather the winged nuts of the Slippery Elm in Canada; their August flight to the rice-fields of Carolina; their Mississippi Valley pilgrimage when the acorns and beech-mast were falling ripe

What a rich, full morning that was. Everything seemed to turn up for them. As they walked over a piney hill, two large birds sprang from the ground and whirred through the trees.

"Ruffed Grouse or 'patridge," as the farmers call

42

8

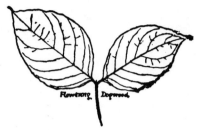
Flowering Dogwood

The Collarless Stranger

them. There's a pair lives nigh aboots here. They come on this bank for the Wintergreen berries."

And Yan was quick to pull and taste them. He filled his pockets with the aromatic plant—berries and all—and chewed it as he went. While they walked, a faint, far drum-thump fell on their ears. "What's that?" he exclaimed, ever on the alert. The stranger listened and said:

"That's the bird ye ha' just seen; that's the Cock Partridge drumming for his mate."

The Pewee of his early memories became the Phœbe of books. That day his brookside singer became the Song-sparrow; the brown triller, the Veery Thrush. The Trilliums, white and red, the Dogtooth Violet, the Spring-beauty, the Trailing Arbutus—all for the first time got names and became real friends, instead of elusive and beautiful, but depressing mysteries.

The stranger warmed, too, and his rugged features glowed; he saw in Yan one minded like himself, tormented with the knowledge-hunger, as in youth he himself had been; and now it was a priceless privilege to save the boy some of what he had suffered. His gratitude to Yan grew fervid, and Yan—he took in every word; nothing that he heard was forgotten. He was in a dream, for he had found at last the greatest thing on earth—sympathy—broad, intelligent, comprehensive sympathy.

That spring morning was ever after like a new epoch in Yan's mind—not his memory, that was a

43

Trailing Arbutus

Spring Beauty

Two Little Savages

thing of the past—but in his mind, his living present.

And the strongest, realest thing in it all was, not the rugged stranger with his kind ways, not the new birds and plants, but the smell of the Wintergreen.

Smell's appeal to the memory is far better, stronger, more real than that of any other sense. The Indians know this; many of them, in time, find out the smell that conjures up their happiest hours, and keep it by them in the medicine bag. It is very real and dear to them—that handful of Pine needles, that lump of Rat-musk, or that piece of Spruce gum. It adds the crown of happy memory to their reveries.

And yet this belief is one of the first attacked by silly White-men, who profess to enlighten the Red-man's darkness. They, in their ignorance, denounce it as absurd, while men of science know its simple truth.

Yan did not know that he had stumbled on a secret of the Indian medicine bag. But ever afterward that wonderful day was called back to him, conjured up by his "medicine," this simple, natural magic, the smell of the Wintergreen.

He appreciated that morning more than he could tell, and yet he did a characteristic foolish thing, that put him in a wrong light and left him so in the stranger's mind.

It was past noon. They had long lingered; the stranger spoke of the many things he had at home; then at length said he must be going. "Weel,

44

10

good-by, laddie; Ah hope Ah'll see you again." He held out his hand. Yan shook it warmly; but he was dazed with thinking and with reaction; his diffidence and timidity were strong; he never rose to the stranger's veiled offer. He let him go without even learning his name or address.

When it was too late, Yan awoke to his blunder. He haunted all those woods in hopes of chancing on him there again, but he never did.

45

I

Behavioral Adjustments to Habitat

Editor's Comments on Papers 2, 3, and 4

2 **Koskimies:** *The Life of the Swift,* Micropus apus *(L.), in Relation to the Weather*

3 **Haartman:** *Territory in the Pied Flycatcher* Muscicapa hypoleuca

4 **Myers and Poole:** *A Study of the Biology of the Wild Rabbit,* Oryctolagus cuniculus *(L.), in Confined Populations: I. The Effects of Density on Home Range and the Formation of Breeding Groups*

The complex of environmental factors surrounding an animal is generally called the *habitat.* Originally the term included only physical factors (temperature, pH, etc.) but soon biotic factors (plants, predators, etc.) were included. Behavioral adjustments to environment or habitat originally focused almost entirely on such simple maintenance activities as feeding, sleeping, and avoidance of predators. Early papers (1910–1920) provided terribly dull reading because they were almost entirely descriptive and are not reproduced here. A very early paper (Adams, 1908), not reproduced here, described birds in habitats but said little about behavior. Early papers mainly illustrate that a separation between behavior and ecology existed in the minds of the investigators. Gradually research showed that behavior had much more complex relations to the habitat, especially through the social organization of the population. The series of papers reproduced here illustrate this research.

As an example, Koskimies (1950) describes the relation of behavior of the swift to weather conditions and its food supply. In this case the behavior is a response to ecological factors and does not influence them in any way. In Koskimies' words (letter of August 1973), "The significance of my 1950 paper on the life of the swift, as viewed today, might be that it succeeded, to a certain extent at least, in elucidating the complex interplay of the many behavioral and bioenergetic adaptations of this highly specialized species."

This paper sets a high standard in the interpretation of ecological data. Koskimies cautiously considers the various aspects of weather that might influence energetic relations and, for example, notes that a correlation with decreasing atmospheric pressure does not mean that pressure *per se* is a causal factor but that other aspects may be responsible for the decline in numbers of insects and hence energy inputs to the Swifts. He also emphasizes the principle that a factor such as temperature has an optimum effect on behavioral or ecological situations. The concept that there is an optimum as well as a minimum and maximum response is important in ecology as well as other sciences. He pursues the explanation of a behavior pattern such as resting to a ecologic or physiologic basis. He had either the insight or the good luck to study a bird that is partially heterothermic. Because the bird's temperatures varied greatly, Koskimies could see energetic relations clearly at the extremes, thus illustrating a principle of scientific research that an extreme or aberrant case may reveal a relationship that is often inconspicuous in an ordinary species.

Behavior begins to receive recognition by the study of territory as demonstrated in Haartman's (1956) paper. The behavior determines the distribution and numbers in relation to a factor (nest holes) of the habitat. Haartman used an experimental approach (adding next boxes) to demonstrate that the number of flycatchers in the woods

14

was limited by the number of nest holes. His data show very clearly that addition of boxes (1941–1945) resulted in an increase that ceased (1946–1950) but again resumed when more boxes were added (1951–1953). Although nest holes would seem to be a purely ecologic factor, Haartman showed that the behavior at the nest hole really determined the number of birds. Thus population was determined by interaction of behavior and an ecological factor.

Most of Haartman's paper is a very detailed description of behavior that is necessary to substantiate the conclusions about defense of nest boxes. An aspect that has not been adequately explored is the similarity of behavior of pied flycatchers and some other birds, such as starlings, that nest in holes. For example, males of both species lead the female to the hole. This convergence in behavior illustrates the principle that an ecologic situation (nesting in holes) can force the behavior of taxonomically unrelated species into similar patterns.

The complexity of the interactions between behavior and ecological factors is illustrated by the studies of rabbits in pens in Australia (Myers and Poole, 1959). Behavior determines the occurrence of a disease (myxomatosis) and also starvation. The observational methods present considerable advance over earlier papers, both in collection and analysis of data, and the maps are especially useful. A complete reorientation of the view of habitat and behavior is now exhibited. Rabbits arrange themselves in a rather rigid "peck order," so the dominant rabbits have first choice of food and burrows. Myers and Poole show how the habitat is parceled out among the rabbits and how starvation may result for certain rabbits due to their low social rank. This orientation is a complete reversal from the early papers, such as Adams (1908) and even Koskimies (1950), in which the habitat was said to control behavior. It must be recognized, of course, that species differ greatly and that the behavior of some truly is governed by habitat conditions.

The ultimate in this developmental sequence is the use of behavioral knowledge for management or control (see Part V).

Reference

Adams, Charles C. 1908. Ecological succession of birds. *Auk 25* (2): 109–153.

Influential Books

Henderson, A. M., 1913. *The Fitness of the Environment.* Macmillan Publishing Co. Inc., New York, 317 pp. Emphasized the view that environment has particular attributes that fit the animal and thus encouraged the view that environment and behavior fit together.
Hickey, J. J. 1943. *A Guide to Bird Watching.* Oxford University Press, New York, 264 pp. Significant presentation of methods for obtaining data about behavior and ecologic factors.

2

Copyright © 1950 by Academia Scientiarum Fennica (Suomalainen Tiedeakatemia)

Reprinted from *Ann. Acad. Sci. Fenn., Ser. A, IV: Biologica*, **15**, 33–40 (Mar. 10, 1950)

The Life of the Swift, *Micropus apus* (L.), in Relation to the Weather

JUKKA KOSKIMIES

G. Influence of the weather upon the quantity of aeroplankton and the food supply of the swift.

As the different investigations concerning the aeroplankton have shown and as is commonly known on the basis of every-day field observations, the weather produces very remarkable fluctuations in the amount and composition of the aeroplankton. GLICK (1939), however, emphasizes especially that it is very difficult — for the present even impossible — to find the factor or factors which are at work in different cases. On the basis of series of material of sufficiently long duration it is always, it is true, possible to get in general outline valid rules on the influence of different weather factors upon the insect fauna in the air. Very often,

Editor's Note: Throughout this volume, a row of asterisks indicates that material has been omitted from the original article.

however — especially when the examination is restricted to single weather factors, for instance temperature, humidity, winds, etc. — one comes across exceptions where the »general rules» do not hold good. As has been emphasized in different connections (cf. KOSKIMIES 1947a), it is not possible to separate the different weather factors when their biological effect is to be examined and regard any one biological phenomenon as being caused by a separate weather factor. The weather as a biological environmental factor must be understood as a whole, in which numerous physical factors (= separate weather factors) in their joint action (= the weather) only, produce the biological effect that can be observed.

Naturally this does not mean that the analysis of separate weather factors is unnecessary when the biological effects of the weather are to be examined. But the separate factors jointly shaping the so-called weather and its different manifestations, only c h a r a c t e r i z e the general weather conditions to which, however, besides, there belong a great number of other factors. If, for instance, the data from an aeroplankton analysis prove that with decreasing atmospheric pressure the number of insects in the air increases, this does not indicate that it is the decreasing pressure which causes the increase in the number of insects in the air. It indicates only that in such weather which is characterized (among other factors) by decreasing atmospheric pressure the number of insects in the air increases.

It is, furthermore, to be noted that from the ascertained correlation between the amount of aeroplankton and a certain weather factor it is still a long way to the causal explanation of the connections existing between these factors. The amount of aeroplankton, for instance, is often not governed by the weather conditions existing at the moment of observation, but by the weather conditions some time beforehand (when the start of the insects from the ground has occurred) and very often also by the development of weather conditions over a long period before the moment of observation (cf. PALMÉN 1944). It is natural that for instance the rain has an entirely different effect on the amount of aeroplankton after a longish dry period than after a continued rain period. Therefore it is likewise understandable that according to GLICK (op.c., p. 93) sometimes, when the weather on the basis of previous experiences seems ideal for the abundant appearance of insects, only few individuals are caught. Correspondingly, in very »bad» weather insects can be caught at times in quite remarkable numbers.

Owing to the extraordinary complexity of the factors involved, wholly reliable and indisputable results can be obtained only with extreme difficulty. The following account attempts to give at least some intimations about the factors that are at work.

Temperature.

It is evident that temperature is one of the most important factors controlling the amount of aeroplankton. Wherever and whenever the temperature is favourable, insects usually appear in abundance.

The temperature on the ground surface (the »ground surface» must here be understood widely, in contrast to the upper layers of the air) naturally produces an effect in the first place upon the insect fauna in the layers nearest the ground surface. As the aeroplankton in the upper layers too arises from the ground surface and is thus dependent on the conditions existing there, in the following the ground surface temperature and its influence on the aeroplankton will chiefly be examined. The fluctuations of the aeroplankton will be followed chiefly in the lowest air layer (up to 100 m.), because the influence of the surface conditions is there more noticeable than in the upper layers and because the conditions in this zone in particular are from the point of view of this investigation most interesting.

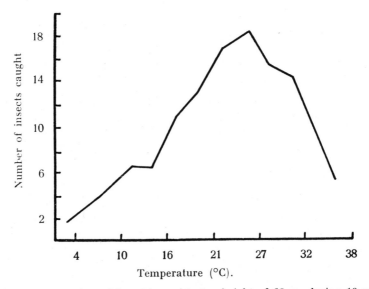

Fig. 3. Average number of insects caught at a height of 60 m. during 10 mins. at different temperatures. After GLICK 1939.

Fig. 3 shows that at a height of 60 m. the aeroplankton is most abundant when the surface temperature is about $+ 25°$ C. This temperature forms a rather narrow optimal zone. Outside it the number of insects rapidly diminishes. Thus even a lowering or rise of $10°$ (to $+ 15°$ or $+ 35°$ respectively) causes a decrease of insects to half the amount observed in optimal conditions.

The dependence of the activity of insects upon the temperature is one of the main problems of research on insect ecology (cf. CHAPMAN et al. 1926; BODENHEIMER & SCHENKIN 1928). Investigations have led to the discovery that the normal activity of different insect species is confined to temperature limits characteristic of each species. In practice the lower limit usually has a greater importance. The lower temperature limit varies remarkably in different species (cf. e.g. PALMÉN 1944, p. 113) being, however, in most cases over + 10 ° C (UVAROV 1931; WELLINGTON 1945). The experimental investigations of PALMÉN (op.c.) concerning, for instance, the beetles, which make the most important component of the swift's food, have shown that only very few species fly at temperatures under + 12 ° C. The lower temperature limit of most species is + 16 — 17 ° C.

These values prove sufficiently that the activity necessary for flying is confined to rather high temperatures. PALMÉN (likewise a number of other authors mentioned by him on p. 116) produces evidence that a condition necessary for the mass appearance of insects is a sufficiently high temperature, in most cases a rise in temperature. Even though a moderately abundant appearance of insects in the air does not require such conditions as lead to the mass appearances described by such authors as PALMÉN, it is evident that the temperature must markedly exceed the average lower temperature limit of insects. If it falls below the limit demanded by insects in the aeroplankton, it follows that a transfer of new insects to the aeroplankton is interrupted. Then there will only be insects in the air as long as those already there are kept in the air by vertical air currents or if horizontal currents carry more from elsewhere. The determination of the temperature limit leading to such a condition is, of course, impossible in the absence of more thorough investigations. But considering the above-mentioned experimental investigations (UVAROV, PALMÉN, WELLINGTON, op.c.) it might be regarded as probable that *even a decrease of the air temperature to + 15 ° C makes so many insects unfit for active flying that their number in the air becomes quite small already.* On the basis of aeroplankton analyses also a temperature of about + 15° may be regarded as the lower limit of the normal appearance of insects in the air.

The greatest abundance of different insect orders and spiders at a height of 60 m. was, according to GLICK, attained at the following temperatures:

Araneida	10 — 13 °
Homoptera	21 — 24 °
Heteroptera	
Coleoptera	24 — 27°
Diptera	
Hymenoptera	30 — 33 °

The »optimum» of the spiders at rather low temperatures can probably be explained as due to their winglessness, so that they are not, of course, to the same extent dependent on the influence of the temperature as the other, in most cases actively flying, forms. In the autumn there also appear numerous young spiders, which are driven with ease into the air and thus cause a seeming »optimum» in the colder autumn weather. The proportion of spiders in the whole aeroplankton is, however, so small (about 6 %) that their possibly greater abundance in cold weather cannot replace the shortage of other groups.

At a height of 300 — 600 m. insects are caught most numerously when the surface temperature has already exceeded the optimum and the number of insects there is on the decrease. This is probably caused, on hot days especially, by the intensive convection (see p. 42 ff.), which tends to transfer aeroplankton from lower air layers upwards. At the same time, owing to the surface temperature exceeding the general optimum, the lower layers do not receive further additions from the ground surface.

On the basis of these facts it is to be regarded as evident that *if the temperature during the breeding season of the swift falls below + 15 ° C, the food deficiency is already quite serious.* The cold causes a greater energy consumption and further increases the food requirements and food deficit. Thus low temperature is often the main factor causing even very serious catastrophes during the breeding season of various birds, especially swifts and swallows (comp. p. 70 ff.).

It has been likewise ascertained that the appearance of insects has a rather distinct upper temperature limit too. Owing to this on sultry days the food supply worsens, particularly in the early afternoon when the convection is most vigorous. The transfer of aeroplankton to the upper air layers leads also to a lengthening of the distances in the hunting flights. The same effect may likewise be produced by a localization of insect appearances, which is also caused by localized convective currents. The effect of high temperature is thus difficult to distinguish from the effect of convection. — At all events, on hot days in the early afternoon there apparently exists a period during which the food supply of the swift is worse. During this time the swifts — it is true, also many other birds — have a distinct rest period.

The daily activity and feeding frequency are highly dependent on the temperature. The effect of temperature is very clearly to be seen at the commencement of activity in the morning. Fig. 4 drawn on the basis of the observations of WEITNAUER (1947) indicates that there exists a very distinct correlation between the morning temperature and the time of the first flight from the nest. The lower the temperature in the morning, the later the swift leaves its nest. At first sight this seems very surprising. One would rather expect the fall in temperature to cause a lengthening of the time of the daily activity, so that a sufficient food supply should be assured.

20

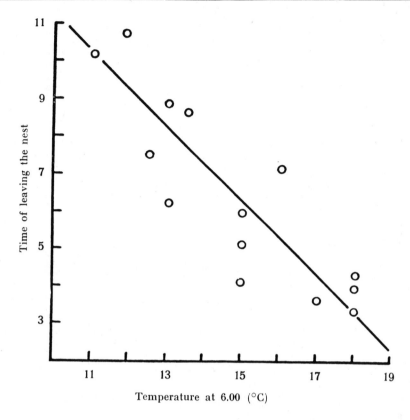

Fig. 4. Beginning of activity of the swift in relation to the morning temperature
according to WEITNAUER's (1947) data.

The above fact may be interpreted as that the swifts do not set out on
their catching flights until the temperature of the open air has become
sufficiently high for the normal flight of insects. It would, therefore, be
very interesting to know the temperature during the first outward flight.
For this purpose the author, assuming that the morning temperature
changes regularly in the direction defined by the temperature values of 6
and 12 a.m. of the respective day, given by WEITNAUER, by extrapolating
has approximately calculated the temperature during each first outward
flight. Counted in this way the temperature at this time varies (with one
exception) from + 13 ° to + 17 ° C, being on the average about + 15 ° C.
It is interesting to note that this value corresponds very well to the average
lower temperature limit necessary for the normal flight activity of insects.
The more thorough explanation of the obvious correlation between the
temperature and the time of onset of activity is a very interesting problem
for future research.

The cessation of activity in the evening takes place at a very defined time (comp. also the observations on p. 13) and quite independently of the temperature (Fig. 5).

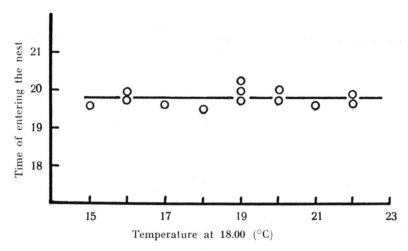

Fig. 5.   Cessation of activity of the swift in relation to the evening temperature according to WEITNAUER's (1947) data.

As the quantity of food in cold weather is smaller than normally, it leads during the nestling period of the young to a lowered feeding frequency. Further, as has been stated, the period of the daily activity on such days is remarkably shortened. Owing to this the total number of feeding visits during cold days is greatly reduced. This fact is also very clearly proved by Fig. 6, which has been drawn on the basis of the observations reported by WEITNAUER (op.c.). Assuming that the feeding frequency decreases continually in proportion to the lowering of the temperature it can be seen that *when the mean temperature of the day is lowered to about $+ 12 \,^{\circ} C$, the feeding frequency is in practice zero.* The optimal feeding frequency (about 30 visits per day) is reached only on days when the mean temperature is close to $+ 20\,^{\circ}$.

CHISLETT (1947 a) and concerning *M. melba* BARTELS (1931) and HÄRRI (1931) report similar observations. According to them the nestlings are fed very seldom during rainy and cold weather so that the young remain in practice without food.

VLEUGEL (1947, p. 224) does not agree with DAANJE's (1944) opinion that the length of the daily activity of the swift is strongly affected by the weather. He bases his view upon the fact that »bij de, door mij nauwkeurig onderzochte soorten (Vink, Keep en Witte kwik) is dit niet het geval» and says that »De grote factor is de lichtsterkte». In fact, the swift differs to a very great extent from most other birds in its relation to external factors.

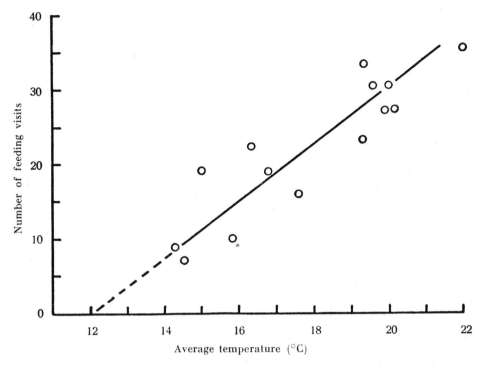

Fig. 6. Feeding frequency of the swift in relation to the average temperature according to WEITNAUER's (1947) data.

It is thus quite obvious how very greatly the length of its daily activity is affected by the prevailing weather, even if the light intensity is by far the more important factor governing he length of daily activity in most other birds.   This is true especially where the light feeding time (»Fresshelligkeit», GROEBBELS 1932) is concerned. It must be recalled that the light feeding time is not in all species identical with the light period of the day. The light feeding time of the swift, for instance, includes only that part of the light period when there are food insects in the air. During a day-long bad weather period the light feeding time and thus the activity of the swift are quite restricted.

* * * * * * *

Literature

BARTELS, M., 1931. Beobachtungen an Brutplätzen des Alpenseglers, Micropus m. melba (L.)—Journ. Ornith., 79, p. 1–28.
BODENHEIMER, F. S. & D. SCHENKIN, 1928, Über die Temperaturabhängigkeit von Insekten.—Zeitschr. vergl. Physiol., 8, p. 1–15.

Literature

CHAPMAN, R. N. et. al., 1926, Studies in the ecology of sand dune insects.—Ecology, 7, p. 416–426.
CHISLETT, R., 1947 a, Swifts at Masham, 1946.—Naturalist, 820, p. 9–10.
DAANJE, A., 1944, De vliegende Gierzwaluw, Apus a. apus (L.).—Ardea, 33, p. 74–84.
GLICK, P. A., 1939, The distribution of insects, spiders and mites in the air.—Techn. Bull. U.S. Dept. Agric., 673, p. 1–150.
GROEBBELS, Fr., 1932–37, Der Vogel, 1–2.—Berlin.
HÄRRI, H., 1931, Die Alpensegler-Kolonie Seengen.—Ornith. Beob., 28, p. 187–190.
KOSKIMIES, J., 1947 a, Tekijöistä, jotka säätelevät lintujen muuttoa.—Luonnon Tutkija, 51, p. 84–86.
PALMÉN, E., 1944, Die anemohydrochore Ausbreitung der Insekten als zoogeographischer Faktor.—Ann. Zool. Soc. Zool.—Bot. Fenn. Vanamo, 10; 1, p. 1–262.
UVAROV, B. P., 1931, Insects and climate.—Trans. Entom. Soc. London, 79, p. 1–247.
VLEUGEL, D. A., 1947, De duur van de vliegdag van de Gierzwaluw, Apus a. apus (L.).—Ardea, 35, p. 222–224.
WEITNAUER, E., 1947, Am Neste des Mauerseglers, Apus a. apus (L.).—Ornith. Beob., 44, Beiheft, p. 133–182.
WELLINGTON, W. G., 1945, Conditions governing the distribution of insects in the free atmosphere.—Canad. Entom., 77, p. 7–15, 21–28, 44–49, 69–74.

3

Reprinted from *Ibis*, **98**, 460–475 (1956)

TERRITORY IN THE PIED FLYCATCHER
MUSCICAPA HYPOLEUCA.

By Lars von Haartman.

Received on 9 December 1955.

The Pied Flycatcher *Muscicapa hypoleuca* was studied in an area in S.W. Finland during the years 1941–1955. The following notes on territorial behaviour represent fragments of this study. To understand the significance of territory nearly all the methods developed by modern field ornithology are needed. I have used these methods in an attempt to arrive at a synthesis.

THE CHOICE OF TERRITORY.

In the Chaffinch, which finds a suitable nesting place in every tree, the male first occupies a territory, the nesting site being chosen afterwards by the female. In the Pied Flycatcher the choice of territory occurs in quite a different way. The male returns before the female and seeks a nesting-hole. If he finds one he inspects it repeatedly, at first hesitantly. Should it prove satisfactory, the male begins to sing and to display other territorial behaviour. In choosing territory, the terrain in which the hole is situated is of relatively slight importance. When nest-boxes were put up in forest types where the species never nests under natural conditions, e.g. a certain barren type of pine forest, with high thin trunks, Pied Flycatchers settled there in large numbers (v. Haartman 1954, cf. also Trettau 1952). Obviously the choice of territory in birds often depends on innate releasing mechanisms with a few key stimuli (v. Haartman 1945 b, cf. Fabricius & Gustafson (1954) on a fish species). A suitable hole seems to be the most important key-stimulus for the male Pied Flycatcher, and perhaps also for many other hole-nesting species. This is easy to understand : nothing guarantees that a certain forest area, which offers plenty of food, also offers a nesting-hole, whereas the reverse is more likely to hold true.

An interesting analogy is found in some territorial fish-species. The male Three-spined Stickleback *Gasterosteus aculeatus* takes up a territory, and afterwards begins to build a nest, starting it here and there simultaneously (Tinbergen 1951). The Stickleback's demands on the nesting-site are so unspecialized that any territory will satisfy him. *Gobius microps* behaves quite differently, nesting under empty shells of the mussel *Mya arenaria*. If the male does not find a *Mya* shell (or a good substitute) territorial behaviour wholly fails to appear (Nyman 1953).

TERRITORIES, POPULATION DENSITY AND EMIGRATION.

As the number of nesting-holes is nearly everywhere smaller than the demand, it acts as an ecological minimum factor, limiting the population density in the Pied Flycatcher, and perhaps other hole-nesting species too. This is clearly seen in areas where the population has multiplied after the putting-up of nest-boxes. Fig. 1 shows the course of events in my study-area. Part of this figure (until 1950—v. Haartman 1951 b) was reproduced by D. Lack. Neglecting the number of available nest-boxes (which was shown in my original graph) he concluded that the number of pairs has reached its maximum in 1945–1950. This could have been a consequence of

FIGURE 1.—Number of nest-boxes where clutches were laid by Pied Flycatchers at Lemsjöholm, 1941–1955. Arrows indicate years in which the number of nest-boxes was considerably increased.

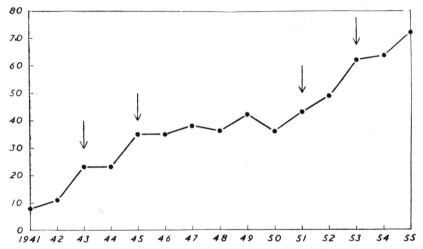

territorial behaviour. However, the premises were not correct, for every time that more nest-boxes were put up, the number of nesting Pied Fly-catchers rose and this occurred even after 1950. Few ornithologists are probably wealthy enough to supply more nest-boxes in an area of 4 sq. km. than the Pied Flycatchers can use. Even restricting ourselves to a much smaller area we have to use an unexpectedly large number of nest-boxes to obtain this effect. Experiments of this sort have been started at Lems-jöholm, but are not yet completed.

Of my ringed Flycatchers nearly all the surviving males, some of the females and a very few first-year birds returned to my study-area (v. Haartman 1949). We may ask whether the non-returning individuals fail to return owing to competition for the nest-boxes, or if they wander off " spontaneously ". Kalela especially (e.g. 1944) has greatly over-emphasized the role of population pressure in the dispersal of birds.

The greater the number of nest-boxes, the weaker, of course, becomes the competition for them.   Table 1 shows the recoveries of ringed Pied Fly-catchers during the years 1941–49 and 1950–54.   In 1951 the number of nest-boxes was greatly increased, which should influence the recoveries of birds ringed in the previous year.

TABLE 1.   *Recoveries of Pied Flycatchers at Lemsjöholm during years with less and more numerous nest-boxes.   The recovery percentages for the birds ringed in 1950–54 are probably somewhat too low, as further recoveries may possibly be obtained during the next few years.*

Years		Ringed	Recovered
1941–49	Young	958	$8\male = 0.84\% \pm 0.42$ $4\female = 0.42\%$
1950–54	,,	929	$15\male = 1.61\% \pm 0.41$ $5\female = 0.54\%$
1941–49	old $\female\female$	204	$27 = 13.3\%$
1950–54	,,	191	$27 = 14.1\%$
1941–49	old $\male\male$	206	$79 = 38.3\%$
1953–54	,,	37	$13 = 35.1\%$

The *Ortstreue* of old males and probably also of old females has not increased, whereas young males returned twice as often as before. The difference is not statistically significant, but is in reality higher than is indicated by Table 1, as only some of the males have been caught in certain of the nineteen-fifties.   The rising number of young males recovered can be explained in two ways.   Either the number of unpaired males (which are difficult to catch) has fallen or old males sometimes chase young ones away from their territories of the previous year (seen in at least one case).   The chased young males perhaps leave the study-area if there are not enough free nest-boxes.   There is, however, no reason to overstress the influence of density upon *Ortstreue* ; most Pied Flycatchers obviously leave the ringing area " spontaneously ".   This seems especially likely in respect of long-range dispersals, the most remarkable of which occurred recently : a female ringed as a nestling in Finland in 1954, nested in Sweden in 1955, 570 km. away.

AGGRESSIVE DISPLAY.

An intruding male Pied Flycatcher elicits the following main reactions in the territory owner :

(1) Most encounters do not result in real fights, but the intruder recedes immediately when the territory owner shows threat display. This consists of fluffing the white feathers (breast, wing-patches, rump).

The front is turned towards the intruder. From this visual angle the threatening male looks enormously enlarged. One wing and also the tail twitches rapidly and continuously, without being raised as high as in the alarm movement, from which this movement is possibly derived. Jumping around makes the male highly conspicuous. The threat is accompanied by a characteristic sound (v. Haartman & Löhrl 1950). Relatively seldom does the male show an *Imponierflug* against the intruder, as the Collared Flycatcher does (Löhrl 1951).   If the antagonist does not flee, displacement pecking is sometimes seen. Often the males (the inferior individual first) jump to bare places on the ground, this being probably the appetitive behaviour of food-catching. Some fights end on fields, which the Flycatchers never otherwise visit.

(2) Alternating with threat display the male shows a reaction which is otherwise addressed to the female : he flies to the nest entrance and even jumps in, singing the excited song.   Löhrl (1951) interprets this performance as meaning in human language " here I nest ". It would seem more plausible to roar to an intruder " here you do not nest ", which, indeed, is the meaning of threat display. Thus, it seems likely that the nest demonstration, when interrupting fights, is a manifestation of ambivalence, the male reacting momentarily as if the antagonist were a female.   A male was seen reacting with clear ambivalence to a stuffed male, attacking it, showing the nesting hole, and trying to copulate.

FIGURE 2.—Threatening male Pied Flycatcher.

(3) Sometimes two males fight violently and persistently for a nest-box.   In some cases at least such fights arose when the owner momentarily left the territory (as Pied Flycatchers often do), and another male had time to inspect the nest-hole and establish itself as territory owner.   When the first male returned a very hard struggle resulted.   The males peck at each other with their bills, fluttering in the air and slowly sinking.   The *Fluchtdistanz* of these combatants is greatly diminished.   Many authors have asked why the males do not fight even more violently than they do.   The answer is possibly that such belligerent males would succumb

more easily to predators. Besides, the interspecific fighting against competitors (which probably is to some extent correlated with " courage " in intraspecific fighting) for a nest-hole not infrequently ends with the flycatcher's death. The competition with the Great Tit is especially dangerous (cf. also Löhrl 1950 on the Collared Flycatcher versus the Tree Sparrow). But, from the standpoint of selection, a living flycatcher is better than a dead lion.

When one male becomes dominant, he pursues the other in a wild flight, uttering excited song (v. Haartman 1947), and a sound otherwise heard in attacks on squirrels, woodpeckers, etc. Alternating with these attacks he demonstrates the nest-hole. These fights sometimes end unexpectedly with the inferior males taking over the territory, when the dominating one has other territories than this (see p. 467), and withdraws to one of them.

Aggressiveness decreases markedly when the female has started to lay. Males feeding young are comparatively indifferent to intruders. Helpers that are strangers sometimes associate with feeding pairs without being driven off (v. Haartman 1953 ; Löhrl 1949 on the Collared Flycatcher).

Females are hostile towards females, and even sometimes towards foreign males, the intensity of reaction being, however, low. Nest demonstrations and violent pursuing flights were never seen in females. The few attempts of two females to settle with the same male in the same nest-box ended in the later arrival being driven off by the first female.

COURTSHIP.

If a female approaches the nest of an unpaired male, the latter shows very characteristic behaviour. He flies in *Imponierflug* to the nest entrance and jumps in, singing an excited song. This is a long, whispering song, ending with some harsh notes " z z z z ". Very likely the same " z z z " is uttered by some of the females, greeting their mates when meeting at the nest during the feeding of the young. If this is correct, its function is probably blocking of aggressiveness and/or escape.

After a while the male looks out. If the female is in the vicinity he withdraws into the nest-hole, otherwise he leaves it. Mostly the female follows the male to the nest entrance and flutters outside. Only when this performance has been repeated often enough, does she dare to follow him into the inside of the nest-hole. As seen with a " nidoscope " (v. Haartman 1952 a), what is going on lacks dramatic point : the male jumps out without more ado, and so does the female, after sitting still for a while. It is by no means rare for the female to refuse the male's invitation, and leave him in search of another mate. As was pointed out by me in 1949, p. 14, analogous nest-demonstration plays a part in the courtship of many bird- and fish-species in which the male chooses the nesting place, commonly a hole, or builds the nest (cf. also Nyman 1953, Wallraff 1953, Armstrong 1955, Creutz 1955).

The female sometimes starts to build very soon after settling in. The nest-demonstration of the male then gradually ceases. Singing suddenly decreases when the female arrives (Table 2). The unmated male Pied Flyctacher is a very indefatigable singer. During a full-day observation (v. Haartman et al. unpublished), a male sang 3620 times, but this was certainly not a top performance. After the arrival of a female the number of songs per day will hardly exceed about 1000. Besides, the songs are the shorter, the longer the males have been mated. This evidence indicates that the main function of song is to attract females, though other functions are not excluded.

TABLE 2. *Number of songs in 10-minute periods uttered by a male Pied Flycatcher before (ordinary type) and after (italics) the arrival of a female. The first female (on 18 May) disappeared after one afternoon, the second mated with the male.*

Time of day	May 18	May 19	May 20	May 21	Time of day	May 18	May 19	May 20	May 21
3–4 a.m.	70	?	77	?	12–13 p.m.	?	?	66	*10*
4–5	89	?	101	?	13–14	86	56	?	*37*
5–6	?	?	?	?	14–15	81	?	44	*0*
6–7	?	90	?	?	15–16	84	?	?	*38*
7–8	?	?	?	?	16–17	72	2	78	?
8–9	96	94	?	*7*	17–18	45	?	?	*0*
9–10	?	?	78	*1*	18–19	0	*10*	?	*37*
10–11	49	4	98	*15*	19–20	38	?	76	?
11–12	79	?	?	*0*	20–21	0	*20*	20	*40*

Copulations take place in the nesting tree or nearby during a few days before the start of egg-laying. In connection with sexual excitement or as real aggressiveness, the male is seen to chase the female in a wild " sexual flight ", very much resembling the pursuing flight performed by two males.

FORM AND SIZE OF TERRITORY.

As population density of the Pied Flycatcher is mostly restricted by the number of available nesting-holes, no other restricting mechanism, e.g. large territories, is needed. Moreover, the display of the male, which includes the demonstration of the nest-hole, restricts his mobility. The appetitive behaviour, consisting of moving around, which characterizes the territorial behaviour of a great many species, is nearly absent in the Pied Flycatcher. Its territories are very small. A Pied Flycatcher's home is his castle.

Most of the singing is done in the immediate vicinity of the nest (cf. Table 3). If the nesting tree has a bare trunk, the male often sings on the roof of the nest-box.

TABLE 3. *Number of song-phrases of a male Pied Flycatcher in* 1941. (*Not whole-day observations.*) *There were two nest-boxes near to each other, and the male sang periodically near each of them.*

	Place song uttered	May 21	May 25	May 29	June 2	June 10
Nest-box 1	In the nesting tree	922	271	118	628	0
,,	In the 2 nearest trees	1960	395	32	85	0
	More distant	1150	65	382	214	1
Nest-box 2	In the nesting tree	0	9	410	0	339
,,	In the 4 nearest trees	0	0	0	0	75

Territorial combats, too, are most frequent near to the nest. Table 4 shows the distance between the encounters and nestboxes of all the males studied, during a certain period of observation. The distance is measured from the territory-owner's, not from the intruder's, nest. If the encounter occurred far away from the nest of both combatants, the distance was measured from both nests.

TABLE 4. *Distance of territorial encounters from nest.*

Metres	0	10	20	30	40	50	60	70
Number of encounters	15	15	6	5	4	4	2	0

Metres	80	90	100	110	120	130	140
Number of encounters	0	3	1	3	0	0	1

Schematically the territory of the Pied Flycatcher can be presented in the manner shown in Fig. 3 A. At least in many open-nesters, the territory seems rather to be of the type presented in Fig. 3 B. Thus, in *Eremophila alpestris* the territorial fights undulate to and fro over an invisible territorial boundary

(Pickwell 1931). For a bird of this type, Mrs. Nice's (1941) definition of territory as a defended area fits especially well. The Pied Flycatcher rather defends a point, the nest-hole. This enables the species locally to breed in astonishingly high densities (e.g. 16 pairs per hectare=7 per acre in Frankfurt a.M. according to Pfeifer & Ruppert 1953 ; for other examples see Creutz 1955 : 249), and to increase enormously after the putting up of nest-boxes in an area (cf. Fig. 1).

FIGURE 3.—Schematic presentation of territorial fighting in the Pied Flycatcher (left) and an ordinary open-nesting species (right).

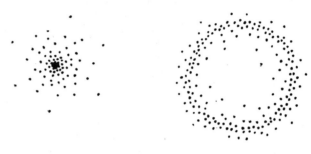

Fig. 3 A.   Fig. 3 B.

The influence of the nest-hole upon the shape of the territory can be illustrated by many kinds of observations :

1. In a certain area three nest-boxes were situated close to each other for a number of years. Males settling here tried to sing round about all three nest-boxes (Map 1) or, if one of them was already occupied, near to two of them. During these years the territories clearly had two or three centres. During the years following the removal of one of the nest-boxes the male never paid any attention to this part of the area (Map 2).

2. The shape of territory can easily be changed if a nest-box is put up near to its boundaries. The male then extends his singing to the vicinity of this nest-box too.

3. As soon as a male left a nest-box one of the neighbours took it over and began to sing in its vicinity (Map 2).

4. When a male possessed two nest-boxes simultaneously, he mostly soon lost one of them. An intruder could take it over almost without a fight when the male was singing round about the other. After this usurpation the first owner hardly ever tried to visit his former domain (Map 3–5).

The female (Map 1) does not restrict herself to the immediate vicinity of the nest-box as the male mostly does. She collects nesting material some ten or even hundred metres from the nest. Once I tried to film the female collecting nesting-material and arranged a table with all she needed directly

MAP 1.—Territory of male 40560 in 1943. Note concentration of singing around
    nest-boxes A, B, and C (in 1941 and 1942 the territories in this area showed the
    same character) .●=singing place.   ♀=♀ seen.   Squares=buildings. Dotted
    lines=roads.   Nest-boxes are indicated with lines and letters. For other
    features of the terrain see map in v. Haartman 1945 : 29.
MAP. 2.—Territory of male 55885 in 1946. After removal of nest-box B, there was
    no singing in this part of the area. When the original male D disappeared,
    male A soon started to sing at D also.

MAP 1   MAP 2

MAP 3.—Territory of male 55810 in 1945, before 24 May.
MAP 4.—Territory of the same male after occupying of nest-box A by male 54381
                        on 24 May.
MAP 5.—Territory of male 54381 after 24 May inclusive.

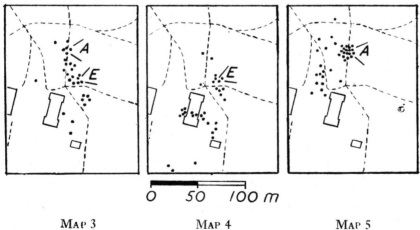

MAP 3   MAP 4   MAP 5

under the nest.   Still she continued to seek for the same kind of material in a forest edge fifty or sixty metres away.   Only when I placed the *non plus ultra* of material, a wad of hairs from the tail of a horse, under the nest, could she be induced to use it.

During cold spells, when nest-building ceased, the female was sometimes seen wandering rather far from the nest, followed by her mostly silent mate.

Both sexes search for food in an area which scarcely coincides with the original territory of the male, the former being much greater (Map 6–7). Feeding areas of adjacent pairs often overlap (Map 8).

POLYTERRITORIALITY.

The male Pied Flycatcher differs from most other Passerine birds in having multiple territories in one nesting season.   I have (v. Haartman 1949) named this phenomenon polyterritoriality (opposite : monoterritoriality). Creutz (1943) and Trettau & Merkel (1943) have already made some observations on polyterritoriality.   In my earlier works I have mostly dealt with polyterritoriality in polygamous males, which may have led some readers to believe that the phenomenon is principally restricted to these.

TABLE 5.   *Polyterritoriality* 1941–1947.

Number of territories during one season	1	2	3	4	5	6
Number of males	79	111	39	6	1	1

Analogies with the male Pied Flycatcher's polyterritoriality are perhaps not altogether lacking among other birds.   Ruiter mentions some observations indicating its occurrence in the Redstart, and Armington describes polyterritoriality combined with polygamy in the Whitethroat.   Polyterritorial behaviour was seen by me in the Greenish and Wood Warblers (v. Haartman 1952 b).

It is by no means always that the male stays for good in his first territory. A territory held before the nesting territory has been named by me a preterritorium (in German : *Vorrevier*).   The territory where the male gets a mate may be named a nesting territory.   Often he leaves the nesting territory for a longer or shorter time and establishes a secondary territory (*Nebenrevier*). If he succeeds in getting a female in this territory too, it could be named a second nesting territory.   This classification of territories perhaps seems complicated, but it is the Pied Flycatcher's fault rather than mine.   Colour-ringing has been absolutely necessary in analysing this behaviour.

Most males have more than one territory.   Table 5 shows the numbers of territories noted during 1941–1947 in ringed males (uncertain cases and males staying only a short time in the study area have been omitted).

34

MAP 6 (left).—Singing places (+and ●) of two males in 1945.
MAP 7 (right).—Excursions of the same males, and their mates during feeding of
    nestlings, showing incongruity between singing and food-seeking areas.

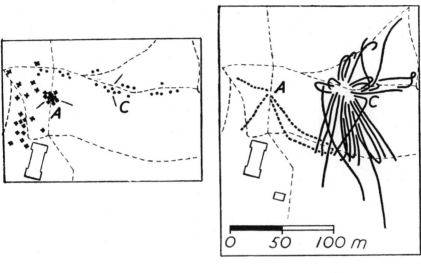

MAP 6   MAP 7

MAP 8.—Overlap of food-seeking areas of two pairs with nestlings.

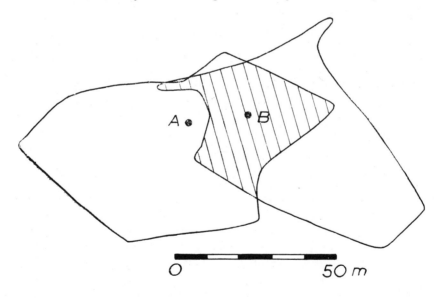

MAP 8

The distance between the territories, measured from nest-box to nest-box (Table 6), is often considerable. As the study area was only 2 km. in diameter and adjacent areas were only superficially investigated, very long distances could not be established. As a matter of fact, the territories were seldom farther than 1 km. from each other. In polygamous males the distance between the two nesting territories was at most a good 400 metres.

TABLE 6. *Distances between territories held by individual male Pied Flycatchers at Lemsjöholm in any one season.* *=cases of monoterritoriality.*

Distance in Kilometres	Number of cases			
	Between most distant territories all types	Between pre-territory and nesting territory	Between nesting and secondary territories	Between first and second nesting territory in polygamous males
Nil*	79	135	105	—
0.1	36	11	30	11
0.2	21	10	14	3
0.3	22	9	13	3
0.4	17	8	10	6
0.5	12	5	3	—
0.6	13	4	3	—
0.7	11	4	4	—
0.8	5	1	3	—
0.9	2	1	1	—
1.0	5	1	2	—
1.1	4	3	—	—
1.2	1	1	—	—
1.3	3	—	2	—
1.4	2	1	—	—
1.5	—	—	—	—
1.6	1	1	—	—
1.7	1	—	—	—
2.2	1	—	—	—
2.9	1	—	—	—

The trapping and ringing of males may sometimes have caused them to leave the territory (cf. Schüz & Löhrl 1952), but in all those cases where emigration took place two or more days after ringing, this cannot be the explanation. If a male does not get a mate he usually emigrates to a new territory (v. Haartman 1949). This does not, however, always happen. Thus, in 1947, the one-year-old male 53031 stayed for 26 days in one and the same territory, without getting a female. In 1948 he returned to the same territory, and again waited several weeks in vain for a mate. The nest-box here was, in fact, so poor that it would hardly have satisfied a female.

During the period 1941–1947 the observed times spent by unpaired males in one territory were as shown in Table 7, the actual times being probably in most cases somewhat longer :

TABLE 7. *Records of territory-occupation by unpaired males.*

Days	1	2–5	6–10	11–15	16–20	21–25	26–30	31–35
Number of males	33	38	20	6	4	4	3	0

The emigration to a new territory probably has significance. If a male does not get a mate, the cause may be unsuitability of the territory or nest-box. The emigration thus gives him a new chance.

At last in the first week of July the unpaired males cease to sing and they disappear. The mated males sometimes stay with their females and broods to the end of the month, although some strongly moulting males leave their broods (v. Haartman 1954 and especially Creutz 1955).

The most enigmatic feature of polyterritoriality is the male's emigration from the nesting territory to a second territory. Most males take up a second territory or at least leave their nesting territory in search of one as the following observations show :—

	1945	1946
Males with probably no second territory	7	12
No second territory established but males left the nesting territory	10	8
Males seen in a second territory	14	14

Emigration to the second territory occurs in general at a very definite moment, in connection with the beginning of egg-laying. The exact time was not always easy to establish, as the males are very silent in their nesting territories during these days. At all events in a few, carefully observed, pairs I have established that the male left his mate before the onset of egg-laying. During 1944–46 emigration took place on the dates shown in Table 8.

TABLE 8. *No. of days before and after laying of the first egg when the male emigrated from his nesting territory. All complete copulations seen up to 1955 are included.*

Date	−6	−5	−4	−3	−2	−1	0	1	2	3	4	5	6	7	8	9
Number of emigrating males	0	0	1	3	6	14	23	5	2	0	1	1	0	0	1	0
Copulations	1	1	2	1	5	3	0	0	0	0	0	0	0	0	0	0

Copulations are most frequent 1–2 days before laying.   Tinbergen (1939) assumed that successive polygamy arises when the female refuses to copulate with her still sexually active mate.   The male Pied Flycatcher's emigration from the nesting territory can hardly be explained in this way, as the abandoned females sometimes copulated with neighbouring males during the absence of their mates.

The cause of taking up a second territory is thus not clear, but its function is obvious.   A male taking up a second territory has a chance to get a second mate and thus obtain more offspring than other males.   He runs no great risks.   Of course, the nestlings die from starvation if the female dies during his absence.   Strange males were seen to copulate with the female when her mate was absent (the paternity of the young thus being somewhat uncertain). I have even once established that an egg-laying female, whose mate was absent, was driven off by a strange male, which occupied the nest-box and showed it to a strange female.

POLYGAMY.

As I cannot add anything important to my earlier papers on polygamy (v. Haartman 1945 a, 1950, 1951 a) the discussion will be very brief.

Excluding promiscuity (as occurring in some Icterids, etc.) there are two main types of polygamy in birds, simultaneous and successive.   The first, when a male has dealings with two females simultaneously, is an exception in the Pied Flycatcher.   Mostly the interval between the egg-laying of the two mates of a polygamous male is at least five days.   This is due to the male's habit of leaving his first mate just before egg-laying.   If he immediately finds a new territory, and gets a second mate, she can at the earliest begin to lay after about 5 days, which is the time the ovocytes require to develop fully.   When the second female starts to lay, the male leaves her and returns to his first mate, aiding her in feeding the brood, whereas the second female mostly has to care for her brood alone.   Only when the second female lays much later than the first, does the male manage to help her too.   As a female alone can to a certain extent compensate for the work done by her mate (v. Haartman 1954), this does not necessarily involve starvation for the brood.

Polygamy is not uncommon in the Pied Flycatcher, some 13% of all females in my study area being mated with polygamous males.   Young males seldom succeed in getting two females.   Some of them do not get mated at all (Trettau & Merkel 1943, v. Haartman 1951 b, 1952 b, Campbell 1955, Creutz 1955) being in some way inferior to the older males in their courtship.

Egg-laying by two females in the same nestbox (cf. especially Creutz 1955) cannot without further investigation be taken as polygamy.   There are often eggs of other birds in a Pied Flycatcher's nest, yet there is no question of accusing the male of a *mésalliance*.

With regard to the number of territories of the polygynous males (v. Haartman 1951a : 265) we may perhaps speak of mono- and polyterritorial polygyny, if a term is really needed.   Creutz (1955) speaks of intra- and extra-territorial polygyny, which is, however, misleading, as the second nesting territory is as much a territory as the first.

Most male Pied Flycatchers are potentially polygamous, i.e. hold a second territory.   That so few of them obtain a second mate depends on the fact that there is no excess of females.   At the time when the males start to wait for their second females, only a few of these are still unmated, i.e. late returners and females which have lost their first brood (a female generally leaves the territory after having lost her brood, a male does not).

Polyandry was established twice.   The female copulated with neighbouring males when her mate left the nesting territory too early.

SUMMARY.

1. The male Pied Flycatcher does not show definite territorial behaviour before finding a suitable nesting-hole.   In species with unspecialized nesting biology, territory is chosen before nest-site.

2. The restricted number of nesting-holes is in most areas the main factor limiting population density in the Pied Flycatcher.   As the number of nest-boxes in an area was multiplied, the percentage of returning birds ringed as young rose, whereas that of old individuals did not rise.   It is concluded that competition for nesting-places forces a certain number of young birds to emigrate.   As the effect is relatively slight, it is concluded that most individuals emigrate for other reasons.

3. Aggressive display and courtship are briefly described.

4. Since overcrowding is mostly effectively prevented by the restricted number of nesting-holes, the Pied Flycatcher does not need large territories.   As the male's courtship includes a repeated nest-demonstration, his absence from the vicinity of the nest-site would be unfavourable.   Hence, the territory is small, the male principally defending the nesting-hole.   The Pied Flycatcher's home is his castle.

5. Most male Pied Flycatchers are polyterritorial.   The distance between territories held by one male in the same season amounted at most to 2–3 km.

6. Polygamy is briefly described.   The pre-requisite of the male's polygamy is in most cases polyterritoriality.

REFERENCES.

ARMINGTON, S. 1951.   Polygami och polyterritorialism hos törnsångaren (*Sylvia communis* Lath.).   Vår Fågelvärld 10 : 26–31.

ARMSTRONG, E. A. 1955.   The Wren.   London.

CAMPBELL, B. 1955.   A population of Pied Flycatchers (*Muscicapa hypoleuca*).   Acta XI Congr. Intern. Ornith. : 428–434.

CREUTZ, G. 1943.   Die Brutbiologie des Trauerfliegenschnäppers (*Muscicapa h. hypoleuca* Pallas).—Ber. Ver. Schles. Orn. 28 : 28–38.

CREUTZ, G. 1953.   Die Brutbiologie des Trauerfliegenschnäppers (*Muscicapa h. hypoleuca* Pallas).   Ber. Ver. Schles. Orn. 28 : 1–10.

CREUTZ, G. 1955.   Der Trauerschnäpper (*Muscicapa hypoleuca* (Pallas)).   Eine Populationsstudie.   J. Orn. 96 : 241–326.

FABRICIUS, E. & GUSTAFSON, K.-J. 1954.   Further aquarium observations on the spawning behaviour of the Char *Salmo alpinus* L. Inst. Freshwater Research, Drottningholm, Rep. 35 : 58–104.

HAARTMAN, L. V. 1945 a.   Some cases of polygamy of the Pied Flycatcher (*Muscicapa h. hypoleuca* Pall.).—Vår Fågelvärld 4 : 27–32.

HAARTMAN, L. v. 1945 b. Zur Biologie der Wasser- und Ufervögel im Schärenmeer Südwestfinnlands. Acta Zool. Fenn. 44.

HAARTMAN, L. v. 1947. Das auslösende Schema, ein Grundbegriff moderner Instinkt- analyse. Ajatus 14 : 5–37.

HAARTMAN, L. v. 1949. Der Trauerfliegenschnäpper. I. Ortstreue und Rassen- bildung. Acta Zool. Fenn. 56.

HAARTMAN, L. v. 1950. Kirjosiepon moniavioisuus. Luonnon Tutkija 1950 :73–77.

HAARTMAN, L. v. 1951 a. Successive Polygamy. Behaviour 3 : 256–274.

HAARTMAN, L. v. 1951 b. Der Trauerfliegenschnäpper. II. Populationsprobleme. Acta Zool. Fenn. 67.

HAARTMAN, L. v. 1952 a. Svartvitia flugsnapparens (Muscicapa hypoleuca) reaktion på sina ägg. Svensk Faunistisk Revy 14 : 84–92.

HAARTMAN, L. v. 1952 b. Über ungepaarte Männchen in Grenzpopulationen der Kleinvögel. Acta Soc. Fauna et Flora Fenn. 69 (1).

HAARTMAN L. v. 1953. Was reizt den Trauerfliegenschnäpper (Muscicapa hypoleuca) zu füttern ? Vogelwarte 16 : 157–164.

HAARTMAN, L. v. 1954. Der Trauerfliegenschnäpper. III. Die Nahrungsbiologie. Acta Zool. Fenn. 83.

HAARTMAN, L. v. & LÖHRL, H. 1950. Die Lautässerungen des Trauer- und Halsband-fliegenschnäppers, Muscicapa h. hypoleuca (Pall.) und M. a. albicollis Temminck. Ornis Fenn. 17 : 85–97.

KALELA, O. 1944. Zur Frage der Ausbreitungstendenz der Tiere. Ann. Zool. Soc. Zool. Bot. Fenn. Vanamo 10 (3).

LACK, D. 1954. The natural Regulation of Animal Numbers. Oxford.

LÖHRL, H. 1949. Polygynie, Sprengung der Ehegemeinschaft und Adoption beim Halsbandfliegenschnäpper (Muscicapa a. albicollis). Vogelwarte 15 : 94–100.

LÖHRL, H. 1950. Zur " Verdrängung " von Meisen durch Fliegenschnäpper. Vogelwelt 71 : 39–41.

LÖHRL, H. 1951. Balz und Paarbildung beim Halsbandfliegenschnäpper. J. Orn. 93 : 41–60.

NICE, M. M., 1941. The role of territory in bird life. Amer. Midl. Natural. 26 : 441–497.

NYMAN, K. J. 1953. Observations on the behaviour of Gobius microps. Acta Soc. Fauna et Flora Fenn. 69 (5).

PFEIFER, S. & RUPPRET, K. 1953. Versuche zur Steigerung der Siedlungsdichte höhlen- und buschbrütender Vogelarten. Biol. Abhandl. 6.

PICKWELL, G. B. 1931. The Prairie Horned Lark. Trans. Acad. Sci. St. Louis 27.

RUITER, C. J. S. 1941. Waarnemingen omtrent de levenswijze van de gekraagde roodstart, Phoenicurus ph. phoenicurus (L.). Ardea 30 : 175–214.

SCHÜZ, E. & LÖHRL, H. 1954. Mehr Strenge gegenüber dem Stoff–gerade in der Ornithologie. Vogelwarte 17 : 1–6.

TINBERGEN, N. 1939. The behavior of the Snow Bunting in spring. Trans. Linn. Soc. New York 5.

TINBERGEN, N. 1951. The Study of Instinct. Oxford.

TRETTAU, W. 1952. Planberingung des Trauerfliegenschnäppers (Muscicapa hypo- leuca) in Hessen. Vogelwarte 16 : 89–95.

TRETTAU, W. & MERKEL, F. 1943. Ergebnisse einer Planberingung des Trauer- fliegenschnäppers (Muscicapa hypoleuca Pallas) in Schlesien. Vogelzug 14 : 77–90.

WALLRAFF, H. G. 1953. Beobachtungen zur Brutbiologie des Stares (Sturnus v. vulgaris L.) in Nürnberg. J. Orn. 94 : 36–67.

Reprinted from *C.S.I.R.O. Wildlife Res.*, 4(1), 14–26 (1959)

A STUDY OF THE BIOLOGY OF THE WILD RABBIT, *ORYCTOLAGUS CUNICULUS* (L.), IN CONFINED POPULATIONS

I. THE EFFECTS OF DENSITY ON HOME RANGE AND THE FORMATION OF BREEDING GROUPS

By K. Myers* and W. E. Poole*

[*Manuscript received March 8, 1959*]

Summary

Adult rabbits in confined populations inhabit a well-defined home range within which they rest, feed, and breed. The average area of home range becomes smaller as rabbit numbers increase, and the home ranges of adult females are smaller than those of adult males.

Rabbits form small groups during the breeding season, usually consisting of two or three males and several females. A strict dominance-hierarchy is established among the males. The more dominant males roam over larger areas than their subordinates and eject foreign males, thus exhibiting territorial behaviour.

I. INTRODUCTION

Very few studies of the movements and social behaviour of rabbits have been made. Most of such work concerns the eastern cottontail, *Sylvilagus floridanus mearnsi* Allen (Dalke and Sime 1938; Allen 1939; Schwartz 1941; Haugen 1942). These workers recorded a wide variation in the extent of home ranges for this species—1·4 to 100 acres for males and 1·2 to 22·5 acres for females—but in each case the home ranges of the males were larger than those of the females. Haugen (1942) also reported a decrease in the extent of home ranges of females during the non-breeding season. Findings of a similar nature have been recorded for *Sylvilagus bachmani* (Waterhouse) (Connell 1954) and *Sylvilagus audubonii* Baird (Ingles 1941). Since all this work was based on the marking and recapturing of individuals in live-traps, no information could be gathered on the social organization of the species concerned.

Little evidence has been obtained on the incidence or extent of home ranges for the European rabbit, *Oryctolagus cuniculus* (L.). Southern (1940) considered that the warren population of 150 rabbits that he was studying in England maintained themselves on an area of 2 acres. Later he described (Southern 1948) "spheres of influence" of adult males and "territorial conservativeness" of adult females, thus demonstrating an areal distribution of individuals within the population itself. He also described for the first time the rudimentary principles of the social organization of a natural rabbit population.

*Wildlife Survey Section, C.S.I.R.O., Field Station, Albury, N.S.W.

The term "range" and "home range" are given synonymous meaning by many workers. In this paper, "home range" is used exclusively to denote the area traversed by each individual in its normal daily activities of food gathering, mating, resting, and caring for young (Burt 1943). The term "territory" is used as defined by Burt (1943) to denote an area defended by its occupant or occupants against competing members of the same species.

The difficulties associated with the observation of natural populations of rabbits have led recently to their study within large enclosures (Carrick 1956; Rowley 1956; Myers 1958; Mykytowycz 1958). Mykytowycz added considerably to Southern's findings with his experimental demonstration of a strict dominance-hierarchy in the social grouping of rabbits.

The data presented here also amplify Southern's observations, and form part of a general study on three confined populations of European rabbits at Thurgoona, New South Wales (Myers 1958).

II. METHODS

Three fenced enclosures (A, B, and C), each $7 \cdot 5$ by $2 \cdot 6$ chains (2 acres) in area, were constructed side by side, with water supply and five groups of shaded logs as described earlier (Myers 1958 and Plate 1, Fig. 2). Two hides, 8 ft square and 10 ft above the ground, were built on the dividing fences to give maximum visibility over the whole area. For night observations each enclosure was illuminated by 22 Philips "Philora" 160-W globes, 15 ft high, in standard reflectors, powered by a 240-V, 5-kVA generator situated near by. The experimental area is shown in ground plan in Figure 1 and as a panorama in Plate 1, Figure 1.

Recognition of individual rabbits was facilitated by the use of chicken wing-tabs inserted in the ears (Phillips, Stephens, and Worden 1952; Thompson and Armour 1954) bearing various patterns in combinations of five different colours (red, blue, green, white, yellow) of "Scotchlite" reflective tape (Richter 1955; Dunnet 1956; Ealey and Dunnet 1956). The patterns used and found most effective are illustrated in Plate 1, Figure 3. Combinations of white and yellow or blue and green were not used as these pairs of colours were difficult to distinguish.

During the day the patterns and colours were readily identified using either Zeiss 15×50 binoculars or a Broadhurst and Clarkson telescope with 3 in. lens and variable magnification. At night individual rabbits were recognized by attaching a 6-V sealed-beam spotlight to the telescope and aligning it to illuminate the rabbit being observed.

The fur on the abdomen and inner thighs of each female was dyed with one or two of the following cold-water dyes (supplied by I.C.I.A.N.Z. Ltd., Dyestuffs Section): methyl violet 2BN200, malachite green AS, rhodamine B500, auromine OS, magenta PN150, victoria blue BN150, chrysoidine YN160. Since the females lined their nests with fur plucked from these parts it was possible to relate each litter to its parent (Myers 1958).

The enclosures were searched systematically each day, with as little disturbance as possible and without handling the rabbits, and the location at which each rabbit was squatting was recorded. Rabbits were handled only once per month during censuses.

The enclosures were stocked on July 16, 1957, with adult rabbits caught in neighbouring paddocks, at the following rates:

Enclosure A, two males (M1, M2) and four females (F1–F4);

Enclosure B, four males (M3–M6) and six females (F5–F10);

Enclosure C, nine males (M7–M15) and ten females (F11–F20).

Observations on their behaviour commenced on October 7, when breeding was in full swing, and continued until December 13. Watches were held on 9 days during this period, mainly between 3 p.m. and 11 p.m. and included one 24-hour period. The average number of hours spent in observation was 50 per enclosure.

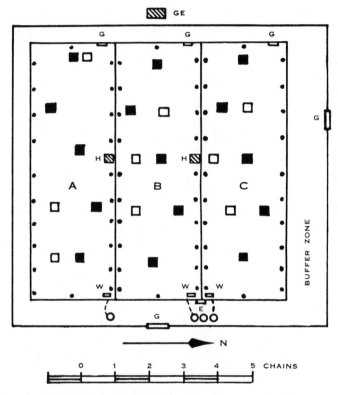

Fig. 1.—Ground plan of experimental enclosures. The dots along the fences indicate the positions of the lights. ■, Log shelters; □, pasture enclosures; ○, water tanks; *W*, drinking trays; *H*, hides; *GE*, generator; *G*, gates; *E*, evaporation tray.

III. RESULTS

The distribution of home ranges, the lie-up positions, and the pattern of the fall of litters for the three enclosures are presented in Figures 2, 3, and 4. For the purposes of plotting each home range, sightings that resulted from experimental interference were ignored. These amounted to less than 5 per cent. of the total average sightings of 60 per animal.

(a) Home Ranges

The average size of home range for the males decreased from 2 acres in A (two males) to 0·87 acres in B (four males) and 0·54 acres in C (nine males). Similarly the average size of home range for the females in the same enclosures decreased from 2 acres in A (four females) to 0·62 acres in B (six females) and 0·39 acres in C (ten females) (see Tables 1 and 2).

In B and C the male home ranges were 40 per cent. larger than those of the females. In A, all animals ranged the 2 acres, irrespective of sex.

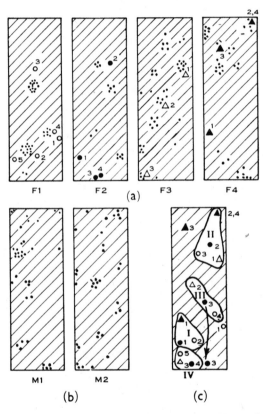

Fig. 2.—(a) Home ranges (cross-hatched), lie-up positions (dots), and burrows (numbers) where litters were born for four females in enclosure A. Numbers also denote sequence of littering for each female.

(b) Home ranges (cross-hatched), and lie-up positions (dots) for two males in enclosure A.

(c) Fall of litters in groups in enclosure A. Litters in area I were born August 28–30; in area II (except No. 4 of F4), September 27–30; in area III, October 27–31; and in area IV, November 29–December 2. Symbols denote individual females as in (a), and numbers denote the sequence of littering for each female.

In enclosure B the female home ranges fell into two distinct groups—F5, F6, F9, and F7, F8, F10 (Fig. 3). The ranges of the females in each group overlapped almost completely, but each group range was discrete. Similarly there were two groups of males, M3, M5 and M4, M6 (Fig. 3).

In enclosure C there were three such groups of females, F18, F19, F20—F12, F13, F17—F11, F14, F15, F16, and three groups of males, M7, M15—M11, M13—M8, M12, M14 (Fig. 4). One group of males (M9, M10) moved over an area that over-lapped all the other group areas.

In enclosures B and C, each female group was overlapped completely by a male group. In C, one male group (M7, M15) also partially overlapped a second group of females (F18, F19, F20).

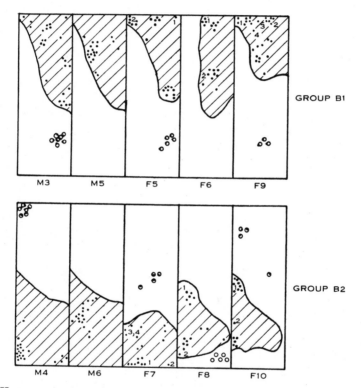

Fig. 3.—Home ranges (cross-hatched), normal lie-up positions (dots), and refuge areas (circles) of six females and four males forming two breeding groups, B1 and B2, in enclosure B. The numbers denote burrows where litters were born and sequence of littering of each female.

(b) Lie-up Positions

The main lie-up, or day-time resting, positions fell within the confines of the home range of each individual. Many rabbits used a secondary, less important lie-up area, often well away from the normal home range (see e.g. M3 and M4, Fig. 3). This was a refuge area for which the animal invariably made when flushed. Since social behaviour was well-nigh absent during the day-time resting period, the possession of a refuge area in foreign territory was not generally resented. In the evening, when social activities recommenced, rabbits using such refuge areas quickly returned to their own home ranges.

The lie-up positions of females were grouped about the burrows in which their young were dropped.

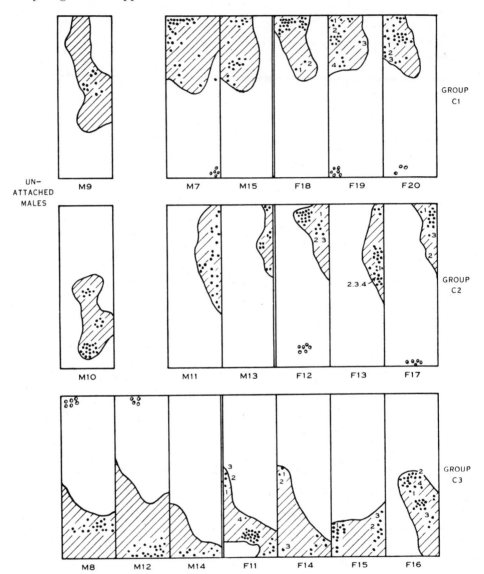

Fig. 4.—Home ranges (cross-hatched), normal lie-up positions (dots), and refuge areas (circles) of ten females and nine males, forming three breeding groups, C1, C2, C3, in enclosure C. Two males were unattached to these groups. The numbers denote burrows where litters were born and sequence of littering of each female.

(c) Litters

All litters were born within the home range of the females concerned. In enclosure C, most females dropped their litters within one small area. Some females used the same burrow for successive litters (F12 and F13, Fig. 4). None of the 35 litters born was deserted, and no nestlings suffered retarded growth rates owing to handling. In enclosure B, the litters were also born within the home range, but breeding burrows tended to be more widely separated (Fig. 3). Of 18 litters born, two were deserted and the kittens in another almost died as a result of handling. In enclosure A, 16 litters were distributed widely throughout the paddock (Fig. 2). Two of these were deserted.

(d) Social Behaviour

The shape and sizes of the home ranges were apparently determined by the social relationships that existed between the individual rabbits in each enclosure.

TABLE 1

AREAS OF HOME RANGE OF TEN RABBITS ON 2 ACRES (ENCLOSURE B)

Group	Males	Area of Home Range (Acres)	Females	Area of Home Range (Acres)
B1	M3	0·84	F5	0·65
	M5	0·81	F6	0·71
			F9	0·63
B2	M4	0·92	F7	0·59
	M6	0·92	F8	0·57
			F10	0·56
Mean	—	0·87	—	0·62

(i) *Enclosure A.*—In this enclosure only a single breeding group existed. Although the females did not confine their activities to a definite range, each drop of litters took place within a small area, demonstrating that they were moving and breeding as a group. This is illustrated in Figure 2(c). The litters in area I were born between August 28 and 30, those in area II (except the fourth litter of F4— see below) between September 27 and 30, those in area III between October 27 and 31, and those in area IV between November 29 and December 2. An exception to the grouping was the third litter of F2, which would have followed the above pattern by being dropped in area III between October 27 and 31, but as a result of experimental interference at the nest (which had been constructed in readiness) was actually dropped, on October 28, near area IV.

F1 was 3 weeks pregnant when introduced into the enclosure and dropped her first litter on July 23, before the other females commenced breeding. She deserted this litter and conceived again a week later with F2 and F4, and dropped with them in area I.

F4 associated with the group at first and, in common with the other females, she had her first three litters sired by M2. When her third litter was due, however, she failed to follow the other females all the way to nesting area III and dropped the litter a short distance from area II (Fig. 2(c)).

After this apparent weakening of group attraction to her she deserted the group completely and returned to her earlier nesting burrow in area II for her fourth litter. Several of her earlier progeny were then living in that region. At the same time, M1 successfully challenged M2 for the right to mate with F4, and it became obvious that a second breeding group was in process of formation; but dry conditions set in immediately afterwards, and reproductive activity ceased.

No fighting was observed among the females.

Throughout most of the period M2 dominated M1 completely. In 37 observed meetings, M2 chased or threatened M1 on 35 occasions and maintained complete

TABLE 2

AREAS OF HOME RANGE OF 19 RABBITS ON 2 ACRES (ENCLOSURE C)

Group	Males	Area of Home Range (Acres)	Females	Area of Home Range (Acres)
C1	M7	0·78	F18	0·37
	M15	0·51	F19	0·45
			F20	0·39
C2	M11	0·50	F12	0·29
	M13	0·17	F13	0·28
			F17	0·19
C3	M14	0·65	F11	0·43
	M12	0·85	F14	0·62
	M8	0·41	F15	0·46
			F16	0·43
Unattached	M10	0·39		
	M9	0·62		
Mean	—	0·54	—	0·39

control over the breeding females. On the last two occasions, however, M1 fought savagely and successfully for F4 when she left the group for her third litter. Her fourth litter was the only one sired by M1 during the whole period.

(ii) *Enclosure B.*—In this enclosure, two breeding groups developed, each consisting of two males and three females—M3, M5, F5, F6, F9 (group B1) and M4, M6, F7, F8, F10 (group B2) (Fig. 3).

One male in each group completely dominated the other (M3 dominated M5, M4 dominated M6). Table 3 lists the outcome of all observed meetings between the four males.

Incursions by males from one group into the territory of the other usually resulted in the intruder being chased out, irrespective of his status in his own group. Thus M3 chased M4 five times and M4 chased M3 three times (see Table 3), the switch in dominance occurring when the chased rabbit was out of his own territory.

No female was observed fighting, save for two or three threats in defence of a nesting burrow.

(iii) *Enclosure C.*—Three breeding groups developed, each consisting of three or four females and two or three males: F18, F19, F20, M7, M15 (group C1); F12, F13, F17, M11, M13 (group C2); and F11, F14, F15, F16, M8, M12, M14 (group C3) (Fig. 4).

The results of all the observed meetings between males in this enclosure are listed in Table 4.

TABLE 3

DOMINANCE RELATIONSHIPS BETWEEN FOUR MALES IN ENCLOSURE B
(Intra-group relationships in *italics*)

Number of Occasions on which Rabbit Chased
Other Rabbits (Reading Down Column)

	Breeding Group and Rabbit Number	B1		B2		Total Submissive Reactions
		M3	M5	M4	M6	
B1	M3	—	*0*	3	2	5
	M5	*37*	—	6	3	46
B2	M4	5	3	—	*0*	8
	M6	2	1	*28*	—	31
Total dominant reactions		44	4	37	5	90

Number of Occasions on which Rabbit was Chased by Other Rabbits (Reading Across Columns)

In groups C1 and C2 one male dominated the other completely. Thus M7 > M15 and M11 > M13 (using the symbol > to signify "was dominant to"). In group C3, where three males were present, a dominance-hierarchy developed, where M14 > M12 > M8.

In each group, as in B, the second-ranking male roamed most or all of the group territory with the dominant buck, playing a large part in its defence, and repeatedly challenging him for the right to mate with females in oestrus. The challenging role of the second-ranking male occurred even when there were only two males in a group.

M7, the dominant male in group C1, also showed dominance over M11, the dominant male in group C2, and maintained control over two of the three females in that group, F12 and F17. The third female, F13, nested outside the range of M7 and mated with M11. More recent observations have shown that such displays represent an intermediate stage in the formation of new groups, when the influence of the dominant male is hanging on, though weakening.

Two males, M10 amd M9, were attached to no breeding groups, but continually attempted to break into one or another, attracted by females in oestrus. M10, a strong buck, dominated the other unattached male, M9. M10 also dominated M13 and M8, members of breeding groups C2 and C3 respectively, but the dominant males in these groups prevented him from displacing the subordinates. There is little doubt that with more females present M10 would have held dominance over a group of his own, a situation which, in fact, came about during the 1958 breeding season.

TABLR 4

DOMINANCE RELATIONSHIPS BETWEEN NINE MALES IN ENCLOSURE C

(Intra-group relationships in *italics*)

Number of Occasions on which Rabbit Chased
Other Rabbits (Reading Down Column)

Number of Occasions on which Rabbit was Chased by Other Rabbits (Reading Across Columns)

Breeding Group and Rabbit Number		C1		C2		C3			*		Total Submissive Reactions
		M7	M15	M11	M13	M14	M12	M8	M10	M9	
C1	M7	—	*0*	0	0	2	1	†	0	0	3
	M15	*14*	—	4	0	1	1	†	0	0	20
C2	M11	11	7	—	*0*	†	2	†	0	0	20
	M13	9	10	*7*	—	†	†	†	2	5	33
C3	M14	1	1	†	†	—	*0*	*0*	0	0	2
	M12	2	1	0	†	*19*	—	*0*	0	0	22
	M8	†	†	†	†	*8*	*15*	—	5	†	28
*	M10	5	4	3	0	7	11	0	—	0	30
	M9	11	14	7	0	5	11	†	13	—	61
Total dominant reactions		53	37	21	0	42	41	0	20	5	219

*Unattached.   †No meeting.

Three males, M8, M9, and M13, were extremely subordinate in their reactions to other males with whom they came into frequent contact (Table 4).

The dominant males, in general, ranged over larger areas than their subordinates (Table 2 and Fig. 4). Meetings between the dominant males of the main groups (C1 and C3) were infrequent.

No dominance-hierarchy was in evidence among the females, although each female was aggressive when near her breeding burrow. Only one fierce fight between females is known to have occurred, when F14 and F15 fought over a nesting site for their third litters. F15 was wounded on the abdomen and retired to another burrow. On several occasions females were observed ejecting females of foreign groups from their areas, which suggests that defence of territory by females may also occur as higher densities lead to increased contact.

IV. DISCUSSION

A preliminary report, combining the work described in this paper and the results of a similar study (being made concurrently at Canberra) of the social and territorial behaviour of a small colony of wild rabbits, has already been published (Myers and Mykytowycz 1958). Mykytowycz (1958) has since presented his observations in more detail. He described a rigid linear ranking order among males and females with bucks dominating bucks and does dominating does, and further stated that the top-ranking buck, in effect, controlled the group. He also described the greater range of movement and the aggressive alertness of the dominant buck, and mentioned briefly the challenging role of the second-ranking male and the hostility of all members of the colony to newly introduced rabbits. With the exception of the evidence relating to a dominance-hierarchy amongst the females, and to the role of the dominant buck in the establishment and control of the colony, these findings parallel our description of the behaviour of a single breeding group. By providing artificial warrens, Mykytowycz demonstrated the strong attraction offered by established warren systems to breeding females. Despite disturbance by handling, his group's activities remained centred about the warrens, in contrast to the movements of the group in our enclosure A. In addition, by giving the females a favoured site to compete for, he found that dominance-relationships developed. At Albury no warrens were allowed to develop. Later observations there have shown, however, that with the influx of new age-classes into the population the dominance of females by females does occur.

The data presented in this paper suggest that group formation is primarily a function of female behaviour. There appears to be a strong desire amongst adult female rabbits to associate in small groups when breeding. This is best demonstrated by the behaviour of the four females in enclosure A, moving and breeding in close company. Male rabbits, on the other hand, display no signs of mutual attraction but merely distribute themselves in relation to the groups of females, and most of the social activities that occur during the breeding season stem from male protection of such groups. The inter-male struggles within each group lead to the establishment of a strict dominance-hierarchy, with the dominant males playing the most active reproductive roles. The protection of each group of females by the dominant males virtually amounts to territorial defence, since foreign males are ejected.

It appears that the size of a breeding group is limited to two or three males and three or four females. Little evidence was obtained to explain the formation of new breeding groups. The behaviour of F4 in enclosure A suggested that one way in which new groups form is when one or more females leave an established group and settle in a new area. Unattached males are attracted to them, and a new social grouping is quickly established. This pattern of events has been confirmed by later observations.

The evidence clearly shows that each rabbit confines its daily activities to a restricted area, the shape and size of which are the result of its owner's adjustments to the behaviour of its companions. The term "home range" is very appropriate in describing such an area.

Since the three enclosures were subjected to the same conditions of experimental control, and there were no shortages of food, shelter, or water (details will be published later), the data presented point to the important effect of density alone on the behaviour of rabbit populations. Increasing density leads to a decrease in the size of individual home ranges, a decrease in the size of group territory, and an increase in the number of breeding groups. Later observations show that with further increase in density the home ranges of all members of the group become synonymous with the group territory itself. Density also appears to affect the behaviour of breeding does. Thus in enclosures A and B the females demonstrated a readiness to move away and desert their young after interference. However, in C—the most densely populated of the three enclosures—no desertions or retarded growth rates in nestlings occurred.

In natural populations the warren itself is undoubtedly the centre of group-activity. That an established warren is very attractive to rabbits is evidenced by Mykytowycz's (1958) observations. In enclosure A any of the four areas of breeding activity would have developed to warren status if left undisturbed. The facts that a large natural warren contains more than one breeding group and that group territories may radiate outwards from the warren have already been hinted at by Southern (1948). The close contact of breeding groups at warrens, together with the presence of other age-classes, points to a more complex pattern of social behaviour in natural populations than that described here for confined, adult rabbits.

V. Acknowledgments

The technical assistance of Messrs. C. S. Hale, E. C. Schneider, and T. Babij, Wildlife Survey Section, C.S.I.R.O., Albury, is gratefully recorded.

The authors also wish to express their appreciation for the material assistance and cooperation given by Mr. Frank Hood, "Kywanna", Thurgoona, on whose property the work was done.

Messrs. F. N. Ratcliffe, J. H. Calaby, and S. J. J. F. Davies, Wildlife Survey Section, C.S.I.R.O., Canberra, read the manuscript and offered helpful suggestions.

VI. References

Allen, D. L. (1939).—Michigan cotton tails in winter. *J. Wildl. Mgmt.* **3**: 307–22.

Burt, W. H. (1943).—Territoriality and home range concept as applied to mammals. *J. Mammal.* **24**: 346–52.

Carrick, R. (1956).—Radioiodine as an indicator of free-feeding activity of the rabbit, *Oryctolagus cuniculus* (L.). *C.S.I.R.O. Wildl. Res.* **1**: 106–13.

Connell, J. H. (1954).—Home range and mobility of brush rabbits in California chaparral. *J. Mammal.* **35**: 392–405.

Dalke, P. D., and Sime, P. R. (1938).—Home and seasonal ranges of the eastern cottontail in Connecticut. *Trans. 3rd N. Amer. Wildl. Conf.*, pp. 659–69.

DUNNET, G. M. (1956).—A population study of the quokka, *Setonix brachyurus* Quoy & Gaimard (Marsupialia). I. Techniques for trapping and marking. *C.S.I.R.O. Wildl. Res.* **1**: 73–8.

EALEY, E. H. M., and DUNNET, G. M. (1956).—Plastic collars with patterns of reflective tape for marking nocturnal mammals. *C.S.I.R.O. Wildl. Res.* **1**: 59–62.

HAUGEN, A. O. (1942).—Life history studies of the cottontail rabbit in south western Michigan. *Amer. Midl. Nat.* **28**: 204–44.

INGLES, LLOYD G. (1941).—Natural history observations on the Audubon cottontail. *J. Mammal.* **22**: 227–50.

MYERS, K. (1958).—Further observations on the use of field enclosures for the study of the wild rabbit, *Oryctolagus cuniculus* (L.). *C.S.I.R.O. Wildl. Res.* **3**: 40–9.

MYERS, K., and MYKYTOWYCZ, R. (1958).—Social behaviour in the wild rabbit, *Oryctolagus cuniculus* (L.). *Nature* **181**: 1515–6.

MYKYTOWYCZ, R. (1958).—Social behaviour of an experimental colony of wild rabbits, *Oryctolagus cuniculus* (L.). *C.S.I.R.O. Wildl. Res.* **3**: 7–25.

PHILLIPS, W. M., STEPHENS, M. N., and WORDEN, A. N. (1952).—Observations on the rabbit in west Wales. *Nature* **169**: 869–70.

RICHTER, W. C. (1955).—A technique for night identification of animals. *J. Wildl. Mgmt.* **19**: 159–60.

ROWLEY, IAN (1956).—Field enclosures for the study of the wild rabbit, *Oryctolagus cuniculus* (L.). *C.S.I.R.O. Wildl. Res.* **1**: 101–5.

SCHWARTZ, C. W. (1941).—Home range of the cottontail in central Missouri. *J. Mammal.* **22**: 386–92.

SOUTHERN, H. N. (1940).—The ecology and population dynamics of the wild rabbit, *Oryctolagus cuniculus*. *Ann. Appl. Biol.* **27**: 509–26.

SOUTHERN, H. N. (1948).—Sexual and aggressive behaviour in the wild rabbit. *Behaviour* **1**: 173–94.

THOMPSON, H. V., and ARMOUR, C. J. (1954).—Methods of marking wild rabbits. *J. Wildl. Mgmt.* **18**: 411–3.

EXPLANATION OF PLATE 1

Fig. 1.—View of experimental enclosures. (Photograph by C. S. Hale and E. C. Schneider.)

Fig. 2.—Single group of shaded logs in enclosure A. (Photograph by K. Myers.)

Fig. 3.—Chicken wing-tabs showing various patterns in combinations of five colours of "Scotchlite" reflective tape. (Photograph by K. Myers.)

RABBIT BIOLOGY IN CONFINED POPULATIONS. I

II
Reproductive Behavior

Editor's Comments on Papers 5 Through 9

Reproduction, including breeding behavior, is basic to ecological understanding of the population for the obvious reason that it produces the animals for the next generation. Information about reproduction in terms of litter or clutch size, seasons, and locations of nests has been available for many species, first from naturalists' observations and then from comprehensive studies. The papers chosen for this part illustrate how reproduction and behavior interact with environmental factors.

A very important early study (not reproduced here) by Brambell and Rowlands (1936) set a new standard by using large numbers of animals, a procedure forgotten since the days of the India Plague Commission around the turn of the century. Earlier work was often based on relatively few observations or records and the data were sometimes biased. From large numbers useful figures on birth rates and other parameters could be derived. The influence of the technique of obtaining data on large numbers of animals was very great and is often overlooked. As long as anatomical and taxonomic investigations were conducted, a few (50–100) specimens sufficed. But when ecologic and reproductive investigations began, it was necessary to obtain as many as 50–100 specimens each month to analyze for age and sex differences and for seasonal changes. The volume of data soon led to the use of punch cards and mechanical sorters and, of course, eventually to computers.

In another early paper Bodenheimer (1949, reproduced here in part) emphasizes the connection of nutrition, reproduction, and ecology, and almost in passing noted that copulation was necessary for ovulation, bringing behavior to the fore. Thus behavior began to intrude into ecological studies of reproduction. The connection to behavior was set by Bodenheimer's conviction that the birth rate was the major aspect of the fluctuations of microtine populations, a conclusion that remains convincing to this day. Had mortality been Bodenheimer's preference, the connection to behavior would have been weak. But reproductive rates are greatly influenced by behavior and in this case by the importance of copulation as a stimulus to ovulation. The existence of post-partum estrus emphasized behavior and also leads to studies of the endocrine system. The studies on nutrition and reproduction produced equivocal results, as is true today.

A very influential book by Darling (1938, reproduced here in part) made a very significant advance by proposing a hypothesis about behavior as an ecological factor and testing it. In addition, the selection chosen for reproduction also shows the nature of methods of collection and analyses of data in 1937. The contribution is measured better by the amount of new information derived from the process of testing the hypothesis than by its correctness, for it turns out that much of the "Darling" effect (earlier nesting in large colonies) resulted from differences in age composition rather than from flocking. The data of Coulson and White (1956, 1958) for kittiwakes showed that small colonies tended to be new colonies and hence have a high proportion of young birds, which generally, among all birds, have a low rate of reproduction. Thus it seems likely that in Darling's gulls the effects of age and size were confounded. Note, however, that the exaggeration of the social factor was done primarily by other persons, who christened it the "Darling effect." Darling himself stated that he did not wish to overemphasize the significance of social factor and that age might be important. But ornithologists were grasping for behavioral principles and here seemed to be one that benefited from a catchy name. Darling's own interpretations are more cautious.

Blanchard (1941) provided a striking advance in the use of reproductive organs as a tool for interpreting the effect of environment on behavior. She adapted histological methods to behavioral research to permit determination of the endocrine basis of seasonal changes. She also collected large numbers of specimens at various seasons and from various areas, and thus could discuss ecologic changes that resulted from behavioral differences. Her work (including several subsequent papers) set a new standard for interpretation of behavioral and anatomical data.

A more complex topic is represented by Lack's (1948) contribution to evolution, which of course has an ecologic basis. Behavior, especially feeding rate, seems to be critical in the evolution of clutch size in birds. The connection between behavior and ecology is more tenuous than in the previous examples, but ecology and evolution through survival are obviously entwined. In this case, behavior through such patterns as feeding rate and renesting have intimate connections to evolution. Under these circumstances it is apparent that evolution depends on behavior, but this dependence has not been prominent in other studies of evolution. Although many studies of evolution of species, especially insects, have used behavior as a tool for separating mating preferences, its relevance to ecology has been meager.

A final example demonstrates the relation of breeding time to season. While many persons described the changes in breeding behavior with seasons in temperate climates and changing lengths of day, Miller (1959) tested many hypotheses derived from temperate regions by considering breeding near the equator. His results produced considerable revision in our ideas of the ecological control of breeding behavior. The work is relatively recent and thus has an influence that is difficult to evaluate, but a significant aspect is the recognition that a species may have reservoirs of behavioral and physiological responses that are not used. Assuming that members of the genus *Zonotrichia* originated in the Northern Hemisphere and there developed, through survival, a response to the long photoperiod, their recent invasion of the equatorial regions left them with a response that lacked a stimulus. Although the birds adapted by breeding all year, presumably the response to a long photoperiod remained, so that those that continued the spread into the Southern Hemisphere could again respond to long days.

Papers of the type chosen here paved the way for research on larger mammals and

57

birds which has flourished since about 1950. Thus the reproduction and breeding of large herbivores, kangaroos, and many game species is now rather completely known, thanks to the influence of the early papers on small mammals and birds.

References

Brambell, F. W. R., and I. W. Rowlands. 1936. Reproduction of the bank vole (*Evotomys glareolus*) Schreber: I. The oestrus cycle of the female. *Phil. Trans. Roy. Soc. London 226:* 71–97.

Coulson, J. C., and E. White. 1956. A study of colonies of the kittiwake *Rissa tridactyla* (L.). *Ibis 98* (1): 63–79.

Coulson, J.C., and E. White. 1958. The effect of age on the breeding biology of the kittiwake *Rissa tridactyla. Ibis 100* (1): 40–51.

An Influential Book

Heape, W. 1931. *Emigration, Migration and Nomadism.* W. Heffner and Son, Cambridge, England, 369 pp. Note the entwining of reproductive physiology (and thus behavior) to the seasons and to behavior.

5

Reprinted from *Problems of Vole Populations in the Middle East,* The Research Council of Israel, 1949, pp. 38, 40–42, 44–47

Problems of Vole Populations in the Middle East

F. S. BODENHEIMER

IV. Physiology and ecology of reproduction of Microtus guentheri.

The just mentioned lack of evidence about any permanently dominant or regulatory role of enemies and diseases upon the numbers of voles makes the study of the physiology of reproduction and fertility in *Microtus guentheri* the more imperative. It may easily prove to be the focal point in the fluctuations of vole populations. We therefore pay increasing attention to this aspect of the problem. The simple data about age of maturity, size and seasonal fluctuations of litters, duration of pregnancy, etc. have been given in Chapter II.

The estrous cycle. Zondek and Sulman (1940) wrote about *M. guentheri*: "The vaginae of the females were morphologically continuously in the characteristic stage of diestrous. Since the other animals kept under the same conditions, but in which no vaginal smears were taken bred rapidly and continuously, it must be assumed that in *Microtus guentheri* mating takes place when the vagina does not show typical cornification... No vaginal cycle with a cornification stage occurs as in white mice and rats. Injections of large doses of estrogonic hormones, however, induce cornification." Yet there was one very fundamental difference between the females from which the vaginal smears were made and the breeding control pairs : the latter were permanently mated, whilst the former were permanently isolated.

Bodenheimer and Sulman (1946) corrected this statement by showing that a real vaginal cycle of cornfication as well as the formation of corpora lutea appears in *M. guentheri* as the natural response to copulation. The Levante vole thus falls into the same group, to which the rabbit, the ferret and the cat were known to belong, where ovulation and the accompanying processes are only produced during the season of sexual readiness. In nature ovulation can only be induced in them during a typical estrous provoked by copulatory stimuli, which obviously are transmitted through the central nervous system.

Editor's Note: Figures 7 and 8 have been omitted, owing to limitations of space.

In the meantime Bodenheimer and Lasch (in preparation) have followed the changes in the vaginal smear from pairing to birth. The vaginal smear shows a typical diestrous during the entire pre-copulation stage. The following changes are observed in the vaginal smear. The numbers in brackets indicate days after pairing in our experiments, not necessarily days after copulation.

Stage	Day	Morphological changes
I.	(1-5)	Increased cornification: initially with a prevalence of stratum granulosum cells, later of stratum lucidum cells.
II.	(1-5)	Massive desquamation with abundant ceratohyaline cells which form the so-called "cheesy substance".
III.	(9)	The appearance of estrous cells indicates the peak of cornfication.
	(1-7) (2-7)	Somewhat earlier sperms only are found in the vaginal smear and subsequently sperms mixed with estrous cells.
IV.	(2-12)	Massive desquamation of stratum lucidum and estrous cells.
V.	(4-16)	Medium cornification with prevalence of stratum lucidum cells.
VI.	(7-19)	Massive leucocyte invasion.
VII.	(14-26)	Typical pregnancy smear: much slime containing autolysing or else deformed cells (swollen or wrinkled).
VIII.	(28)	Disappearance of the typical pregnancy smear with birth. A number of stratum granulosum and stratum lucidum cells appear.

The following table (again the data do not indicate exact days after copulation, but only after pairing) correlates the vaginal cycle with the processes in the ovaries and the cornua uteri. (r: right, l: left, e: embryo:

Day after pairing	Vaginal smear	Corpora lutea	Cornua uteri
5	desquamation	None	normal
9	a prop formed by sperms and desquamation	Folliciles in ovulation	Serosa with many capillaries
11	pregnancy smear	r:2,l:3 corp.lut.	r:2,l:3 e (21 mg, 3.5 mm long)
14	dto.	r:1,l:3 c.l.	r:1,l:4 e (12 mg, 2.8 mm)
17	dto.	r:1,l:4 c.l.	r:0,l:2 e + 1 resorption (137, 107 mg, 9.8 mm)
21	dto.	r:3,l:1 c.l.	r:3,l:1 e (61 mg, 7.3 mm).

The first implantation of an embryo was observed at the 12th day after pairing (not necessarily after copulation). The bursting of the ovarial follicle was observed from the fifth day after pairing, the typical corpus luteum from the 14th day onwards. It can be recognised a few days earlier, but has not yet its final colouration.

Seasonal changes in the gonads. The size and structure of the gonads undergoes in *Microtus guentheri* a number of characteristic seasonal changes. These are compiled in the accompanying table. We first discuss briefly these changes in our laboratory voles at Jerusalem. The testes were studied by A. Yanovsky. The total weight of the testes of males about 50 days old shows two peaks: in late autumn and in late winter, with a deep low in mid-summer. The average width of the tubuli follows this seasonal trend approximately but not too closely. The differences are much smaller than in the weight, being mainly made up by changes in the mass of the interstitional tissue. The intensity of sperm production as measured by the number of sperms in the ocular field follows rather closely the change of total weight. From May to July the testes are in the abdominal cavity, the rest of the year in half to complete descensus. The seasonal changes in the total weight agree rather closely with those in the total weight of the testes. For comparison we add the seasonal trend of the toal weight of the thyroidea, which is definitely opposite to that of the gonads with peaks in June and December, and is lowest in November and February.

The seasonal trend of total weight of testes of freshly trapped

61

males in Ganigar agrees with that of our laboratory voles at Jerusalem. Comparing these conspicuous changes in the total weight as well as in the histological structure of the gonads (see fig. 8), we are astonished to find that the seasonal fertility of our laboratory voles (see table p. 40) reflects so slightly the magnitude of these changes. This is however not correct under all circumstances. At Tel Aviv, for instance, the voles kept in the Biological-paedagogical Institute are almost sterile in summer (Dr. H. Mendelssohn i. l.). Tel Aviv is, of course, much hotter in summer than Jerusalem. And in our field observations at Ganigar the breeding season was restricted from November to May on non-irrigated fields. It is quite probable that the complete sterility of the females during summer and most of the autumn was a result of the exceptionally dry year, as in other years we have obtained a certain amount of pregnant females during the entire year. But there is no doubt that fertility in nature is much lower in summer in normal years than in late autumn and in late winter, when it reaches its peak. In moderate climates fertility is interrupted in the genus *Microtus* in both sexes in winter, or at least much reduced. Israel seems to be the first country where an aestival interruption or reduction of fertility has been established.

Influence of food on fertility. A series of experiments was conducted in order to find out how far difference of basic nutrition influences fertility in the Levante vole. The index to the different diets indicates the number of young born per pair and 1000 days mated life:

Standard nutrition	73	Carrot	18
Mixed grains, liver oil, lemon	34	Half of standard nutrition	0
Mixed grains, lemon	13	Meat, wheat, water	0
Mixed grains, water	43	Meat	0
Wheat, water	18	Standard food, light 24 hours	32
Mixed grains, carrot, liver oil	7	Dto., light 6 hours	58
Wheat, barley, carrot	75	Dto., dark 24 hours	70
Carrot, lemon	14	Dto., thermostat 20°C	42
Carrot, lemon, liver oil	5	Dto., thermostat 30°C	22
Carrot, water	35	Dto., thermostat 35°C	0

Seasonal trend in fertility and connected glands in Microtus guentheri

	M o n t h												
Laboratory voles (Jerusalem 1943/45)	V	VI	VII	VIII	IX	X	XI	XII	I	II	III	IV	V
Total weight of testes mg.	145	81	55	129	92	174	250	156	101	219	190	170	147
Average width of tubuli (micra)	41	35	37	36	33	45	44	34	40	49	44	46	39
% tubuli with 10-50 sperms	38	41	25	7	6	48	62	30	71	36	51	56	56
% tubuli with over 50 sperms	3	0	0	0	0	35	0	0	4	47	3	31	7
Descensus testiculorum	0	0	0	1	1	½	¾	½	1	½	1	½	0
Weight of ovaries mg.	15.4	12.3	14.0	14.4	15.8	20.1	23.8	13.2	16.6	25.5	25.7	20.0	—
Weight of thyroidea mg.	8.9	9.9	9.6	9.2	8.9	8.2	7.6	9.4	7.1	6.5	6.8	8.1	—
Field voles (Ganigar 1947/48)													
Weight of testes (non irrigated fields)	162	22	20	30	60	130	255	364	322	418	365	222	—
Dto. (on irrigated fields)	—	31	—	—	—	300	344	495	481	678	—	—	—
% of pregnant females (non-irrig.)	41	0	0	0	0	0	18	58	17	7	72	47	—
Dto. (irrigated fields)	—	31	—	—	—	0	25	59	58	75	—	—	—
Average no. of embryos (non irrig.)	8.8	—	—	—	—	—	5.5	5.2	4.6	10.0	9.2	8.8	—
per female (irrig. fields).	—	5.0	—	—	—	—	4.5	8.5	7.2	10.0	—	—	—

Unfortunately nutrition with green herbs only was omitted at that time, as we were not then aware of the importance of this food in nature. In summer it is available only on irrigated fields. Without going into detail the extended series of experiments show clearly that differences in nutrition find their expression in differences in fertility. The same is true for light and temperature. Illumination (24 hours per day) for a number of generations had a definitely inhibiting influence on fertility, whilst complete darkness for a number of generations had not the slighest influence on it. Differences in relative humidity had no appreciable effect upon fertility, but constant high temperatures had a very inhibitive effect.

To return to nutrition : Apart from basic food a number of special foods exist which exert a definite influence on fertility. Thus, Dr. H. Mendelssohn of Tel Aviv , told us that he produced immediate fertility in his voles in summer by feeding them with hemp seeds. And F. Sulman isolated an estrogenic substance from the rootstock of mandrake. In 1934, the Friedmans discovered the presence in certain plants, such as alfalfa, corn and oats, of a gonadotropic factor which is active in the estrous rabbit. Research by Friedman and Mitchell (1942) has tended to show that the concentration of this factor in the plants aforementioned and probably in other and wild plants as well, is subject to seasonal and longer "cycles". In a few cases the gonadotropic effect of plant extracts was even stronger than that of human pregnancy urine. Bradbury (1944) however pointed out that the effective principle in plants is not primarily gonadotropic, but neurotoxic. The plant extract proved to be without gonadotropic effect in *Rattus*. Rabbits and rats treated with large doses of the substance developed toxic clinical symptoms which culminated in the death of the animals from pulmonary edema. It seems probable therefore that the stimulation of gonadotropic pituitary secretion by the neurotoxic plant- extract is a secondary effect of central brain stimulation. Injection of copper, cadmium, picrotoxin, etc. are known to produce similar effects. In the light of these findings, the presence in *Microtus* of a rabbit-type of estrous cycle assumes a primary ecological importance. This is the reason why plant extracts with a gonadotropic effect on rabbits are inactive in rats and mice (Bradbury 1944). We have been unable so far, to obtain any strong gonadotropic effect from alfalfa plant extracts as tested by injections into rabbits. But the number of experiments in this direction was rather limited, as war conditions made the acquisition of rabbits for scientific experiments an almost impermissible luxury. When comparing the vaginal smear of virgin female voles, which were fed with an addition of alfalfa to the normal diet, with those fed only on the standard diet, we sometimes obtained preestrous in the former, but never in the latter groups. And this result is fully confirmed by the comparison of the testes (total weight) from non-irrigated and irrigated fields at Ganigar (see table p. 44).

We have called the Friedman-effect gonadotropic, because its symptoms in the proper dosage are similar to those obtained by *raised* activity of gonadotropic hormones. This means the effect of the secretions of the anterior lobe of the pituitary glands, such as can be obtained from human pregnancy urine, from pregnant mare serum, or directly from the extract of the anterior lobes of the pituitary. All three of these sources were either fed or injected into voles, in order to study the efffect of intensive gonadotropic hormonal activity on the fertility of *Microtus guentheri* (Bodenheimer and Sulman 1946). The following table shows the increase of litter size following administration of gonadotropin (75 females by injection, 65 females by ingestion):

Microtus guenteri

Litter size (control equal to 100)

	Control	Gonadotropin injected	Gonadotropin ingested
Average size of litter per pregnant female	100	131	114
Percentage of births within 30 days after pairing	100	144	120
Average size of litter per all females together	100	178	166

The last horizontal column expresses the result of the combined effect of raised ovulation and of raised readiness to copulate. The gonadotropic effect obtained by ingestion or injection of pituitary anterior lobal hormones is, in the magnitude of a fertility, one and a half to twice that obtained in the control pairs.

It should be borne in mind that in the initial phase of all vole outbreaks observed by us the average litter size was 6 to 14, as against 3 to 8 in normal years. Also during the early phase of a vole outbreak the percentage of pregnant females is considerably increased. These facts by themselves suggested the action of a gonadotropic factor in the initial mechanism of a vole outbreak, which is responsible for an increase in ovulation as well as for a greater readiness of the female to copulate. This mechanism may be initiated by a number of extero-stimulations, such as acting through food or through central nervous stimulation, either directly or indirectly upon the anterior lobe of the pituitary, but it is not restricted to cyclic outbreaks. Elton (1931 p. 673) has, for instance, shown that in *Apodemus* the number of embryos per pregnant female is highest at the peak of the breeding season, being smaller before and after the peak. This hints at the presence of a very similar gonadotropic lability as in *Microtus*. Apart from the increase of the number of embryos per litter and the closer coincidence of births there is still one other mechanism which may have a great effect on vole outbreaks : the lengthening or the shortening of the main breeding season, probably via rainfall through conditions of vegetation. This possibility will be discussed in the following chapter.

* * * * * * *

VII. Bibliography

F. S. BODENHEIMER and F. SULMAN, The estrous cycle *Microtus guentheri* and its ecological implications. Ecology. 27, 1946. p. 255–256.

J. T. BRADBURY, The rabbit ovulating factor of plant juice. Am. J. Physiol. 142. 1944. p. 487.

C. ELTON, The health and the parasites of a wild mouse population. Proc. Zool. Soc. London 1931. p. 657–721.

M. H. and G. S. FRIEDMAN, A gonad-stimulating extract from Alfalfa meal. Proc. Soc. Exper. Biol. and Med. 31. 1934. p. 842.

M. H. FRIEDMAN, and J. W. MITCHEL, Variation in the yield of gonadotropic material from green plants in relation to the season of growth and the pH of the fresh juice. Endocrinology. 29. 1941. p. 172–178.

B. ZONDEK and F. SULMAN, Vaginal cycle of *Microtus guentheri*. Proc. Soc. Exper. Biol. and Med. 43. 1940. p. 86–88.

66

6

Reprinted from *Bird Flocks and the Breeding Cycle*, Cambridge University Press, London, 1938, pp. 61–78

Bird Flocks and the Breeding Cycle

F. F. DARLING

The largest colony in 1936 of 84–90 herring gulls began laying eggs on May 7th and finished on May 23rd, an inclusive period of 17 days. The second colony of 30–34 birds started later on May 12th and finished on June 3rd, a period of 23 days. The third colony of 20 birds began later still on May 18th and finished on June 12th, a period of 26 days. The fourth colony of 4 birds made nests in the few days following May 26th, but they never laid eggs.

The larger colony of lesser black-backed gulls—
72–80 birds—laid their first egg on May 15th and
the last egg was laid on June 7th, a period of 24 days.
I should point out that the first egg was sucked by
one of the crows and no more eggs appeared until
May 21st, after which date they were laid in rapid
succession. If that solitary first egg is ignored, the
laying period for the whole colony was 18 days, a
strictly comparable period with that of the large
colony of herring gulls. The second colony of 18
birds began laying on May 23rd and finished on
June 12th, a period of 21 days.

The breeding season of 1937 showed several
remarkable differences from that of 1936. Herring
gulls and lesser black-backed gulls were more
numerous by approximately 50 per cent. This
increase in itself would naturally make for more
young reared, but each female bird influences the
total by the possible variation in the number of eggs
laid each season. The average number of eggs per nest
in the large herring gullery in 1936 was 2·1; in 1937 it
was 2·9. The average number of eggs per nest in the
larger lesser black-back gullery was 2·2 in 1936 and
2·9 in 1937. The figures are closely comparable
between the two species and doubtless point to an
improvement in the environmental complex in 1937.
The number of infertile eggs was also less in the
latter year.

The herring gulls began laying on May 9th at

the largest and second largest gulleries. The first colony finished on May 25th, an inclusive period of 17 days and the same length of time as last year. By a system of pegging it was possible to note the course of egg-laying through the period and a modal curve can be constructed showing, as might be expected, the rise sharper than the fall. The gullery was searched every second day from the time the first eggs appeared until several days after the last ones were laid. The figures for plotting were based on the number of nests at each search which contained eggs for the first time.

May 1937	9th	11th	13th	15th	17th
New nests with eggs	3	9	17	14	11

May 1937	19th	21st	23rd	25th	17 days
New nests with eggs	7	4	no search	last egg laid	189 eggs in 65 nests

The second herring gullery, where eggs also first appeared on May 9th, was searched in the same manner, but the results have been confused by several pairs of lesser black-backed gulls nesting with them. The gullery was in such a position that it was not possible to see with certainty which nests belonged to each species. As a general rule the lesser black-backed gull makes a larger and more grassy nest than the herring gull, but the general rule could not justifiably be applied to the particular case for overlapping of nest types was common. The eggs of the two species are practically alike. I can vouch for the fact that two herring gulls in this group began laying on May 27th and finished on May 30th, which extends the egg-laying period of the colony to 22 days at least. The colony contained a minimum of 18 herring gull nests. This intrusion of another species has entirely upset the possibility of estimating the rearing percentage in 1937 and it has not been attempted. Lesser black-backed chicks are usually darker in their immature plumage than those of the herring gull, but the criterion is not sharp enough for application in a particular case.

The third gullery was also invaded by lesser black-backed gulls in 1937 and the laying period of the herring gulls was masked in consequence. Laying began on May 11th and one herring gull at least finished laying on June 1st, which gives a minimum period for the colony of 22 days. There were at least 9 herring gull nests.

The fourth colony which contained 4 birds in 1936 had grown to 6 birds in 1937 and each of the three pairs laid a full clutch of three eggs. They laid between May 13th and May 27th, an inclusive period of 15 days. This period for the smallest colony is less than for the largest one. But, following up the general tendency postulated, for small colonies to spread ovulation over a longer period than large ones, there must come a limit to its expression, because the smallest possible unit of one pair would normally take only 5 days to lay the clutch, i.e. three eggs laid on alternate days. The breeding of these three pairs points further to the improvement of 1937 as a breeding season over that of 1936.

No nests were robbed in the large herring gullery in 1936 or 1937, but two suffered at the second gullery in the early part of the 1937 season. One bird laid a fresh clutch before the end of the 22-day laying period and the other did not nest again.

Goethe (1937) gives dates of onset of laying in his colony of herring gulls which are much the same as I have noted, but he found some of his birds laying as late as the latter half of June. His colony was an immense one of 5000 to 6000 pairs, but he points out that different parts of the colony began laying at different times, so I imagine that the whole number was not one colony but a number of socially discrete colonies which might show similar sequences to those I have found in the much smaller colonies of Priest Island.

The lesser black-back colonies yielded data in 1937 comparable with those of the year before. In the larger colony of approximately 120 birds laying began on May 17th and finished on June 8th, an inclusive period of 23 days. The modal curve which can be plotted from the figures is flatter than the one for the large herring gullery. The same principle was followed in searching the gullery every second day, but the weather broke rather badly in the last

DBF 5

week of May and, because of the greater time taken in searching this widespread colony, I could not bring myself to put the birds off their eggs on days of high, cold winds. I do not place the same reliance on the modal curve, as such, of this larger lesser black-back colony, because the nests were much more difficult to find and some were found when two eggs were present already.

May 1937	17th	19th	22nd	25th	28th	30th
New nests with eggs	3	5	6	9	8	7

June 1937	2nd	4th	6th	8th	23 days
New nests with eggs	6	5	5	2 both full	163 eggs in 56 nests

The second lesser black-back gullery of 30 birds began laying 2 days later than the larger one, on May 19th, and finished on June 12th, an inclusive period of 25 days. The figures relating to the course of laying in this gullery are absolute, for the area covered by them is much less than that of the big colony and all the nests are on a steep slope above a lochan, from the opposite side of which the whole colony can be watched.

May 1937	19th	21st	23rd	27th	30th
New nests with eggs	1	1	2	2	2

June 1937	2nd	5th	8th	11th	12th	25 days
New nests with eggs	3	2	1	1	fin- ished	42 eggs in 15 nests

The above figures show in almost all cases that the larger colonies not only start laying earlier, but the time taken by the whole colonies to lay their crops of eggs is shorter than in the colonies of lesser numbers. The same fact emerges from comparison of the second herring gull colony with the third in 1936. The figures are too striking to allow of coincidence and it is suggested that the most obvious interpretation is to be found in the total value of visual auditory stimulation for each pair in the larger flocks, compared with the total amount of such stimulation in the smaller flocks. If such a theory is correct, the observation is easily explained of the two pairs of gulls which reached the stage in the reproductive cycle of nest building but not of laying eggs.

I do not wish to overestimate the significance of the social factor in the breeding cycle but to point to its existence. It is reasonable to imagine that when other environmental conditions are of an optimum character, the social factor would play a lesser part, and when they are bad, it may become crucially important. My 1937 figures, for example, are not so striking as those of 1936. The spring of 1936 was a bad one and I think that sea food for the gulls was scarce because they fed so largely on the sown oats of the mainland crofts. There were comparatively few oat casts on the gulleries in 1937 and the crofters told me the gulls had troubled them much less this

5-2

year. It was noticed, instead, that in 1937 the gulls went out to sea a mile or two in great flocks and fed to repletion on small fry, probably herring, which was present at the surface of the sea this year in profusion. The weather of 1936 was unusually bad in the spring; that of 1937 was unusually good from April 1st.

Let us leave these phenomena for the moment and note the later effects of this shortness or protraction of the egg-laying phase. The total period of hatching of the eggs in the colony corresponds closely with the length of time taken in egg-laying, and it is at that point we leave the psycho-physiological state of the parents and follow the fate of the offspring.

On Priest Island the chicks of the lesser black-backed and herring gulls are preyed upon principally by the great black-backed gull and the heron. The largest herring gullery, for example, had a pair of the big gulls nesting on the western and eastern side of it in 1936, and in 1937 one pair of great black-backs nested in the middle of the herring gullery. There were three more pairs on the island, nesting in close proximity to the other gulleries. The herons come over from the mainland where these birds are very common, and as they frequently stay on the island overnight, I imagine they are non-breeding birds, probably from the large heronry on the Fionn Loch eighteen or twenty miles away.

I have observed that these two predators on the

gull chicks take them only while they are in the down stage, i.e. for the first fortnight of their lives. Once the chicks are growing quill feathers, the predators appear to leave them alone and the chicks are more or less immune from attack until they are fledged. The peregrine falcon takes toll of them afterwards, but the fate of the chicks after the fledgling state is outside the subject of the essay. I have also noticed that the herring gulls themselves rarely peck any of the chicks when the down stage is past.

The herring gulls and lesser black-backed gulls do not mob the great black-back, though the separate pairs on their nesting territories may object when the big gull intrudes. This toleration exists probably because the big bird appears not unlike themselves, and because it has mixed with them in the gullery throughout the season. The heron is mobbed on sight with violence, whether on mischief bent or not. The result is that the great black-back takes such chicks as it can find whenever it wishes and the heron takes those it can snap by stealth.

A steady toll of the chicks is taken while they are in the down stage. If this period is a comparatively short one for the whole chick crop of the colony, the percentage taken will be less than where the down period of the chick crop is extended. The state of affairs may be represented diagrammatically as in the figure overleaf.

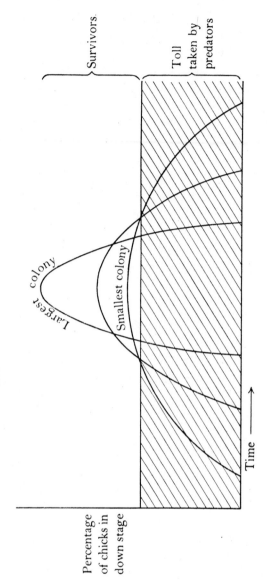

Diagram illustrating how the survival rate is influenced by the "spread" of time in which the eggs of a colony of herring gulls are laid

The chicks become more active the older they grow and the observer finds it difficult to estimate the number remaining in the colony. If he walks about a herring gullery counting the chicks, he is sure to miss some of them which are squatting against the rock; and in the lesser black-back colonies the chicks are hiding in the heather and bracken. A reasonably accurate but not an absolute count can be made when the chicks are just fledged and are at the cliff edges taking their first flights; or, in the case of the lesser black-backs, when they are on the lochans learning to fly by rising off the water into the wind. It is unsafe to count much later than that time because the young birds range about the island and are missed. Here are the figures recorded, and expressed as percentages of the chicks hatched:

1936

Herring gull colony, No. 1. 84–90 birds.
> 40 nests; 8 more without eggs; 84 eggs; 72 hatched; 35 chicks reached fledgling age, $= 48·6$ per cent.

Herring gull colony, No. 2. 30–34 birds.
> 12 nests; 2 more without eggs; 26 eggs; 22 hatched; 8 chicks reached fledgling age, $= 36·4$ per cent.

Herring gull colony, No. 3. 20 birds.
> 7 nests; 1 more without eggs; 16 eggs; 14 hatched; 3 chicks reached fledgling age, $= 21·4$ per cent.

Herring gull colony, No. 4. 4 birds.
> 2 nests; no eggs laid.

Lesser black-backed gull colony, No. 1. 72–80 birds.
    36 nests; 10 more without eggs; 78 eggs; 72
    hatched; 40–45 chicks reached fledgling age,
    = 55·5–62·5 per cent.

Lesser black-backed gull colony, No. 2. 18 birds.
    7 nests; 3 more without eggs; 15 eggs; 14 hatched;
    8 chicks reached fledgling age, = 59·1 per cent.

The comparative rearing figures for some of the gulleries in 1937 are not given because of the mixing of the two species in some of the herring gull colonies, to which confusing occurrence allusion has been made already. The following figures from four of the colonies are of comparable accuracy and value with those given above. It will at once be remarked that the rearing percentage of the large herring gullery is extremely low. This is owing to a catastrophic northerly gale which happened on June 28th–29th. There was extremely heavy rain as well and the enormous waves broke over the greater part of the gullery which, it will be remembered, is at the north-west corner of the island and two sides of its triangular shape are sea cliffs. I went over this gullery on the morning of June 29th and found many chicks drowned and more washed over the edge of the cliffs. Many chicks at the second herring gullery suffered a like fate. Such a gale as this was must be a rare occurrence in late June.

This incident of the gale and heavy rain had one interesting consequence for the observer of the gullery. Some of the birds were totally bereft of their young and in five or six days' time, on July 3rd and 4th, I heard much more noise in the gullery than had been made since incubation commenced. There was a recrudescence for a few days longer of the early stages of courting behaviour among these birds. The first and second types of display outlined in the Appendix were common and the third type was attained in a few instances. Beyond this point the birds did not go and no new nests were built or eggs laid. Doubtless regression of the gonads had gone too far to allow of the cycle being wholly repeated at this time.

1937

Herring gull colony, No. 1. 130–150 birds.
> 65 nests; 4 more without eggs; 189 eggs; 181 hatched; 76 reached fledgling age, =41·4 per cent.

Herring gull colony, No. 4. 6 birds.
> 3 nests; 9 eggs; 8 hatched; 1 chick reached fledgling age, =12·5 per cent.

Lesser black-backed gull colony, No. 1. 120 birds.
> 56 nests; 26 more without eggs; 163 eggs; 155 hatched; 75–90 chicks reached fledgling age, =48·4–58·0 per cent.

Lesser black-backed gull colony, No. 2. 30 birds.
> 15 nests; 4 more without eggs; 43 eggs; 41 hatched; 22 chicks reached fledgling age, =53·6 per cent.

The wide-nesting habit of the lesser black-backs, once they have developed a habit of communal courtship, is itself an advantage towards a higher survival rate among the chicks. Let us consider the environmental factor in relation to the social interactions. The lesser black-backed gull nests a fortnight later than the herring gull. When the lesser black-back lays its eggs the ground may be relatively bare—almost as bare as that of the herring gullery, which produces no luxury of foliage at any time; and it has been shown that lesser black-backs' eggs thus uncovered and at times unprotected are lost to the hooded crow. The incubation period is 23–25 days and when the lesser black-back chicks hatch, they become noticeably more active than the herring gull chicks. I have often seen them swimming on the lochans while still in the down stage, and it will be obvious that they will have less chance of intruding upon other nesting territories than if they were members of a close-nesting group. By the time the lesser black-back chicks hatch, the ground has become covered by bracken and other herbage and the chicks disappear among it. Obviously a predatory enemy has greater difficulty in finding a member of the chick crop when seventy chicks are on thirty acres with plenty of ground cover, than when the same number are on one acre with but little ground cover.

The chick crop suffers loss from the adult gulls of the same colony to some extent, and this also occurs while the chicks are in the down stage. When the chicks are newly hatched they do not move far from the nest, but if they do, they are likely to be pecked and may possibly be eaten by neighbouring gulls. Even their own parents do not brood them if the chicks have wandered far from the nesting-site, and examples are on record of parents killing their own chicks when both are away from the nesting terri-tories. The parent-child relation seems to strengthen as the days go by, and when the chicks are normally wandering about the gullery in half-fledged con-dition they are found and fed by their parents and are but little persecuted by the other adults. Loss of this kind is heavier in gulls which follow a close-nesting habit. A few die thus in the herring gulleries, I cannot say just how many, but hardly any in the wider-nesting lesser black-back colonies. Kirkman (1937) has devoted a whole chapter to what he calls "The Massacre of the Innocents" in a large black-headed gull colony. Nesting is closer in this species than in the herring gullery and Kirkman shows that losses are considerable. He observes that disturbance scatters the chicks and the journey back to their own nest is of a particularly hazardous nature because of the pecks the chicks may receive from adults other than their parents.

To return to the species I have watched, we see that the larger colonies have been more successful than the small ones; that one species by an extended nesting range, though with a no less close sociality, is more successful than another; and that when colonies are very small, they may be wholly unsuccessful.

The greater success of larger colonies does not appear to me to lie in any vague factor of mutual protection, but in the nearer approach to synchronization of the breeding cycle throughout the colony. Howard has shown us that synchronization is essential for reproduction on psycho-physiological grounds and that that state is partly brought about by the stimulation of courtship ritual. I suggest that in birds social at the breeding season, the importance of synchronization extends to the ecological field of survival rate. That state is brought about by the same means of courtship display which synchronizes pairs of birds, but, in addition, the numbers of the flock are a potent factor because of the quantitative and cumulative value of their presence, general activity and courtship display. My observations show that as well as better synchronization of the breeding cycle, the onset of egg-laying is slightly earlier in the larger flocks. I do not suggest any particular survival value to rest in this phenomenon, but I think it supports the contention that the size of the flock has an influence on the breeding cycle and that sociality cannot be ignored as an exteroceptive factor.

It has been suggested to me that the different flocks observed were different age groups and that young birds might be expected to embark upon the breeding cycle at a rather later date. Unfortunately, ages cannot be recognized after the adult plumage is reached and the suggestion cannot be followed up; but I do know that the largest colony of herring gulls contains a few birds each year that are just reaching adult plumage. I do not think there are grounds for believing that these birds flock in age groups, and such an objection to the theory put forward in this essay would be pure conjecture.

It would be interesting to know the threshold of distance over which the visual and auditory stimulation may exert an influence which doubtless varies in different species. Three of the herring gulleries on Priest Island were four hundred and five hundred yards apart along a line as the crow flies. The fourth colony was a mile distant on the south side of the island and quite isolated. The lesser black-backed gull colonies were also four or five hundred yards distant from each other and not nearer than that to the herring gulleries. My opinion is that the stimulation of voice and presence does not extend beyond a hundred to two hundred yards, and the stimulation provided by posture and display to a much shorter distance, but this is only an opinion which I cannot back with positive evidence. The possible threshold of numbers will be treated in the next chapter.

The observations made of the gulls and the description given of their behaviour must bear the greater share of the burden of my thesis. Their behaviour and history during the breeding season in one set of conditions have been watched more closely and for a longer and more consecutive period than I have been able to watch any other species of birds. How surely it has been borne upon me that the glimpses of minutes, hours, days or even weeks, which a life of bird watching as a hobby have given, are inadequate for an interpretation or solution of the deeper problems of evolution, natural selection and survival in the bird world! We need time, time, time and a sense of timelessness. Our pictures of behaviour must be detailed in time equally with those of space.

I now proceed to describe in rather less detailed fashion the phenomena of communal courtship, sometimes with and sometimes without social nesting, in a few other species of birds. The gulls do not indulge in many patterns of communal courtship though they nest in colonies, but some birds execute intricate figures in company and present another aspect of the same principle of social stimulation towards the integration of the breeding cycle. As far as possible those of Priest Island and of northern and western Scotland will be described, for I know them and their conditions more intimately.

7

Reprinted from *Univ. Calif. Publ. Zool.*, **46**(1), 57–62 (1941)

The White-Crowned Sparrows (*Zonotrichia leucophrys*) of the Pacific Seaboard: Environment and Annual Cycle

BARBARA D. BLANCHARD

RATE OF DEVELOPMENT OF THE TESTES

The rate of testis development, from the first change from inactive condition to breeding, is analyzed for 1936–37 and 1937–38 in table 7. In both years the rate is slow at first, then rapid, then slow again. This tends toward the sigmoid curve characteristic of growth-rate graphs. Although this curve is composite, there are sufficient items to give us a close approximation of the curve of testis enlargement for one individual.

ANNUAL VARIATIONS IN GONAD CYCLE

STAGE 5

Variation in date.—I can establish the fact that identical histologic stages in the testis show year-to-year differences in date. The point where one or more gonads filled with primary spermatocytes in synapsis occur (Stage 5) is the earliest point in the cycle which is represented by sufficient material to be comparable in all four years. A number of gonads have been sectioned and investigated at, and close to, this point, as shown in the graphs (figs. 15–19). It is unfortunate that in 1936 only four gonads were collected at this point, but, as will be shown later, these are consistently related to the rest of the cycle and establish the point with reasonable accuracy. The approximate dates for the appearance of one or more gonads in Stage 5 are as follows: in 1936, January 12; in 1938, January 27; in 1934, January 28; in 1935, February 8; and in 1937, February 15.

In 1935 the first gonad in Stage 5 occurs on February 1; but, since its volume is considerably smaller than the first gonad in Stage 5 collected in other years, I allowed several days for it to reach a comparable volume.

Variation in both rate and starting point.—The fact that Stage 5 comes at different dates each year means one of three things: that there is a variation in the time of earliest change (when interstitial cells appear) only, or in the subsequent rate of development only, or in both. Comparison of the two years 1936–37 and 1937–38 (figs. 18 and 19) shows conclusively that the interval from first change to Stage 5 must be widely variable in length. In 1937 and 1938, for example, birds with testes of almost identical average volumes (1.44 mm.³ and 1.49 mm.³) were taken on January 6–8 and on January 9, respectively; yet the rates of subsequent development diverged so sharply that Stage 5 was not reached until February 15 in 1937, whereas in 1938 it was

TABLE 7

RATE OF DEVELOPMENT OF TESTES

Period of Fourteen Days	Number of Testes	Volume Range (mm.³)	Volume Average (mm.³)	Percentage of Increase over Previous Two Weeks
		1936–1937		
December 28–January 10.....	7	0.41– 2.36	1.29	....
January 11–24..............	5	0.80– 2.09	1.48	14.7
January 25–February 7.......	5	1.83– 4.00	3.11	110.0
February 8–21..............	4	2.22– 9.13	6.00	92.9
February 22–March 7........	5	2.67– 41.10	12.01	100.1
March 8–21................	5	8.11–114.60	66.99	457.7
March 22–April 4...........	3	116.30–140.49	125.40	87.2
April 5–18.................	1	....	152.75	21.8
		1937–1938		
November 4–17.............	3	0.54– 0.84	0.72	....
November 18–December 1...	1	....	0.17	....
December 2–15.............	2	0.59– 1.01	0.80	....
December 16–29...........	5	0.88– 1.27	1.08	....
December 30–January 12....	3	0.95– 2.03	1.19	37.2
January 13–26..............	5	0.88– 3.93	2.32	53.6
January 27–February 9......	3	8.03– 16.08	12.40	434.9
February 10–23.............	3	8.03– 42.08	19.93	60.7
February 24–March 9........	2	51.32– 70.71	61.02	206.7
March 10–23...............	3	107.00–126.66	117.31	92.2

reached by January 27. The contrast is still more marked if we compare the average volumes for these dates. On February 15, 1937, the average testis volume of four birds was only 6 mm.³; on January 27, 1938, the average testis volume of two birds was 10.21 mm.³ If we interpolate, we find that a testis volume of 6 mm.³ would have been reached in 1938 by January 25, 21 days earlier than in 1937. The rate of testis volume increase averaged only .12 mm.³ per day in 1937; that of an identical segment in 1938 averaged .41 mm.³ per day, a rate over three times as fast as in 1937.

Variation in starting point, on the other hand, is evident as between 1935–36 and three other years, though by less direct and satisfactory evidence. If the time of beginning had been the same in 1935–36 as in 1936–37 (that is, not before January 1), the period from first change to Stage 5 (which we know took not less than 46 days in 1936–37, not less than 30 days in 1934–35, and not less than 25 days in 1937–38) would have been compressed to 12 days in 1935–36. This is very nearly proof by *reductio ad absurdum* that the time of first change is variable.

Thus the Nuttall sparrow fails to show the "eminent regularity" from year to year in date of incidence and rate of recrudescence of the testes which many authors assume to be generally characteristic of the cycle in wild birds (for example, Rowan, 1929, p. 175) and which appears to be assumed by Bissonnette and Chapnick (1930).

FIRST COPULATIONS

The annual differences in testis development are reflected in behavior. The median dates for first copulations of the banded pairs show annual variations in the same direction and to about the same degree as do the dates of Stage 5,

TABLE 8

FIRST COPULATIONS

Year	Number of Pairs	Median Date	Extremes
1936.....................	3	March 4	March 3–5
1934.....................	1	March 11	......
1935.....................	4	March 27	March 19–April 11
1938.....................	3	April 3	April 2–10
1937.....................	6	April 5	March 7–April 14

except for 1938 when, as explained below (p. 74) extraordinary weather conditions on the verge of nesting time prevented copulation from taking place until long after the testes were fully developed, and threw the year out of sequence.

The difference between 1936 and 1937 is substantiated by differences in dates of first copulation of the same pair (XIII) :

YEAR	FIRST COPULATION	DIFFERENCE	MEDIAN DATE FOR POPULATION	DIFFERENCE
1936.............	March 4		March 4	
		38 days		32 days
1937............	April 11		April 5	

FIRST EGGS

The dates for first eggs have been graphed for all five years (fig. 26). The years are arranged not in chronological sequence but from earliest to latest, according to the dates of the reproductive stages. There are wide annual differences between the median dates of first eggs, always in the same direction and of the same order of magnitude (again with the exception of 1938) as those already described for Stage 5 and first copulation.

These differences are substantiated by the fact that I have first egg dates for a single pair for three years and for two other pairs for two consecutive years. These show year-to-year variations, always in the same direction as, and usually of a similar order of magnitude with, the population as a whole.

These females were adults except for female I, which was a first-year bird in 1934.

TABLE 9

FIRST EGGS

Year	Number of Females	Median Date	Extremes
1936...................	7	March 11	March 3–17
1934...................	6	March 22–23	March 21–26
1935...................	6	April 4	April 3–18
1938...................	5	April 7	March 31–April 14
1937...................	9	April 13	March 25–April 20

TABLE 10

FIRST EGGS: CONSECUTIVE YEARS FOR THE SAME PAIRS

Pair	Year	Date of First Egg	Difference	Median Date for Population	Difference
I...........	1934	March 21		March 22–23	
	1935	March 31	10 days	April 4	12–13 days
	1936	March 11	20 days	March 11	24 days
II–IV........	1935	March 24		April 4	
	1936	March 12	12 days	March 11	24 days
XIII........	1936	March 12		March 11	
	1937	April 13	32 days	April 13	31 days

DISCUSSION OF ANNUAL VARIATION

We find that the intervals between the median dates for the attainment of Stage 5 and the median dates for first copulation for 1935, 1936, and 1937 show a maximum variation of only five days, which is very slight, considering that the material is insufficient to give perfect average dates.

YEAR	APPROXIMATE DATE OF STAGE 5	MEDIAN DATE OF COPULATION	INTERVAL
1935...................	February 8	March 27	47 days
1936...................	January 12	March 4	52 days
1937...................	February 15	April 5	49 days

As has been said, the extraordinary late conditions in 1938 upset the behavior phenomena of the latter part of the cycle.

The interval between Stage 5 and the first egg for each year is, again with the exception of 1938, equally consistent, with a maximum variation of only five or six days.

It follows from this that intervals of a similar order of magnitude are to be found from year to year between the time of appearance of the first gonads in Stage 5 and the median dates for first copulations and first eggs.

YEAR	APPROXIMATE DATE OF STAGE 5	MEDIAN DATE OF FIRST EGGS	INTERVAL
1934	January 28	March 22–23	53–54 days
1935	February 8	April 4	55 days
1936	January 12	March 11	59 days
1937	February 15	April 13	57 days

Thus, up to and including 1937, there was every reason to believe that intervals of a similar order of magnitude were to be expected from year to year between the time of appearance of the first gonads in Stage 5 and the median dates for first copulations and first eggs; in other words, that the parts of the reproductive cycle subsequent to Stage 5 were relatively, if not absolutely,

TABLE 11

STAGE 5, FIRST COPULATIONS AND FIRST EGGS

	Pairs of Years		Difference between Pairs of Years	Maximum Variation in Differences
	1934	1935		
Stage 5	Jan. 28	Feb. 8	11 days	
First copulation	....	....	......	2 days
First eggs	Mar. 22–23	Apr. 4	12–13 days	
	1935	1936		
Stage 5	Feb. 8	Jan. 12	27 days	
First copulation	Mar. 27	Mar. 4	24 days	4 days
First eggs	Apr. 4	Mar. 11	23 days	
	1936	1937		
Stage 5	Jan. 12	Feb. 15	34 days	
First copulation	Mar. 4	Apr. 5	33 days	2 days
First eggs	Mar. 11	Apr. 13	32 days	
	1934	1936		
Stage 5	Jan. 28	Jan. 12	16 days	
First copulation	....	....		5 days
First eggs	Mar. 22–23	Mar. 11	11–12 days	
	1934	1937		
Stage 5	Jan. 28	Feb. 15	18 days	
First copulation	....	....		3–4 days
First eggs	Mar. 22–23	Apr. 13	21–22 days	
	1935	1937		
Stage 5	Feb. 8	Feb. 15	7 days	
First copulation	Mar. 27	Apr. 5	9 days	2 days
First eggs	Apr. 4	Apr. 13	9 days	

invariable in duration. Indeed this still appears to be true, in so far as testis development is concerned, for the record-breaking rains and accompanying cold which occurred following January 27 in 1938 had no perceptible effect on the growth curve of the testes, which continued in a manner strictly characteristic of the previous years. When, however, full breeding condition had been attained (March 13), it was accompanied by none of the usual parallel phenomena of behavior, which were delayed about three weeks.

It was interesting to observe that, in spite of the physiological condition which had been reached, the birds were not bound to the wheel of their reproductive cycle to the extent of being forced to breed under conditions impossible for nesting and rearing of young and that, although the hidden physiological development of the male is little, if at all, susceptible, the behavioristic end-results were modifiable by external conditions even at so late a stage. These circumstances will be discussed more fully in connection with the climatic correlates of the cycle.

CONCLUSION

Since, then, the annual differences in time of occurrence are each year alike for these later stages and events in the reproductive cycle, it follows that some common cause must exist, of external origin and annual variation. Since in four out of five years the intervals from Stage 5 to copulation and egg laying have been shown to be of the same length but to fall between different dates each year, it follows that the factors determining the annual variations must complete their operation before the attainment of Stage 5—that is, before the testes reach a volume of 5 mm.[3]

The fact that the time of first change is variable indicates that at least part of the period when the hypothetical external factors are operative must come before the first change. On the other hand, the fact that the length of the interval from first change to Stage 5 is variable indicates that at least part of the period when the external factors are operative must come after first change.

I have to depend upon the meteorological correlations, as treated in the ensuing section, to find out both how long before the time of first change the external factors may begin to influence the time of commencement and the early rate and also how long after first change the influence of such factors may continue to be operative. It cannot apparently continue, in any perceptible degree, after the occurrence of Stage 5 unless, as in 1938, abnormal conditions render the concomitant behavior physically impossible. That is hardly, however, the type of modifying agency of which I am in search.

* * * * * * *

Literature Cited

BISSONNETTE, T. H., and CHAPNICK, M. H.
    1930. Studies on the sexual cycle in birds. II. The normal progressive changes in the testis from November to May in the European starling (*Sturnus vulgaris*), an introduced, non-migratory bird. Am. Jour. Anat., 45: 307–343, 7 pls.

ROWAN, W.
    1929. Experiments in bird migration. I. Manipulation of the reproductive cycle. Seasonal histological changes in the gonads. Proc. Boston Soc. Nat. Hist., 39: 151–208, pls. 22–32.

$$8$$

Reprinted from *Evolution*, **2**(2), 95–110 (1948)

NATURAL SELECTION AND FAMILY SIZE IN THE STARLING

David Lack

Edward Grey Institute, Dept. Zoological Field Studies, Oxford

Received February 16, 1948

Introduction

It is extremely difficult to measure natural selection in the field because, when one variant is at a marked disadvantage compared with another, it is usually too rare to provide adequate data, while when both variants are common, their selective value is closely similar (except where a heterozygote is at an advantage compared with either homozygote). It might seem especially rash to attempt to study such a problem in the case of clutch-size in birds, when the genetic factors controlling it are unknown. However, clutch-size possesses one characteristic which facilitates such an investigation, due to the fact that it directly affects the number of offspring. The point in question can be made clear by a hypothetical (and over-simplified) example. Suppose that a species consists of two hereditary types, one laying 5 and the other 6 eggs in a clutch. If both types are common, both presumably have an almost equal selective value, and each gives rise to about the same eventual number of descendants, which means that their offspring should survive in the ratio of 6 : 5, a difference which might be large enough to demonstrate in the field.

A preliminary review of the clutch-size problem has been published elsewhere (Lack, 1946, 1947–8). It was there pointed out that the limitation of clutch-size to a comparatively small number of eggs cannot be adequately explained on purely physiological grounds. The view that it is adjusted to the mortality of the species is also untenable. The alternative hypothesis was put forward that it is ultimately selected by the number of young which the parents can raise, the latter being determined by the available food supply. This view, so acceptable *a priori* to the population-geneticist and so difficult, apparently, for the traditional ecologist, was supported by much circumstantial evidence, chiefly relating to seasonal and regional trends of variation in clutch-size.

In the present paper, an attempt is made to provide direct evidence, to test whether, in fact, the young from broods of large size are at a disadvantage compared with those from broods of small size. It might be expected that, for broods above the average size, proportionate mortality among the young would rise as brood-size increased. When this is so, a point is quickly reached where an increase in the number of eggs is offset by the increased mortality, so that there is no increase in the number of young raised. One might reasonably hope to find evidence for this turning point in nature. However, one could not expect to go beyond it, to the point when productivity actually declines with increasing clutch-size, as deleterious mutants tend to be eliminated, hence individuals with disadvantageously large clutches would tend to be very rare. It should be added that even the turning point might be obscured if there were extensive non-genetic modifications in clutch-size, as shown in the parallel case of litter-size in mammals (Lack, in press).

THE RAW DATA

The European Starling (*Sturnus vulgaris*) was selected for study, as the data were more extensive and less complex than for other species, which will be treated in later papers. The data used here come from: (i) Institute for Plantenziektenkundigen Dienst, Wageningen, Holland, for clutch-size and nesting success; (ii) Vogelwarte Sempach, Switzerland, for brood-size and post-fledging survival; (iii) Edward Grey Institute, Oxford, for clutch-size, in MSS by R. Carrick, A. Whitaker, late F. C. R. Jourdain, and the British Trust for Ornithology Hatching and Fledgling Enquiry; (iv) *British Birds* and British Trust for Ornithology Marking Scheme, for brood-size and post-fledging survival. The writer is particularly indebted to Dr. Kluijver in Holland and Dr. Schifferli in Switzerland for generously putting their extensive data at his disposal, when it was found that the British data were insufficient.

CLUTCH-SIZE

In Britain, Holland and Switzerland, the Starling lays its first clutch in April or early May. A smaller number of layings occur in late May and in June, these being due partly to one-year-old birds breeding for the first time, partly to older birds whose first layings were destroyed, and partly to older birds which have already raised one brood in the season. In this paper, all clutches laid in April, and

TABLE 1. *Clutch-size*

Clutch-size	Number of clutches in		
	N. England	Holland	
	April	April	June
1		1	2
2	1	10	5
3	7	41	28
4	29	273	80
5	47	669	64
6	12	447	10
7	6	131	4
8	3	17	
9		1	
10		2	
Total	105	1592	193
Average	4.9	5.2	4.3

Note. Average clutch-size for May layings in Holland was 4.9.

all broods with young old enough to band in May, have been classified as "early" broods, and they have been separated from later layings, which are on the average somewhat smaller in size. The latter point is shown for clutch-size in Holland in table 1, and for brood-size in all three regions in table 2. The differences are statistically significant. In general, 5 is the commonest size of an early laying, and 4 of a late clutch.

The Starling also shows a regional variation in clutch-size. As in many other passerine species, average clutch-size is somewhat smaller in England than in central Europe at the same latitude (Lack, 1947–8). This is shown for clutch-size in England compared with Holland in table 1, and for brood-size in England compared with Holland and Switzerland in table 2. The average differences are rather smaller than those associated with the season of the year, but are statistically significant.

PARENTAL FEEDING RATE

The hypothesis of this paper rests on the assumption that the parent Starlings bring less food to each nestling in a large than a small brood. This was found by

TABLE 2. *Brood-size*

Brood-size	Number of early broods in			Number of late broods in		
	N. W. England	Holland	Switzerland	N. W. England	Holland	Switzerland
1	?	30	65	?	34	44
2	66	98	164	31	77	96
3	166	189	426	66	145	254
4	311	395	989	57	194	391
5	295	425	1235	25	124	285
6	69	196	526	4	29	73
7	6	41	93		14	7.
8		3	15			
9			2			
10			1			
Total	913	1377	3516	183	617	1150
Average	4.1	4.4	4.5	3.3	3.7	4.0

Notes.   (i) The English average has been adjusted to allow for the fact that broods of 1 are not recorded, by assuming that the ratio of broods of 1 to broods of 2 is the same as in Switzerland and Holland.

(ii) The average for all 2186 English broods banded in May was 3.9 and for 538 later broods it was 3.0—these figures are rather smaller than those for N.W. England only.

(iii) The Swiss averages are: for May: 4.5, first half of June: 4.1, second half of June: 3.8 and July: 3.5, based on 3516, 439, 580 and 131 broods respectively.

TABLE 3. *Feeding frequency and brood-size (after Kluijver)*

(i) 1930

Days after hatching	Number of feeding visits per day for	
	brood of 4	brood of 6
3	208	227
4	247	235
5	251	306
6	303	306
7	345	476
8	300	525
9	388	430
10	358	448
11	240	472
12	401	483
13	395	485
14	358	440
15	366	408
16	421	366
17	378	393
Total in 15 days	4959	6000
Total per nestling per day	83	67

(ii) 1931

	brood of 2	brood of 6
Total of feeds per day	224	390
Feeds per day per young	112	65

Note. The first set of records is adapted from Kluijver (1933), table IV, p. 57, and the second set from his table III, p. 55.

Kluijver (1933) to hold in the Starling, as summarised in table 3, but further data on this point are desirable. However, a similar result was obtained for most other species in which this problem has been investigated by Moreau (1947). To summarise these data: with a brood of larger size, the parents increase the number of their feeding visits, but the increase is not sufficient to offset the larger number of young, so that each nestling is fed less often in a large than in a small brood.

MORTALITY IN THE NEST

Owing to the above fact, it was expected that the young would show, on the average, a proportionately higher mortality in large than small broods. The nesting data for the Starling in Britain did not suggest this, but were too few for analysis. Fortunately, extensive nest records are available from Holland since 1922 in the form of the number of eggs laid and the number of young alive on the observer's last visit. These are analysed in table 4, and show decisively that the above expectation is not realised.

TABLE 4. *Nesting success (egg–fledgling)*
in Holland

Clutch-size	Number laid		No. fledged	Per cent fledged	
	All broods	All broods except total failures	All broods	All broods	All broods except total failures

1. April layings

Clutch-size	All broods	All broods except total failures	All broods	All broods	All broods except total failures
3	126	105	86	68	82
4	956	864	727	76	84
5	2975	2770	2324	78	84
6	2298	2184	1784	78	82
7	798	756	620	78	82
8	120	112	87	73	78
others	55	53	34	—	—
Total	7328	6844	5662	77	83

2. Late layings

Clutch-size	All broods	All broods except total failures	All broods	All broods	All broods except total failures
3	162	144	125	77	87
4	888	800	642	72	80
5	1345	1210	949	71	78
6	516	486	340	66	70
7	245	203	159	65	78
others	73	71	46	—	—
Total	3229	2914	2261	70	78

Notes. (i) "Total failure" refers to broods in which no nestling fledged. It is better to omit losses of entire broods, as these are due mainly to predation and are erratic.
(ii) Number "fledged" means number of young alive in nest on observer's last visit; each nest was visited at least weekly.

As interest is in mortality due to malnutrition, losses of entire broods are better omitted from analysis, since they are due mainly to predation, including human destruction. The right-hand column of table 4 shows that, in early layings, the proportion of eggs which give rise to fledged young is approximately the same for clutches of 3, 4, 5, 6, 7 and 8 eggs. For late layings, the answer is similar; (except for an apparent lower success for clutches of 6, which is attributable merely to random sampling, as a similar drop is not found for clutches of 7). When the losses of entire broods are included in the analysis (fifth column of table 4), the results for early layings again show no variation with clutch-size, and the apparent variation in late layings is doubtless due to random sampling, the losses of entire broods being very erratic in their incidence from place to place and year to year. Nestling mortality was found to be independent of brood-size in the four other small passerine species in which this point has been studied (Lack, 1946, and unpublished).

NESTLING WEIGHT

The fact that nestling mortality is similar for broods of different sizes was first discovered for the Robin (*Erithacus rubecula*) (Lack, 1946). To investigate the problem further, nestling Robins were weighed daily from hatching to flying. This unexpectedly revealed that there were extremely marked individual variations in the weight of the nestlings (Lack and Silva, in press; Lees, in press). Lees (in litt.) has found similar large variations in the other small passerine species which he has weighed. In such species, the nestling can remain alive, and develop its feathers normally, although badly under weight, and the chief effect of undernourishment on a nestling is not death in the nest, but a reduced weight when it leaves the nest. The only comparative data so far available on the weights of nestling Starlings in broods of different sizes were obtained in 1947 by R. Carrick

TABLE 5. *Nestling weights of Starling*
Obtained in England in May 1947, by R. Carrick

Age in days	Average weight per nestling in grams in		
	Brood of 2	Brood of 5	Brood of 7
4	28.8	—	21.4
6	48.3	45.1	38.2
15	88.0	77.6	71.4

Note. The limits of weight on day 15 were for brood of 2: 87.5–88.5; for brood of 5: 72.5–83.0; and for brood of 7: 66.0–77.0.

for three broods of different sizes, all raised by fully adult parents, all in the same locality, and all starting incubation on the same date. The results, set out in table 5, fit the view that the average nestling weight decreases as brood-size increases, but many more data are needed before this point can be considered established.

Post-Fledging Survival in Switzerland

Those young which are below normal weight on leaving the nest might well have a reduced chance of subsequent survival. Moreover, any differences in their fledging weight associated with size of brood would tend to be intensified in the period immediately after leaving the nest, when the young are still dependent on their parents for a time; hence the greater the number of young, the less often each will tend to be fed. This means that the possible adverse effects of a large brood-size on survival can be revealed only if the data are extended to cover the period after leaving the nest.

Soon after leaving the nest, young Starlings join with their parents to form large flocks, and it would seem impossible to study the further fate of individual broods in the wild. However, the essential data can be obtained indirectly from bird-banding, which was originally started to study migration, but has since thrown light on many other points of biological interest. For the Starling, the Swiss data are particularly extensive, and most observers recorded the number of young separately for each brood banded.

Should there be, in the period shortly after leaving the nest, a heavier mortality among the young from large than from small broods, then proportionately more individuals from small than large broods should survive to maturity. The proportion of young Starlings from each brood-size alive at the time of leaving the nest is easily (if laboriously) obtained by adding up the numbers of all banded young, separately for each brood-size. The pro-

portion of successful survivors from each brood-size is revealed by adding up the number of individuals from each brood-size recovered after they have attained a certain age. The minimum age for such 'successful survival' was provisionally fixed at three months after leaving the nest, and the results indicate that, in fact, the main mortality due to any undernourishment in the nest has occurred before this age is reached. N. B. It does not matter whether the birds recovered after reaching this age were found dead or alive since the point of interest is whether they successfully attained this age. In fact, most of the Swiss birds were recovered dead, the Dutch alive, while the British sample included many of both types.

The brood-size of a recovered Starling is, of course, found by turning up the band-number concerned in the original banding data. Only about 2% of the banded nestling Starlings are later recovered. The rest die undetected. The assumption is made here that those recovered more than three months after they leave the nest are a random sample of the population remaining alive. (The individuals recovered within the first three months come in a different category, and will be discussed later.)

The results set out in table 6 are highly suggestive. In early broods, 5 is the commonest size of family. The recovery-rate for birds more than three months out of the nest is about the same for those coming from broods of 3, 4 or 5 young, in each case being around 2%. On the other hand, as soon as the brood-size exceeds 5, the recovery-rate falls. For individuals from broods of 6 it is only 1.7%, and for those from broods of 7–8 it is only 1.4%. If, now, the proportion of individuals recovered is multiplied by the original number of young in the brood (as in the right-hand column of table 6), it is seen that, on the average, a brood of 5, 6 or 7 young gives rise to the same number of recoveries, about 0.1 per brood. If these recoveries are representative of the population,

TABLE 6. *Post-fledging survival in relation to brood-size.*
For Swiss Starlings banded as young

Brood-size	No. of broods banded	No. of young banded	Young recovered more than 3 months after fledging		Relative productivity (previous column × brood-size)
			Number	Percentage	
colspan-1. Early broods (banded in May)				
1	65	65	0	—	—
2	164	328	6	1.8	3.7
3	426	1278	26	2.0	6.1
4	989	3956	82	2.1	8.3
5	1235	6175	128	2.1	10.4
6	526	3156	53	1.7	10.1
7	93	651	10	1.5 ⎫	10.2
8	15	120	1	0.8 ⎭	
9	2	18	0	—	—
10	1	10	0	—	—
Total	3516	15757	306	1.94	
2. Late broods (June, July)				
1	44	44	1 ⎫		
2	96	192	8 ⎬	2.3 ⎫	5.8
3	254	762	14 ⎭		
4	391	1564	35	2.2	8.9
5	285	1425	25	1.8	8.8
6	73	438	6	1.4	8.2
7	7	49	—	—	—
Total	1150	4474	89	1.99	

this must mean that an increase in brood-size between 5 and 7 young is approximately counterbalanced by an increase in proportionate mortality among the young, taking place in the period shortly after they leave the nest. (Broods of 9 or 10 young are too scarce for a reliable estimate of their mortality. This scarcity suggests that they are at a selective disadvantage, with productivity below that of broods of smaller size, but this view could be tested only by the experimental addition of young to nests, so as to make up a large series of broods of 9 and 10 young artificially.)

The above data fit the hypothesis advanced earlier, viz. that when brood-size rises above the average, mortality also rises, so that productivity is not increased.

Unfortunately, the proportion of birds recovered is so small that, despite the large numbers banded, the totals recovered are too small for the apparent differences in recovery-rate to be statistically significant. One could not expect bigger differences in the recovery-rates than those actually obtained, since, according to theory, the productivity per brood should be about equal for a brood of 5, 6 or 7 young, as in fact it is. (Had the young from broods of 6 or 7 survived less well, broods of 6 and 7 would have been less productive than broods of 5, and so would presumably have been extremely scarce.) Hence the lack of a significant result is due to the small number of recoveries, and not to the size of the difference found. For a significant result, about twice the present number of recoveries is needed, on the assumption that they are recovered in the same proportions as in table 6. The Swiss are now resuming the intensive banding of Starlings, which was stopped in 1937, but it may be another decade before sufficient further data have been obtained.

Turning to the second half of table 6, in late broods 4 is the commonest size of family, and the recovery-rate falls from 2.2% in broods of 4, to 1.8% in broods of 5, and to about 1.4% in broods of 6. Once again, if the proportion of individuals recovered is multiplied by the number of young in the brood, it is seen that broods of 4, 5 and 6 young are about equally productive. Hence the results from late broods corroborate those from early broods. Unfortunately, the data are again too few for the result to be statistically significant.

In both early and late broods, the decline in survival-rate appears to commence as soon as brood-size rises above the commonest size found in nature (as theory requires), and it is interesting that this brood-size is one smaller in late than in early broods. Table 6 shows that the young from late broods have a similar *total* recovery-rate to those from early broods, i.e., 1.99% cf. 1.94%, but this equality is achieved only because the average size of late broods is smaller by one. For broods of 5 young or more, the survival-rate is lower in late than early broods. This strongly suggests that the reduction in average clutch-size in late as compared with early layings is an advantageous adaptation, and it may be suggested that the food supply available for nestling Starlings is smaller late than early in the season. Kluijver (1933) showed that, though the food is highly variable, the young of early broods are fed mainly on dipterous larvae, those of late broods mainly on Orthoptera. The latter possibly provide a less available or less nutritious diet than the former, but this point requires investigation, as does the regularity of the above difference in diet.

An independent check on the results in table 6 is provided by those banded Starlings found dead within three months of leaving the nest. As those in their third month might already have survived the period in which death occurs through previous undernourishment as a nestling, it is better to restrict analysis to the first two months after leaving the nest. As before, only a small proportion of the banded young which die in this period are later found and reported. These dead birds might be a random sample of the population alive at the time, in which case one would expect the recoveries to be in roughly the same proportions as the birds originally banded. Alternatively, the sample might include some of those juveniles dying as a result of undernourishment, in which case one would ex-

TABLE 7. *Recoveries of birds dead within 2 months of leaving nest*

Brood-size	For Swiss Starlings banded in May			
	No. banded	No. found dead in		Per cent found dead less than 2 months old
		1st month	2nd month	
1	65	—	—	
2	328	—	—	
3	1278	—	2	0.21
4	3956	2	8	
5	6175	14	8	0.35
6	3156	9	5	
7	651	1	2	0.43
8–10	148	—	—	
Total	15757	26	25	0.32

Notes. (i) The differences in recovery-rate are not significant.

(ii) Recovered in 3rd month: 2 from broods of 4, 6 from broods of 5, 1 from broods of 7.

(iii) Recoveries from late broods: in first two months, 3 from broods of 2, 5 from broods of 3, 4 from broods of 4 and 4 from broods of 5: in third month, 1 from brood of 3, 1 from brood of 4, 2 from broods of 5, and 1 from brood of 6.

pect to find a relatively high proportion of individuals from broods of sizes above the average. On neither view, however, would one expect to find a relatively high proportion of dead individuals from broods of small size, such as occurred in table 6.

The data for early broods in table 7 indicate that, during the first two months after leaving the nest, there is a higher recovery-rate of dead young from broods of large than small size, i.e., the reverse of table 6. This therefore supports the view that the young from broods of larger size are at a comparative disadvantage in the period soon after leaving the nest. Unfortunately, the recoveries are again too few to give a result which is statistically significant. (The data for late broods are too few to be worth analysing, but are added as a footnote to table 7.)

BRITISH AND DUTCH DATA

Many fewer Starlings have been banded, and a smaller proportion recovered, in Britain than in Switzerland. The results for those recovered after they are three months old are set out in table 8, and, in general, support those from Switzerland in table 6, though once again they are too few to show whether the apparent differences in recovery-rate are statistically significant. For early broods in Britain, 4 is the commonest size of family, and the recovery-rate for individuals from broods of 4 is 1.9%, whereas for individuals from broods of 5–7 it falls to only 1.6%.* In late broods, the commonest size of family is 3, and the re-

---

* The apparent higher survival rate for British Starlings from broods of 6–7 than from broods of 5 is presumably due to random variation in a small sample. It might be added that in 1933 one man, H. J. Moon, banded 11 broods of 6 young each, from which 5 individuals were later recovered, a very high proportion. For all other years and all other brood-sizes, Moon's recovery-rates were normal. Omitting his 1933 data, the proportion recovered from British broods of 6 is only 1.6%, a much more reasonable figure.

TABLE 8.   *Recoveries of British banded Starlings*

Brood-size	No. of broods banded	No. of young banded	Young recovered more than 3 months old	
			Number	Percentage
1. Early broods (May)				
2	255	510	7 ⎱	1.5
3	455	1365	22 ⎰	
4	734	2936	55	1.9
5	596	2980	44	1.5 ⎱
6	133	798	17 ⎱	2.1 ⎰ 1.6
7	13	91	2 ⎰	
Total	2186	8680	147	1.7
2. Late broods (June, July)				
2	120	240	3	c.1.3
3	191	573	19	3.3
4	165	660	14	2.1
5	54	270	1	c.0.3
6	7	42	—	—
7	1	7	—	—
Total	538	1792	37	2.1

Note. The apparent higher recovery rate from late than early broods is probably due to chance, as it mainly results from one worker's trapped recoveries. When his recoveries are omitted, the proportions recovered are 1.3% for May and 1.5% for late broods.

covery-rate is smaller from broods of 4 and 5 young than from broods of 3.

The recovery-rate for Starlings in their first two months after leaving the nest is much higher in Britain than in Switzerland; hence despite the smaller totals banded, there are enough British recoveries from early broods for a significant result. The data in table 9 show that, in the first two months after leaving the nest, the recovery-rate is significantly higher for dead young from broods above the average in size than for dead young from broods of smaller size. This confirms the results in table 7, and supports the view that the young from broods of above average size experience a heavier mortality than those from broods of smaller size,

TABLE 9. *Recoveries of birds dead within 2 months of leaving nest. For British Starlings banded in May*

Brood-size	No. banded	No. found dead in		Per cent found dead less than 2 months old
		1st month	2nd month	
2	510	3	1 ⎫	
3	1365	3	2 ⎬	0.42
4	2936	9	2	
5	2980	18	6 ⎭	
6	798	3	2 ⎫	0.78
7	91	1	— ⎭	
Total	8680	37	13	0.58

Notes. (i) The difference in recovery-rate for smaller compared with larger broods is significant, $\chi^2 = 4.8$.

(ii) Recoveries in 3rd month: 1 from brood of 2, 1 from brood of 3, 4 from broods of 4, and 2 from broods of 5.

(iii) Recoveries from late broods: in first 2 months, 4 from broods of 2, 6 from broods of 3, 6 from broods of 4, and 5 from broods of 5: in third month, none.

during the first two months after they leave the nest.

A few data are also available from Holland, all obtained from broods banded in the last week of May, 1932 and 1933, by H. N. Kluijver. The high rate of recovery is due to the fact that most recoveries were of individuals trapped alive at nest-sites in a later year. In Holland, as in Switzerland, 5 is the commonest brood-size, and the data in Table 10 show that successful survival to an age of at least 3 months is proportionately much higher for young from broods of 5 or less, than it is for young from broods of more than 5 young. The difference is statistically significant.

To sum up the results in tables 6–10, in Switzerland, Britain and Holland, the differences in survival-rate among the young from broods of different sizes support the hypothesis advanced *a priori,* although in only two cases (tables 9, 10) are the differences statistically significant. However, there are altogether seven in-

TABLE 10. *Recoveries of Dutch Starlings banded in May*

Brood-size	No. of broods banded	No. of young banded	Young recovered more than 3 months old	
			Number	Percentage
2	3	6	— ⎫	
3	5	15	2 ⎪	
4	16	64	6 ⎬	10
5	45	225	24 ⎭	
6	25	150	5 ⎫	
7	3	21	1 ⎬	3
8	1	8	— ⎭	

Note. The difference in recovery-rate for smaller as compared with larger broods is significant, $\chi^2 = 6.7$.

dependent analyses, five for young attaining an age of at least three months (viz. Swiss early and Swiss late broods, British early and British late broods, Dutch early broods), and two others for young dying in the first two months (viz. Swiss early and British early broods). All seven analyses give an answer in accordance with the view that, for broods above the average in size, juvenile mortality is higher; hence, taken together, they strongly suggest that this view is correct.

THE REGIONAL DIFFERENCE

It will be noted that in early broods the commonest number of young is 5 in Switzerland but only 4 in Britain, while in late broods it is 4 in Switzerland but only 3 in Britain. Lack (1947) showed that there was a widespread tendency for the average clutch-size of passerine birds to be larger in Central Europe than in Britain at the same latitude, and argued that a general trend suggests an adaptation. The data in tables 6 and 8 fit in with this suggestion, since, in early broods, the survival-rate begins to decrease with a brood of more than 5 young in Switzerland, but with a brood of more than 4 young in Britain, while in late broods the decrease commences with a brood of more than 4 young in Switzerland but with a brood of more than 3 young in Britain.

This presumably means that more food is available for young Starlings in Switzerland than in Britain, but there are as yet no data on this point. In the parallel case of the Swift (*Apus apus*) in the two regions, there is suggestive evidence that not only average clutch-size but also food supply is greater in Switzerland than in Britain (Lack and Arn, 1947).

Lack (1947) also showed a general tendency in passerine birds for average clutch-size to be higher in the north than the south of the European range. There are not yet sufficient data on this point for the Starling (but see note to table 2).

ADAPTIVE MODIFICATIONS

Although clutch-size might be regarded as a form of polymorphism, the discontinuities are due to the impossibility of laying fractions of an egg, and the hereditary control is probably influenced by a number of genes, as in the case of other quantitative characters, such as body-size. The frequency distribution of clutch-sizes suggests a roughly normal curve.

A further complication is the existence of non-genetic, adaptive modifications. In birds, three important advantageous modifications in clutch-size have been demonstrated, connected with (i) season of year, (ii) age of bird and (iii) differences in feeding conditions from one year to another (Lack, 1947).

(i) In the Starling, the seasonal variation has already been discussed, and there is suggestive evidence that the smaller size of late broods is adaptive. Similarly in the Alpine Swift (*Apus melba*), late clutches are smaller than early clutches, and it has been shown that the survival-rate from broods of large size is smaller in late than it is in early broods (Lack and Arn, 1947).

(ii) Kluijver (1933, 1935) has shown that Starlings breeding for the first time, when about a year old, tend to lay clutches of smaller average size than do older parents. A similar tendency is found in other passerine and near-passerine species (Lack, 1947; Lack and Arn, 1947),

and it is probably general. It seems reasonable to consider this as another advantageous modification, which has become part of the hereditary constitution of the species through natural selection, due to the fact that parents breeding for the first time are less efficient at collecting food for nestlings than are experienced parents, so that the optimum brood-size is somewhat smaller for young than for experienced parents. This supposition has not as yet been tested for any species, and should be investigated.

This second modification means that, unless first-year parents can be separated from the rest, the results of an analysis such as that in table 6 will be somewhat obscured. For some of the broods of small size will be those raised by first-year and inexperienced parents, and their young may have a smaller chance of subsequent survival than have the young from broods of rather larger size raised by experienced parents. In the Starling, however, unlike some other species, Kluijver (loc. cit.) has shown that many of the first-year birds breed later in the season than the older birds, so that the first broods of the latter are usually separated from the former by date alone. Probably most of the early (i.e., May) broods whose survival was analysed in tables 6 and 8 were raised by experienced parents.

(iii) A few passerine species show marked, and others small, differences in average clutch-size from one year to another. A variation of this nature has presumably become a characteristic of the species only when, on balance, it is advantageous. The advantage is obvious in the case of African passerine species which have larger clutches when the rains are heavy (Moreau, 1944), and in the case of rodent predators which have larger clutches when there is a vole or lemming plague (Lack, 1947). There is also suggestive evidence that, in the Alpine Swift, the years of larger average clutch-size are also the years with high nesting success among broods of large size (Lack and Arn, 1947).

Kluijver's data for the April layings of the Starling in Holland are analysed in table 11 for each year separately. The low averages for 1922 and 1925 are possibly doubtful, as in a co-operative study the accuracy of the observers tends to increase with time. However, even omitting the early years, average clutch-size varies between 5.06 in 1931 and 5.55 in 1930, and an analysis of variance shows that the differences in mean are greater than can be attributed to chance alone. There therefore seems to be some annual variation in the size of the first layings of the Starling. The laying time varies from one year to another, associated with the general earliness or lateness of the spring, but the data in table 11 show that this is not the major cause of the clutch-size variation. For the present, the latter cannot be explained, nor is it known whether it is an advantageous mod-

ification. It should be added that it is much smaller in extent than in some other species. Extensive adaptive modification have a highly obscuring effect on the type of analysis undertaken in table 6, since small clutches tend to be laid when conditions are unfavourable for the survival of the young, and large clutches when they are favourable. In an extreme case, with much phenotypic and no hereditary variation in clutch-size, one would expect proportionate survival to be very similar from clutches of all sizes. This point has been discussed in mammals (Lack, in press) and will be discussed later for other species.

Table 11 shows that each individual Starling does not necessarily lay a clutch of the same size in each year, and this is confirmed by data from H. N. Kluijver on the layings of marked individuals in successive years, set out in Appendix I. The data are restricted to first broods, late broods being omitted. Layings of individuals known to be only one year old are also omitted, though some may have been included inadvertently, as most of the birds were of unknown age when banded. Analysis shows that the variance between clutches of the same individual is smaller than the variance between clutches of all individuals, the difference being statistically significant at a probability of .01. Hence, despite some annual variation, each individual has a tendency to lay a clutch of similar size each year (after the first). This accords with the view that clutch-size has a hereditary basis. Data given by Nice (1937) for the Song Sparrow (*Melospiza melodia*), and by Weitnauer (1947) and Lack and Arn (1947) for Swifts (*Apus* spp.), suggest a similar conclusion for these species, but more detailed observations are needed.

TABLE 11. *Annual differences in average clutch-size*

In Holland

Year	No. of nests	Mean clutch-size	S.E.	Mean date of laying of first broods
1922	29	4.79	0.19	—
1923	51	5.29	0.14	22.IV
1924	52	5.37	0.14	28.IV
1925	82	4.76	0.11	20.IV
1928	85	5.28	0.11	30.IV
1929	12	5.42	0.29	4.V
1930	106	5.55	0.09	22.IV
1931	196	5.06	0.07	26.IV
1932	116	5.41	0.09	29.IV
1933	245	5.09	0.06	25.IV
1934	228	5.27	0.07	23.IV
1935	138	5.39	0.09	24.IV
1936	239	5.39	0.06	25.IV
1937 & 42	13	5.00	0.28	—
All years	1592	5.24		

Notes. (i) Variance ratio = 4.69/1.00 at 13/1578 degrees of freedom, which is significant. The result is still significant if the first four years are omitted from consideration.
(ii) The mean date of laying was calculated for that of all clutches laid before May 10th each year.

RECOVERIES FROM THE SAME BROOD

Table 12 shows the number of cases in which two members of the same brood were later recovered, this being confined to cases in which both individuals attained

TABLE 12. *Recoveries from same brood*

Brood-size	No. of broods banded	No. of recoveries over 3 months old	No. of cases of 2 in same brood
		Switzerland	
2	260	14	1
3	680	40	—
4	1380	117	7
5	1520	153	10
6	599	59	3
7	100	10	—
8–10	18	1	—
		Britain	
2	375	7	—
3	646	28	1
4	899	39	2
5	650	29	2
6	140	12	—
7	14	1	—
		Holland	
2–3	8	2	1
4	16	6	—
5	45	24	4
6	25	5	1
7–8	4	1	—

Notes. (i) For Britain, recoveries of birds trapped are omitted.
(ii) In Switzerland, one pair was due to trapping in the same place.
(iii) In Switzerland, there were 6 further cases (omitted in table 12) in which one of the pair was recovered in its 2nd month of life.

an age of at least three months. The number of such pairs recovered is higher than would be expected by chance, and the departure from random is significant. This result is not attributable merely to a tendency for birds in the same brood to stay in the same flock, as the two recoveries were usually from widely separated localities, for instance one from Algeria in winter and the other from Switzerland in summer. A similar tendency is also apparent among British Starlings. To test the latter, all trapped recoveries were omitted, as these come from a few restricted localities, which increases the chance of finding two in the same brood.

The above result could be explained through a tendency for either all or none of the members of each brood to survive the dangerous first three months of life. Such a tendency might occur if the parents fed each nestling in the brood about equally, so that either all or none of them tended to be undernourished at the time of leaving the nest. This point could be studied further by a detailed analysis of nestling weight in the Starling. It will be shown later that, in Song Thrush (*Turdus ericetorum*) and Blackbird (*T. merula*), the frequency with which two individuals from the same brood are recovered is about what would be expected by chance, and there is no tendency of the type shown in table 12. In the Great Tit the same tendency is found as in the Starling, but in this species nearly all the recoveries are due to trapping in restricted localities, so that this result might merely reflect the tendency for individuals from the same brood to stay in the same flock.

POPULATION TURN-OVER

It is seen from table 2 that the Starling has a lower reproductive rate in Britain than in Switzerland. While in both countries the Starling has been increasing in numbers in the last fifty years, the increase has probably been sufficiently slow to mean that the average number of young born each year is very nearly equal to the average annual mortality. The higher reproductive rate of Swiss than of British Starlings should therefore mean that the Swiss birds have a higher annual mortality than the British. This point can be tested by analysing the age at death of each banded bird later recovered, as set out in table 13. This indicates that the mortality in the first year is 73% for Swiss but only 66% for British Starlings, while in later years it is 62% for Swiss and only 55% for British birds. These differences are statistically significant, and therefore support the above hypothesis. However,

TABLE 13. *Annual mortality of banded Starlings in Switzerland and Britain*

Year	Number found dead in	
	Switzerland	Britain
1st	433	261
2nd	97	75
3rd	45	26
4th	8	23
5th	7	6
6th	1	3
7th	—	—
8th	—	1
12th	1	—
Total	592	395
Mortality in first year	73%	66%
Average annual mortality after first year	62%	55%

Notes. (i) For Starlings banded as young, the first year is reckoned from date of ringing to May 31st of following year, and subsequent years from June 1st–May 31st.

(ii) Average annual mortality is given by $\dfrac{D2 + D3 + D4 \cdots}{D2 + 2D3 + 3D4 \cdots}$ where D2, D3, D4 = no. of young found dead in 2nd, 3rd, 4th, $\cdots$ years after ringing.

(iii) The mortality data are analysed in more detail elsewhere (Lack and Schifferli, in press).

it is possible that the small percentage of recovered Starlings is slightly biassed as to age, as discussed in greater detail by Lack and Schifferli (in press). It is scarcely necessary to add that, in a balanced population, a higher mortality-rate is an inevitable result of a higher reproductive rate. The types of danger to which Swiss and British Starlings are respectively subjected are a purely secondary consideration.

CONCLUSION

The data in this paper suggest that, as brood-size increases, the mortality among the young also increases, in such a way that the commonest brood-size found in nature is also the size with optimum productivity. This result appears to hold under the differing seasonal conditions of first and late broods, and the differing regional conditions of Switzerland, Holland and Britain. The facts therefore support the view that clutch-size is determined by natural selection acting through the survival of the young.

However, while the position may be regarded as satisfactory thus far, many difficulties remain. The most important gap in knowledge, and one that will be difficult to fill, is the way in which clutch-size is inherited. Knowledge of this is essential before a really adequate formulation can be achieved. Another difficulty in the present analysis is the persistence of many clutches of 4 in early layings of the Swiss Starling. The survival of individual young from broods of 4 appears to be no better than the survival of individual young from broods of 5; hence the productivity from broods of 4 is lower than that from broods of 5 (see table 6). This being so, one would expect the tendency to lay clutches of 4 to be eliminated by natural selection, whereas in fact it is common. One possibility that may be kept in mind is that it may be advantageous to have a somewhat variable clutch-size, particularly as feeding conditions are not identical from one year to another.

Finally it should be noted that the Starling was selected for this first paper because the results are clear-cut. This is attributable to the possibility that, in this species, the only important non-hereditary modifications in clutch-size are largely sorted out by date of laying. The survival data for other species, treated in later papers, are not so clear-cut. The present data should also be compared with those already published for the Alpine Swift (*Apus melba*), in which the basic problem is the same, and a parallel answer has been obtained. The details, however, are very different, as in Swifts undernourishment of the nestling causes death in the nest, instead of afterwards (Lack and Arn, 1947).

SUMMARY

1. In the European Starling (*Sturnus vulgaris*) the clutch varies from 1–10 eggs. The average size of clutch and brood is smaller in late than early broods, and smaller in England than in Switzerland (tables 1 and 2).

2. Each nestling is fed less often in a brood of large than one of small size (table 3), and the few data on nestling weights support this view (table 5).

3. The percentage of eggs which produce fledged young is closely similar for broods of all the ordinary sizes found in nature (table 4).

4. Among Starlings which survive at least three months after leaving the nest, the recovery-rate is lower for young from broods of above average size than it is for young from broods of average size or smaller (tables 6, 8 and 10). Correspondingly, for Starlings dying within two months of leaving the nest, the recovery rate is higher for young from broods above average size than for young from broods of smaller size (tables 7 and 9). These differences hold for both early and late broods, and for Switzerland, Britain, and Holland. In only two cases are the differences statistically significant, but all of the seven possible analyses give concordant answers.

5. These results imply that, in the first two months after leaving the nest, the mortality-rate is higher among the young from broods of large than small size. As a result, broods with one or two more young than the average give rise to about the same number of descendants as broods of average size, i.e., broods of these sizes are about equally efficient.

6. The smaller size of late than first clutches, and of British than Swiss clutches, is considered to be advantageous, since in each case the mortality-rate rises as soon as brood-size exceeds the commonest size found in nature.

7. Annual differences in average clutch-size are small (table 11). Had there been extensive non-genetic modifications, the apparent rise in mortality-rate with brood-size would have been obscured.

8. Each individual tends to lay a clutch of similar (but not necessarily identical) size in different years (Appendix I).

9. Two individuals from the same brood are recovered rather more frequently than would be expected by chance (table 12).

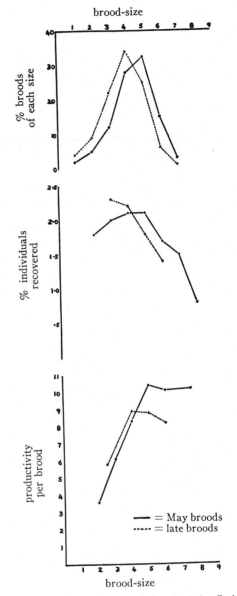

FIG. 1. Brood-Size and Survival in Swiss Starlings Based on Data in Table 6.

10. Swiss Starlings appear to have a higher average annual mortality-rate than British Starlings (table 13). This is an inevitable consequence of the higher reproductive rate of the former.

ACKNOWLEDGMENTS

The writer is particularly indebted to the following persons who have greatly helped during the preparation of this paper: H. N. Kluijver for the loan of many unpublished nesting data from Holland, A. Schifferli for permission to analyse the banding records of the Vogelwarte Sempach, and for hospitality while doing so, E. P. Leach for placing British ringing records at his disposal, E. T. Silva, assistant at the Edward Grey Institute, for extensive help in recording and analysing the raw data, R. Carrick for some unpublished data on clutch-size and nestling weights, and for helpful comments on the manuscript, D. J. Finney for considerable help on the statistical aspects of the work and advice throughout, Professor A. C. Hardy, Dr. E. B. Ford and R. E. Moreau for criticising the manuscript.

LITERATURE CITED

KLUIJVER, H. N. 1933. Bijdrage tot de Biologie en de Ecologie van den Spreeuw (*Sturnus vulgaris vulgaris* Linn.). Med. Plantenziektenkundigen Dienst te Wageningen, no. 69.

KLUIJVER, H. N. 1935. Waarnemingen over de levenswijze van den Spreeuw. Ardea, 24: 133–166.

LACK, D. 1946. Clutch and brood size in the Robin. Brit. Birds, 39: 98–109, 130–135.

LACK, D. 1947-8. The significance of clutch-size; Pts. I and II. Ibis, 89: 302–352; Pt. III, 90: 25–45.

LACK, D. (in press.) The significance of litter-size. J. Anim. Ecol., 17: in press.

LACK, D., and H. ARN. 1947. Die Bedeutung der Gelegegrösse beim Alpensegler. Ornith. Beob., 44: 188–210.

LACK, D., and A. SCHIFFERLI. (in press.) Die Lebensdauer des Stars. Ornith. Beob., 45: in press.

LACK, D., and E. T. SILVA. (in press). The weight of nestling robins. Ibis, 91: in press.

LEES, J. (in press.) Weights of robins, nestlings and adults. Ibis, 91: in press.

MOREAU, R. E. 1947. Relations between number in brood, feeding rate and nestling period in nine species of birds. J. Anim. Ecol., 16: 205–209.

NICE, M. M. 1937. Studies in the life history of the Song Sparrow, I. Trans. Linnaean Soc. N. Y., 4: 108–111.

WEITNAUER, E. 1947. Am Neste des Mauerseglers, *Apus apus apus* (L.). Ornith. Beob., 44: 146.

APPENDIX I. *Successive layings of same female in different years*
(*H. N. Kluijver*)

	A. In three successive years			B. In two successive years	
	1932	1933	1934	Layings of	No. of cases
(i)	6	6	6	3 and 6	1
(ii)	6	6	6	4 and 5	2
(iii)	6	6	6	4 and 6	2
(iv)	6	7	6	5 and 5	3
(v)	6	7	7	5 and 6	5
(vi)	7	7	8	6 and 6	10
				6 and 7	1

Notes: (i) Late layings, and layings known to be due to first-year birds have been omitted, but some of the latter may have been included inadvertently.

(ii) In Part B, the smaller laying did not necessarily come in the earlier of the two years.

APPENDIX II. ANNUAL DIFFERENCES IN SURVIVAL

An analysis of survival-rates in each year separately did not yield results of interest. However, as this point is important in other species, the data are set out below for May broods in Switzerland for those years with a large number of recoveries. For other years in Switzerland, for all late broods, and for all British data, the recoveries were too few for this to be worth while.

Annual differences for May broods in Switzerland

Brood-size	No. of young banded in					No. of young recovered in				
	1933	1934	1935	1936	1937	1933	1934	1935	1936	1937
1	9	6	10	8	7	—	—	—	—	—
2	52	28	62	36	44	1	1	—	—	1
3	153	171	183	147	174	1	7	3	1	3
4	460	500	644	496	548	11	8	15	14	8
5	820	695	1240	645	795	24	21	30	13	8
6	432	324	534	258	444	9	6	5	7	4
7	49	98	112	63	77	1	1	1	1	1
8–10	8	16	49	27	8	—	—	—	—	1
Total	1983	1838	2834	1680	2097	47	44	54	36	26

Reprinted from *Proc. Nat. Acad. Sci. USA*, **45**(7), 1095–1100 (1959)

REPRODUCTIVE CYCLES IN AN EQUATORIAL SPARROW*

By Alden H. Miller

MUSEUM OF VERTEBRATE ZOÖLOGY, UNIVERSITY OF CALIFORNIA, BERKELEY

Read before the Academy, April 29, 1959

The male avian reproductive system undergoes cycles of activity in which there are periods of reduced, nonfunctional gonads alternating with those when the active testes are 150 times the size of the resting stage. Such cycles occur in essentially all wild species that have been investigated. Control of the cycle by the gonadotropic hormones of the pituitary has been demonstrated, a control that parallels the especially well-documented situation in mammals of pituitary regulation of reproductive periodicity.

In every instance in birds in which manipulation of the cycle has been attempted by experimental light treatment, it has been found that greater day length stimulates the pituitary-gonad mechanism. On the other hand there are periods following maximum activity, and also periods in the maturing of individuals, when the mechanism is refractory to the normally stimulating effects of light increments. These seemingly necessary periods of rest and organization have a strong tendency to alternate with recrudescent phases and constitute the basis for ascribing an innate rhythmic attribute to the mechanism. The general conclusion has been drawn that the annual reproductive cycles of species of temperate and high latitudes is the result of this general rhythmic tendency coerced by seasonal photoperiodism so that the onset of the breeding season is precisely timed in adaptation to the average annual development of favorable ecologic conditions for the species; the breeding effort that ensues thus can achieve maximum success.

As one examines species living at progressively lower latitudes, where seasonal photoperiodism is less extreme, many instances of the influence of factors other than day length become apparent. Thus even at latitude 38° some species have breeding seasons during the shortest days of the year in adjustment to rainfall patterns and the related plant and insect food sources.[1] In a few instances at least the actual or proximate stimuli to recrudescence have been traced to factors such as flowering of plants or availability of plant growth[2] or to the onset of rains. These stimuli other than light are best revealed in desert areas where they are variable in timing from year to year and where recrudescence therefore cannot possibly be keyed to any phase of photoperiodism.

Nevertheless the prevailing influence of photoperiods is detected, through critical correlations, in an equatorial direction as far as latitude 10°N where most species breed during the longer days of the year following the vernal equinox.[3] This is true both in wet and dry belts and among species that are partly influenced by seasonal moisture conditions.

In equatorial latitudes, between 5°N and 5°S, the photoperiod fluctuations are so reduced as to have no physiologic effect on reproductive rhythm. One may expect, then, (1) that innate cyclic tendencies will be expressed in some species without the coercion of photoperiodism, (2) that such species may as species show continuous breeding because the normal variability in length of the presumptive individual cycles would put these individuals out of phase with respect to one another, and (3) that other species that inhabit areas with strong rainfall cycles will be coerced into exhibiting coordinate, restricted breeding seasons by evolving sensitivities to some one of a variety of proximate stimuli arising directly or indirectly from rainfall.

As a result of previous work in Colombia, South America, at equatorial latitudes in an area with weak rainfall cycles, it was found that in eight of ten sample species breeding condition in males could be encountered in all segments of the year.[4] But we discovered also, as evidence from equatorial Africa had already partly shown,[5] that adult males of these species undergo testis regression to a point fully equivalent histologically to the state of winter dormancy of north-temperate birds. It was thus evident that individuals underwent some type of cycle even in the continuously breeding species.

The purpose of the investigation of 1958 in Colombia was to ascertain in such a continuously breeding bird the number of cycles and the duration of each phase of the cycle in a given individual through a twelve-month period.   The species selected for study was *Zonotrichia capensis*, a congeneric relative of the North American Sparrows of the genus *Zonotrichia* on which much of the experimental work on photoperiodism and breeding physiology has been conducted.[6, 7, 8]   To achieve significant results on *Zonotrichia capensis*, free-living individuals marked for identification needed to be followed by means of observation and by periodic recapture to ascertain the occurrence of stages of testis enlargement and regression.

Methods and Procedure.—A research station was established in an area of dense population of *Zonotrichia capensis* on the crest of the Western Andes, 6,500 feet elevation, west of the city of Cali, Colombia, at latitude 3°30′N.   This research center was maintained from February 8, 1958, to January 23, 1959.   In the course of the year 160 individuals were captured and individually color marked on an area of four acres.   The most critical results were derived, however, from the subsequent history of the first 50 birds captured in the initial three weeks.   These latter consisted largely of resident adult and postjuvenal individuals, in which there was good survival, of about 78 per cent, through the ensuing ten or eleven months.

Breeding level in males is regarded as a stage of testis development in which mature sperm are present either in bundles or free in the lumen of the seminiferous tubules.   If sperm are not free, it may safely be assumed that they can be so freed within five to ten days under the stimulus of courtship and nesting behavior of the pair and that the male is therefore in potential breeding condition.   In the regressed testis spermatogonia and primary spermatocytes are present but only a few of the latter show synapsis of the chromosomes.   By correlation studies of the linear dimension of the testes and microscopic sections carried out on birds collected in areas adjacent to that occupied by the living marked population, it was determined that testes 6 to 9 mm in length are in potential breeding condition and that testes 1.5 to 2.5 mm in length were in inactive state.   Testes 2.5 to 5.0 mm in length if they are firm or turgid are in various stages of the maturation divisions which implies a prior period of inactivity.   The decline from active state is very rapid and the testis in this transition is soft and usually distinctively yellow.

Further correlation studies among the 80 birds autopsied showed that a cloacal gland development represented by external diameters of 5 to 8 mm of the cloacal protuberance invariably reflected a testis in potential breeding condition.   Moreover, birds that were engaging in steady territorial defense song always possessed an active testis.   The converse situations, however, did not hold.   Reduced cloacal glands and absence of song may at times accompany a fully enlarged testis.

A critical procedure was the recapture of individuals and the performing of a laparotomy whereby the internally situated testis could be measured with calipers. Following closure of the abdominal wall and recovery from anesthesia the bird, within an hour, could be returned to the field, where it would at once resume its singing and nesting behavior if it were in an active state.   This operation was performed as many as three or four times on some individuals without any ill effects and it was particularly important in establishing periods when the testes were in regressed or early recrudescent stages.   The uncertainties of survival in the wild and of recapture at the proper periods for laparotomies were factors on which we gambled but with which we had good fortune.

Age determination of individuals was made by observing presence and progressive disappearance of the ventrally streaked juvenal plumage. All identifiable remnants of such plumage are lost by 80 days of age. Up to eight months of age, and occasionally to ten months, evidence of immaturity persists in the skull in the form of progressively reduced areas of translucent bone in the frontal region; these "windows" are later closed through full separation of the two layers of this bone by an air space and supporting trabeculae. This skull area may be repeatedly examined by operating without anesthesia to slit the thin overlying skin and expose the bone; suturing with scotch tape permits immediate return of the bird to the field. Age in the period from three to nine months can thus be gauged approximately by measuring the size of the frontal "window."

By correlation of all evidence derived from operations, external features of the plumage and the cloaca, and behavior involving song, copulation, and actual participation in nesting, the reproductive state could be ascertained in a number of males over a ten to twelve months' period.

Results.—The evidence indicates a full reproductive cycle, involving one high and one low point, of five to seven months' duration, the normal length being six months. Case histories of 29 postjuvenal and adult males that reflect status over a period of six months or more in no instance offer the slightest suggestive evidence of appreciably longer cycles. Moreover, the like records of 27 females in no instance show sustained breeding activity for more than six months without rest periods.

The positive record of cycles in males consists of 17 timed cycles attested by laparotomies for the low points and by laparotomies and/or positive cloacal or behavioral evidence for the alternating high points. There are five additional cycle records for which the evidence for the low points is indirect, consisting of conspicuous cessation of song and/or heavy involvement with molt.

The most decisive records are those derived from five males in which two complete cycles were traced, thus clearly demonstrating the double cyclic situation for the annual period. The records of these are summarized in Table 1. Herein the

TABLE 1

RECORDS OF CYCLES OF FIVE WILD MALES OF *Zonotrichia capensis*

Individual	High	1st Low Point	High	2nd Low Point	High
466	C	Mar. 2	L, C, N	Aug. 25	C, N
472	. . .	Mar. 20 (Apr. 7)	L, C, N	Aug. 26	C, N
480	L, C	Mar. 16	L, C, N	Oct. 20	C
484	L, C	Mar. 20 (Apr. 8)	C, N	Nov. 3	. . .
806	. . .	Mar. 8	C	Sept. 8	C

critical low points are determined by laparotomies. If the operation revealed a testis in middle stages of recrudescence, the low point was extrapolated from this and the actual date of the operation placed in parentheses. The fully resting testis state is apparently maintained for only two or three weeks at most and thus serves as a convenient reference point. The alternating periods of full breeding potential are attested in these birds by laparotomies (L), and/or greatly enlarged cloacal protuberances (C), and participation in nesting (N).

The length of the sustained breeding condition or plateau was determined rather closely in ten adult birds. These are best expressed as minimum durations based on positive evidence. They range from three to six months, but eight of the ten are in the range of 3 $1/2$ to 4 $1/2$ months. Maximal possible periods might be estimated by extrapolation from intervening lows but would in no instance be more than two or three weeks greater than the proved minima.

In immatures undergoing their first testicular enlargement, the plateau may be shorter. In one instance it was proved to be only 2 $2/3$ months and similar short durations were suggested purely by behavioral evidence in several other immature individuals. Such short plateaus were never found in adults.

The interval of rest could be measured only as the gap between the end of one high plateau and the beginning of the next, owing to the rapidity of regression and the briefness of actual rest (less than a month) and recrudescence. To ascertain and time the progress of these rapid changes would have necessitated more frequent captures and operations than the circumstances and the recovery of the birds would permit. The maximum time for the regression and recovery interval ascertained from six histories ranged from six weeks to three months, the latter probably longer than real; the normal period was two months.

The age in months at which young males first attain breeding condition was found to be as follows in seven individuals with adequate records: 5, 6, 7, 7 $1/3$, 8, 9 $1/3$, 11 $1/3$. There is a very small chance that in the last two instances a very brief recrudescence and subsidence could have occurred short of nine to eleven months and been undetected but it could not have been of such duration as actually to have permitted nesting. Although we were not fortunate enough to find any of these particular males actually engaged in nesting, two females at calculated ages of 4 $1/2$ and 9 $1/2$ months had laid clutches of eggs. In the first instance if we allow for some delay in initiating or completing the postjuvenal molt beyond the normal chronology, the age of breeding might have been five months, but not more.

To be detailed in another paper are full data on weather and on the statistics which show when the largest numbers of individuals are at breeding level. Briefly it may be indicated here that the modal occurrences of breeding fall in May, June, and July and in December, January, and February. These months represent the ends of the wetter seasons and the early parts of the ensuing drier periods; the heaviest rains are in April and May and in late October, November, and December (Table 2). Thus the peaks of breeding are related to and in some

TABLE 2

WEATHER RECORD AT 4 KM NW SAN ANTONIO, 6,500 FT, VALLE, COLOMBIA
Monthly averages of maximum and minimum temperature (°F) and total rainfall,
February 8, 1958 to January 23, 1959

	Minimum	Maximum	Rainfall, in.
January	57.9	70.0	1.1 (23 days only)
February	58.5	71.5	3.065(20 days only)
March	58.6	72.9	5.51
April	58.2	71.4	7.37
May	58.7	72.1	8.05
June	58.6	73.8	1.13
July	57.8	72.5	1.29
August	57.6	73.4	3.47
September	57.3	72.7	1.16
October	56.9	70.7	7.60
November	56.7	68.6	4.95
December	56.7	68.4	6.49
			$\overline{51.23}$ (1 yr. less 15 days)

way induced by the rainfall cycle but not by the weak traces of photoperiodism, for it is to be observed that one peak spans the summer solstice (for $3 \frac{1}{2}°$N latitude) and the other the winter solstice. It should be re-emphasized, however, that birds in breeding state and successful nesting occur in every month of the year.

Conclusion and Summary.—In equatorial species of birds that exhibit potential breeding state at all periods of the year, individual males undergo cycles with maxima and minima equivalent in degree of difference in physiological level to those of their temperate zone relatives. In one such species, *Zonotrichia capensis*, the first of its kind to be studied intensively, the cycles of the individual average six months in duration and two complete cycles are manifest each year. The high plateau of breeding potency in males normally lasts four months and the intervening regression, rest, and recrudescence of the testis occupies two months.

The six-month cycle in *Zonotrichia capensis* is an expression of innate cyclic tendency. It is uncoerced by small variations in photoperiod and is only incompletely controlled by the seasonal occurrence of rainfall, which however does appear to induce a majority of the population to engage in breeding in the late parts of each of the two wetter periods of the year.

Young males attain the breeding plateau at five to eleven months of age. Young females may actually engage in nesting when five months old.

The variable age of attainment of breeding potential and the individual variability of the innate cycle of the adult males is responsible for the production of some young in every month of the year.

* This investigation was supported through a Guggenheim Fellowship and sabbatical privileges of the University of California, with assistance from the Universidad del Valle, Cali, Colombia, and especially from Professor F. Carlos Lehmann V. of the faculty of that institution. Grateful acknowledgment is made to these institutions and parties.

[1] Williamson, F. S. L., *Condor*, **58**, 342–366 (1956).

[2] Marshall, A. J., and H. J. de S. Disney, *Nature*, **180**, 647–649 (1957).

[3] Skutch, A. F., *Ibis*, **92**, 221 (1950).

[4] Miller, A. H., *Acta XI Congressus Internationalis Ornithologici, Basel*, **1954**, 495–503 (1955).

[5] Moreau, R. E., A. L. Wilk, and W. Rowan, *Proc. Zool. Soc. London*, **117**, 345–364 (1947).

[6] Wolfson, A., *Sci. Monthly*, **74**, 191–200 (1952).

[7] Miller, A. H., *Condor*, **56**, 13–20 (1954).

[8] Farner, D. S., in *Recent Studies in Avian Biology*, ed. A. Wolfson (Urbana: University of Illinois Press, 1955), 198–237.

III
Social Behavior

Editor's Comments on Papers 10 Through 18

This topic perhaps shows the greatest influence of behavior as an ecological factor because the organization of animals into a social system determines the availability of resources. Indeed the presumed survival value of social behavior is that the resources are so partitioned that some individuals survive even when resources are limited.

An early study (Whitman, 1919) described some aspects of this type of behavior and is noteworthy for its detail of description and lack of concern for ecological factors. This work, although often overlooked, greatly influenced persons working two decades later. I vividly remember that Mrs. M. M. Nice urged me to study Whitman. Although published in 1919, the observations were made from 1895 to 1906. Whitman used his results in many lectures at Wood's Hole and at several universities. Incidentally, Whitman was one of the imaginative scholars that were attracted to the University of Chicago at its burst of growth around 1900. One must assume that Whitman knew Robert Ezra Park, who began the study of "human ecology" at that time.

Chapman (1935) produced a remarkable study that showed how a very complex set of behaviors was adjusted to the habitat conditions of a dense tropical forest. The influence of this paper lies partly in the development of techniques, but primarily in the underlying awareness that behavior was an ecological factor in the nesting success of the species. The influence of the paper derived partly from Chapman's prestige. In the early 1930s the study of behavior was suspect or at least perplexing to the establishment in ornithology, which was still largely systematic and evolutionary. But Chapman, whose credentials were the best, showed that observing birds was both respectable and productive, even though, with his sense of humor, Chapman listed a cushion as an essential item of equipment. Of greater significance in the science of behavior is the fact that Chapman's subject, the manikin, showed unusual territorial behavior and thus ranked with some species described by Elliot Howard, such as the ruff and black grouse, in his influential *Territory in Bird Life.*

The classic research of Nice (1937) on song sparrows describes territorial behavior. The section is reprinted to illustrate a system of reporting data that influenced a generation of students. It also presents her ideas on the effects of weather, thereby relating behavior and factors of the environment. While the description of territorial behavior could be considered separately from the ecologic ramifications, Nice was aware of these relations and collected detailed data on them (Chart VIII). At that time (1930–1935) the scientific community was hardly aware of the significance of length of day on reproductive processes, including song, although Rowan (1927) had observed changes in gonads but in the context of migration rather than of song. Thus Chart VIII may well represent the relation to length of day rather than to temperature. Note that Nice examined the percent of sunshine instead of the length of day, showing how embryonic were the ideas about photoperiod at that time. Mrs. Nice's influence affected the study of bird behavior more than of ecology, but because her data were so convincing she was able to establish respectability for behavior and field research that spread to ecological research.

Two reprints of my own are included to illustrate several points. Very detailed observations formed the basis for description of complex behavior in one species in a rather quantitative way. Ecological factors permeate the study but rarely dominate the behavioral analysis. The descriptive study (1940) contains some reference to ecological factors, such as the effect of rain on the breeding season, a topic that to this day has not been resolved. The paper clearly shows the influence of Chapman and Nice in the manner of description and of interpretation. The quantitative approach is rather good for that time, and the inclusion of information on reproductive condition was an innovation. The list of references indicates the publications that dominated the topic at that time. In the comprehensive analysis of phylogeny, the behavior is considered to be the direct consequence of ecological conditions that are conducive to flocking. Thus behavior and ecologic factors became entwined. In particular, the viewpoint is adopted that no single cause produced a unique effect but that several conditions were suitable and conspired to produce communal nesting. Other conditions for other species also produced communal nesting. Indeed, this approach resembles in many ways the ecosystem view now popular for energetic relations. Another noteworthy aspect is that the British Guiana observations could only be interpreted after the studies in Cuba clearly showed territorial behavior, because in British Guiana the populations were too sparse to produce fighting sufficiently frequently that the observer would be able to see it. Thus the significance of density of population and the *appearance* of behavior became recognized and permeated later thinking about rodents. The summary connects behavior and ecology at every stage in the presumed evolution of phylogeny.

Turning to mammals, a somewhat different set of influences developed. Since most small mammals are difficult to observe, the data on behavior must be obtained by indirect means such as trapping. Thus Blair (1940) derived information on behavioral aspects such as movements and colonization. These conclusions were then applied to ecological questions such as distribution of resources. From the behavioral viewpoint an interpretation developed eventually from the observations that mammals have a home range. Since an individual after it settles in its range will stay there usually for the rest of

115

its life, its behavior becomes predictable and learning is useful. Thus if a mammal learns that food or a predator is in a certain place, the behavior will not be random searching (like an ameba!) but will follow patterns that are based on learned distribution. In this sense learning behavior has survival value, in contrast to random wandering, where learning is useless because discovery depends upon chance encounters. Blair did not mention this interpretation, but studies by him and others set the stage for two decades of observations of home range.

A study that is frequently quoted for its excellent description of complex behavior should also be quoted for its superior description of the effects of behavior on the habitat (King, 1955). The sections reproduced here emphasize these relations. An unusual viewpoint that now is generally accepted is that the animals may greatly influence the plant population. Although this idea was implicit in the old-fashioned rodent-poisoning campaigns on grazing land at the turn of the century, data to support the idea were not available, and the general conclusion was that the habitat controlled the population without reverse effects, which are now known to be frequent. Superimposed on this effect is the social organization, which organizes the distribution of animals into local groups that, by intensive grazing, alter the vegetation. Thus ecology and behavior are united. Another aspect should be noted. Enough was known about behavior (social rank, territory) to apply these concepts to learning about prairie dogs. Until this time the animal was used to learn about a concept, but now the concept could be used to learn about an animal. According to Dr. King (letter of August 1973) this paper "supported nascent concepts of social control of population size, beneficial environmental conditioning, and cultural heritage of spatial distribution." This paper has been very influential by its example of thorough description of the kind presented in purely ecological papers written much earlier (see Vorhies and Taylor, 1922; and others).

In Koford's paper on the vicuna we now see clearly that the marriage of behavior and ecology factors is complete, although unnoticed. This study sets a new level of excellence in its description of relationships. Koford (letter of August 1973) states that this report is "apparently the first clear example of year-round bird-like territoriality (defense of area) in mammals." A notable aspect is the thorough documentation of maintenance behavior (feeding, urination, sleeping, etc.) as a necessary basis of social behavior. Without such information the major features of social behavior could not have been described in meaningful terms. The results of Koford's observations of territorial relations (not reproduced here) have been noted many times and are an excellent and early example of mammalism territorial complexity. But, although excellent, the results by this time (1957) were another example of territorial behavior. Its great influence lies in the connection of behavior with the ecology of the harsh Andean region.

The paper by Stewart and Aldrich (1951) sets the stage for the use of behavior to examine relations, not of individuals, but of populations to ecologic factors. The significance (Stewart, letter of August 1973) is that "the indications are that for many species a reservoir of mobile nonterritorial males was present and these birds rapidly refilled evacuated territories." A major part of the influence of this paper is its jump to the population level. Again we see that behavior (territorial) is a *tool* for the study of ecology (populations). It was known that the birds had territories, not from studies on all the species, but from general principles. Some field work was done to demonstrate some of the details, and then the existence of territorial behavior became a means to study movement and mortality rates. These ecological factors act at the population rather than the individual level, and thus our knowledge of the ecology of the species is extended again.

References

Rowan, W. 1927. Experimental migration. *Nature 119:*351–353

Vorhies, C. T., and W. P. Taylor. 1922. Life history of the kangaroo rat (*Dipodomys spectabilis spectabilis*). *U.S. Dep. Agr. Bull. 1091*, 40 pp.

Influential Books

Darling, F. F. 1937. *A Herd of Red Deer.* Oxford University Press, New York, 215 pp. Important for details of descriptions.

Howard, Elliot. 1920. *Territory in Bird Life.* Murray, London, 308 pp. A comprehensive theory of behavior.

10

Reprinted from *The Behavior of Pigeons* (Carnegie Inst. Wash. Publ. 257), Carnegie Institution of Washington, Washington, D.C., 1919, pp. 3–14

The Behavior of Pigeons

C. O. WHITMAN

CHAPTER I. THE MATING PERIOD.

The mating period refers to that part of the activities of the reproductive cycle which begins with courting and copulation and ends with the laying of the first egg. It embraces such activities as courting, copulating, and the hunting, acceptance, and construction of the nest. In the majority of cases its duration is six or seven days. The quest for a nesting-place begins one or two days after the initiation of the cycle and extends over a period of two to three days. The final acceptance of the nest-site is indicated by the beginning of nest-building. The construction of the nest continues for the rest of the period and may extend for several days into the incubation period. Copulation occurs intermittently throughout the entire mating period. These statements are based upon a very detailed record of the bronze-wing pigeon,[1] which covers a number of cycles. Excerpts from this record are herewith given in sufficient fullness to establish the above conclusions.[2]

First Cycle—Dec. 10, 1905. The male goes to the nest-box and calls. He attempts copulation. On the 13th copulation was successful. It was attempted on the previous days, but there was no certainty of its completion. On the 14th the nest was accepted and there was some nest-building. The 15th and 16th were occupied with nest-building and copulation, and the first egg was laid on the 16th.

Second Cycle.—On Jan. 1, 1906, the male was ready to begin a new cycle and kept attempting to entice the female. The young bird died on the 7th. Courting and attempts at copulation occurred on the 8th, and the first successful copulations were noted on the 9th. The quest for a nest was begun on the 10th, and this activity, interspersed with copulations, continued until the 13th. Nest-building was noted on the 13th and the 14th. The 15th and the 16th were occupied with nest-building in the forenoon and copulations in the afternoon. The first egg was laid on the afternoon of the 16th.

Third Cycle.—On Jan. 23, both birds were courting, but there was no contact. On the 25th they were interested in courting and copulation. Courting was continued without contact on the next day, when the egg was deserted. The quest for a nest was begun on the 27th, and the nest was accepted on the 30th. Nest-building began at once and successful copulations were noted. No records were made for Jan. 31 and Feb. 1. Nest-building was noted on the 2d, and the egg was laid on the day following.

Fourth Cycle.—Courting and attempted copulation by the male were begun on Feb. 21. The young died on the following morning, and the day was occupied by courting, copulation, and hunting for a nest. The quest for a nest was continued throughout the 23d and 24th, and the final acceptance of the nest was noted on the 25th. Nest-building was recorded on the 26th and the egg was laid on the 28th.

---

[1] The author remarked to one of his students (W. C.) that the bronze-wing pigeons (*Phaps, Ocyphaps*), coming from the Southern Hemisphere, at first bred in Chicago in winter, but that they gradually changed their cycles until they bred in summer.—EDITOR.

[2] In this volume the largest print represents the words of the editor. The author's material has been printed in a medium-sized type. The small type has been used for the tables, notes, and other incidental matter. With this typographical distinction the reader will be enabled to identify the manuscript material.

Fifth Cycle.—The pair was separated on March 16. They were brought together during the mornings of bright days in order to secure photographic records of their mating behavior. Both were ready to mate and copulated whenever brought together. On Apr. 26 they were paired again and courting began immediately. Copulations and acceptance of the nest were recorded on the 28th; and on May 3 the first egg was laid. (Excerpts from R 33.)

THE BREEDING SEASON.

The manuscripts contain but few statements as to the normal breeding-season for the various species of pigeons. Table 1 has been compiled from the various

TABLE 1.—*The breeding season.*

Species.	Time of first egg of season.	Time of last egg of season.
Passenger (*Ectopistes*)	....	8/29/96
" "	3/17/97...............	8/28/97
" "	7/14/98...............	8/ 7/98
" "	2/24/99...............	....
Geopelia humeralis...................	1/13/06 (kept in house)...	3/10/06
Bronze-wing (*Phaps*)...............	12/16/05 (kept in house)..	5/16/06
White-face (*L. picata*)...............	4/11/07...............	6/ 2/07
Fantail (common pigeon)	....	9/22/98
Crested (*Ocyphaps*)	7/25/96...............	
" "	2/ 4/97...............	5,22/97
" "	2/23/98...............	....
Mourning-dove (*Zenaidura*)..........	3/24/97...............	8/ 4/97
" "	....	8/ 8/98
" "	4/26/97...............	8/18/97
" "	3/15/98...............	7,30,98
" "	4/ 3/99...............	7,29,99
" "	4/13/00...............	7,12,00
Band-tail (*C. fasciata*)...............	2/11/03...............	....
Wood-pigeon (*C. palumbus*)..........	1/ 8/01...............	....
Satinette (common pigeon)...........	3/22/98...............	
Common pigeon......................	2/28/97...............	8/12/97
European turtle (*T. turtur*)..........	5/31/97...............	6/ 2,97
Blond rings (*St. risoria*)...............	3/12/95...............	12/23/95
" "	5/12/96...............	12,25/96
" "	2/ 8/96...............	
" "	1/28/97...............	6/ 4/97
" "	1/ 1/97...............	4/21 97
" "	3/ 5/97...............	6/ 8,97
" "	1/26/06...............	
Zenaidura ♂ × white ring ♀	1/ 8/97...............	3/10 97
" "	4/19/97...............	5/ 1 97
" "	6/16/98...............	
White fantail ♂ × ring-dove ♀	1/ 2/97...............	8/ 2,97
Common ♂ × ring-dove ♀	5/ 3/96...............	12/22/96
"	1/23/97...............	8/22/97
Jap. tumbler ♂ × ring ♀	7/23/97...............	12/15/97
"	1/20/98...............	2/21/98
Common ♂ × blond ring ♀	1/12/98...............	....
Archangel ♂ × blond ring ♀	2/18/97...............	8/31/97
Ectopistes ♂ × ring ♀	2/24/97...............	6/29/97
Ectopistes ♂ × bl.-wh. ring hyb. ♀	1/26/97...............	8/15/97
"	3/24/98...............	....
Homer ♂ × blond ring ♀	3/11/97...............	5/ 2/97
"	3/24/97...............	6/28/97
Common ♂ × Jap. turtle ♀	6/ 5/97...............	8/ 5/97
Hyb. ring ♂ × homer ♀	9/ 6/97...............	12/20/97
"	6/16/98...............	12/18/97
Blond ring ♂ × white ring ♀	1/ 9/98...............	2/12/98
Red ring [1] ♂ × blond ring ♀	11/27/96...............	....
	12/20/00 (kept in house)..	8/30/01

[1] Most of the species listed here, besides several others, have been drawn in color; these illustrations are given in Volumes I and II of these works.—EDITOR.

breeding-records, giving the dates for the first and last eggs for the year. Too much reliance may not be placed upon these data. Frequently the birds were not together early in the year, being mated late in the season, and as a consequence the date of the first egg is often not a true index of the beginning of the season. In some cases the records state that the birds were kept in the house, and this special care may have accentuated, in some pairs, the early development of the breeding impulse. As to the location of the birds, within or out of doors, there is no indication in many of the records. Frequently the pairs were separated and remated during the season for experimental purposes and, from the manuscripts dealing with behavior, no clue can be obtained as to the normal end of the season.[1] Further, experimental control and quasi-domestication (see following topic) may profoundly influence the duration of the season. Besides the table referred to above, the following comment is given:

During 1898 not an egg was laid by the passenger-pigeons until July 14. One was laid on July 15, and the remaining pairs soon became active. Why they began so late I do not know. They were brought to Woods Hole, May 25, and should have begun to lay in April at the latest. Possibly I fed them too much corn or too much in general. I ceased giving them corn about the end of June and lessened the amount of seed given. I do not know whether the laying was brought about by the change, but think it probable. In the season of 1899 the first egg from these birds was laid on Feb. 24, and the second on Mar. 1. Perhaps the late and small work of the previous year led to an earlier activity this year. The birds have been in the pen outside all winter, and it is the coldest winter experienced here for many years.

A mourning-dove × ring-dove pair (*Zen. 5-C*) failed to fertilize eggs after August, the end of the regular season for the mourning-dove. Eggs were laid September 5 and 7, after the mourning-dove ceases normally to coo or care for eggs. (R 19, R 11.)

"Fancy pigeons generally show an inclination to mate together some time in the month of February; but much depends upon the temperature, as in very severe weather they will sometimes show no signs of doing so until March, whilst if it be mild some birds, if allowed, would go to nest in January."[2]

PROLONGATION OF BREEDING SEASON IN WILD PIGEONS.[3]

A pair consisting of a mourning-dove and a ring-dove[4] was mated in April 1899. Three sets of eggs were produced and incubated by the end of August. Four sets of eggs were laid between Sept. 8 and Dec. 23, but all failed of development. The male continued to sit faithfully until the first week in November, but after that time he lost interest and performed his incubation duties very irregularly. It is noted that "this pair, although remarkably successful in fertilizing and hatching during the normal season, failed after August, *i.e.*, at the end of the regular season for the mourning-dove."

Another similar pair was also mated in April of the same year. Six young were raised by Nov. 1. A pair of eggs was laid on Nov. 15 and 17 and these were successfully incubated by the male up to Dec. 17, when they were removed. Another set of eggs was produced on Dec. 23 and 25 and the male exchanged faithfully until the removal of the eggs on

---

[1] A large number of complete and reliable data on this topic may be found in Volume II. The designations of pairs of birds—*e.g.*, *Zen. 5-C*—which are often appended in this volume, will enable those who care to do so to make further examination of the entire breeding record, as this is given in the previous volume.—EDITOR.

[2] Fulton and Lumley's Book of Pigeons, London, 1895, p. 35.

[3] For a further reference to this phenomenon, see wood-pigeon, Vol. II, table 87.—EDITOR.

[4] In these designations the name of the male is written first, that of the female last. The same order is used in referring to hybrids.—EDITOR.

2

Jan. 6, 1900. The male sat faithfully in October 1900, but two succeeding sets in November and December received no attention from him. In 1901 the sitting continued until December. It is noted that "the mourning-dove normally ceases to coo or care for eggs after Sept. 1. This late and regular sitting is remarkable, showing that the normal period can be prolonged as a result of domestication. Notice that this 'prolongation' is in a case where the male is unusually successful in fertilizing the ring-dove's eggs." (Excerpts from R 11.)

LENGTH OF MATING PERIOD.

The mating period[1] has a duration of 5 to 9 days; in the majority of cases its duration is 6 or 7 days. Apparently, no species differences exist. The average time for all cases after May is slightly less than that for the extremely early part of the season, but the difference is so slight that no confident assertion can be made. Table 2, giving exact data for several species, has been compiled from the various

TABLE 2.—*Length of mating period.*

Species.	Date of first egg.	Length of period.
Bronze-wing	12/16/05	6 days after mating.
"	1/16/06	8 days after first courting.
"	2/ 3/06	8 days after desertion of eggs.
"	2/28/06	6 days after death of young.
"	5/ 3/06	8 days after pairing.
Geopelia humeralis	1/31/06	8 days after breaking eggs.
" "	2/31/06	10 days after breaking eggs.
" "	3/ 8/06	7 days after desertion of eggs.
" "	4/15/06	7 days after death of young.
Blond ring ♂ × white ring ♀	4/ 9/96	6 days after mating.
Band tail	2/11/03	9 days after beginning nest.
Fantail	8/19/96	6 days after removal of egg.
"	9/22/96	7 days after ceasing incubation.
Zenaidura	4/21/97	9 days after mating.
"	4/26/97	5 days after breaking egg.
"	5/27/97	9 days after leaving young.
"	7/14/97	17 days after breaking egg.
"	4/ 9/98	7 days after breaking egg.
"	4/22/98	9 days ca. after breaking egg.
"	7/ 3/98	7 days after removal of young.
Zenaidura ♂ × white ring ♀	1/ 8/97	7 days after first copulation.
"	3/ 7/97	6 days after removal of egg.
"	3/18/97	6 days after removal of egg.
"	4/19/97	7 days after mating.
"	6/16/98	6 days after removal of eggs.
"	6/26/98	6 days after removal of eggs.
Zenaidura ♂ × blond ring ♀	11/25/99	8 days ca. after removal of eggs.
"	12/21/96	6 days after removal of eggs.
Zenaidura ♂ × ring dove ♀	12/23/99	6 days after removal of eggs.
Ectopistes	7/29/96	9 days after pairing.
"	8/29/96	6 days after first billing.
"	3/17/97	6 days after first courting.
"	4/ 4/97	8 days after removal of egg.
"	4/22/97	5 days after removal of egg.
"	4/28/97	5 days after first courting.
"	5/23/97	7 days after removal of egg.
"	5/30/97	12 days after hatching of young.
"	6/24/97	7 days after pairing.
"	8/16/97	7 days after removal of egg.
"	7/20/98	7 days after removal of egg.
"	7/27/98	6 days after breaking egg.
"	7/25/98	6 days after removal of egg.
"	7/31/98	7 days after removal of egg.
"	8/ 7/98	

[1] That is, the portion of each reproductive cycle during which copulation occurs.

records. The length of this period for females of various kinds, and variously mated, was given the following tests:

On Oct. 16, 1897, I removed the eggs from the following four pairs: (1) a male common dove and a female Japanese turtle; (2) a male archangel and a female ring-dove; (3) a male hybrid between a common pigeon and a ring-dove and a female black Japanese tumbler; (4) a male ring-dove and a female homer. On Oct. 23, or *7 days* after the removal, all of the four females laid again. Thus a Japanese turtle-dove (*Turtur orientalis*) a ring-dove (*Streptopelia risoria*), a tumbler (*Columba gyrans*), and a homer (*Columba tabellaria*) all took the same time.

On June 10, 1898, the eggs were removed from four pairs which had laid between June 1 and 5. The four pairs were: (1) a male mourning-dove and a female white ring-dove; (2) a male mourning-dove and a female ring-dove; (3) a male hybrid and a female ring-dove; (4) a male mourning-dove and a female ring-dove. A second set of eggs was produced by each of these four pairs in *six days* after the removal of the eggs. The first egg in each case was laid at about the same hour (3^h 16^m to 5^h 05^m) in the afternoon of June 16; and the second egg at also about the same hour (7^h 50^m to 8^h 05^m) in the morning of June 18, after an interval of one day and two nights. The conditions of temperature and food were here the same for all. The time required for producing new eggs is sometimes only 5 days, but in colder weather it is sometimes 7 or 8 days.[1]

On June 20 I removed the eggs from the following pairs: (1) a male mourning-dove and a female white ring-dove; (2) a male hybrid and a female ring-dove; (3) a male mourning-dove and a female blond-white hybrid; (4) two blond-white hybrids. These had laid between June 16 and June 19. Eggs were produced in three cases on June 26, *six days* after the removal. In the fourth case, the female persisted in sitting for several days after the removal, and the first egg was dropped on July 2, after a 12-day interval.

A pair of European turtle-doves (*Turtur turtur*) laid eggs on June 3 and 5 and the young were removed on June 26. The first egg of the succeeding cycle was laid on July 2, or *six days* after the removal of the young. (Summarized from R 19.)

Fulton, in his Book of Pigeons, p. 37, states that:

"Most pigeons[2] lay in from 8 to 16 days after being mated, some being what are called much 'freer breeders' than others. If a hen does not lay within three weeks of mating she should be separated from the cock for about 8 days and then returned to him, which will usually produce eggs."

INITIATION OF THE CYCLE.

The male usually takes the initiative in beginning the cycle, and the female may succumb to his advances at once or she may resist for some time. The female may also take the initiative and make advances, the male in the meantime being on the defensive. Either bird may thus remain in a state of readiness for a considerable period of time. If a pair, when ready, are prevented from mating by being separated, both birds may remain in a state of readiness for a month or longer. The primary sexual impulse seems to arise relatively automatically, and the succeeding activities of the cycle must normally await the reciprocal sexual activities of the two birds. Since the primary impulse may arise at different times in a pair of birds, this ability for either to remain in a state of readiness for some time is the means by which the two series of activities become synchronized with each other.

---

[1] After breaking up a nest the birds will often start a new cycle of copulations within half an hour.—EDITOR.

[2] Common or "fancy" pigeons are referred to here. The several species concerned in the preceding paragraphs and studied by the author, are more popularly known as "doves" of various kinds.—EDITOR.

The hypothesis that the sexual impulse arises relatively automatically with each bird does not imply that the two birds do not stimulate each other to any extent. While synchronization is normally effected, yet exceptions occur and the cycle may progress with one bird lagging a few days behind the other. Neither will a bird *always* remain in the first stage of readiness until the succeeding activities of the cycle are released by the reciprocal activity of the mate. Occasionally the cycle may progress to completion without the coöperation of a partner. These statements are illustrated by the following excerpts from the manuscripts.

A pair of *Geopelia humeralis* were put together on Dec. 23, 1905. On Dec. 26 it was noted that the male had every day shown some interest in nesting, but that the female did not seem to be ready. On Jan. 9, 1906, it was noted that the male had kept up his endeavors to interest his mate in a nest, but that so far she had remained indifferent. The first egg was laid on Jan. 21. Allowing 7 to 8 days as the normal time between the beginning of mating and the laying of the first egg, there was a period of about 3 weeks in which the female remained indifferent to the advances of the male. In the third cycle for this pair the female resisted the male's attentions for several days, while in the second and fourth cycles both birds seemed to be ready simultaneously.

In a bronze-wing pair the male took the initiative in all five cycles. The period of feminine resistance varied from 1 to 7 days. On the last cycle the pair was separated on March 16, 1906, just as they were beginning a new cycle. After a separation of 15 days they were brought together for a short time and courting began immediately, both birds seeming to be ready to continue the cycle. They were tested from time to time up to April 28, and both exhibited some interest in mating. They were now left together permanently and the new cycle started at once. Both birds thus remained in a state of more or less readiness for a period of 45 days.

A male mourning-dove (*Zen. 3*) was paired with a white ring-dove (*W 1*) on December 20. The female was eager to mate, and tried all possible ways to ingratiate herself in the favor of the male. He refused to have anything to do with her until 12 days later. On this date (Jan. 1), the female advanced to within a few inches and began billing her wings, bowing, cooing, and casting amorous glances. He soon began to be attracted, billed his wing, and showed that he was yielding. When she touched his head with her beak he repelled her by pecking because of fear. After these maneuvers were continued for 15 or 20 minutes he finally had the courage to take her beak in his, and thereupon the consummation followed quickly. I am quite certain that this was the first time that a union had taken place. On Jan. 5, *W 1* is still doing most of the courting. She is wonderfully persistent, taking no discouragement from any rebuff. On Jan. 20 I saw the male courting the female for the first time. This pairing was never successful. Although unions occurred, the eggs did not develop for lack of fertilization, and the male took no part in incubation. After 3½ months (of trial) he was given another female.

A young male fantail was paired on Aug. 13 with an impure fantail female which had previously mated and had been incubating eggs with another female. Two days later this female took the initiative and attempted to mate with the male. He was quite bashful at first, and although pleased with the advances, appeared to fear to let her come very near. He fought her off the stand and would not let her remain a moment in the cot. During the day he became more enamoured and by degrees suffered her to approach him, but always retreated at the moment she offered her beak, or else attacked her as if not yet quite sure that she could be trusted. This behavior continued throughout the day. On the following day, they were both coquetting and billing, and seemingly fully mated.

Three days later (Aug. 19) the first egg was laid, but the male did not begin to assume his duties of incubation until two days after the second egg was dropped.

The female of a blond and white ring-dove pair was nearly ready to lay and was inclined to mate. The blond male was savage and would not tolerate her presence; I was therefore forced to keep her in a small cage within his cage. On the following day he was still intolerant, but began to yield a little; he went to the box and called. She at once responded, went to him, tried to take the nest, and began cooing in her turn. The male endured this for a time and then drove her off. He next went to the opposite nest-box and repeated the same behavior with her. The white ring was patient and took all his abuse, and by night the blond finally became reconciled to her and condescended to sit beside her. (Excerpts from R 29, R 33, C 7/7, C 7/33, C 7/15.)

That females may take the initiative in mating is also evident from the fact that two females may mate and complete the cycle of nest-building, production of eggs, and incubation. (See topic on Pairing of two females, Chapter III.)

The completion of the cycle without the stimulus of a mate is illustrated by the following references: "Indeed a young and rank hen (domestic pigeon) will often lay, like a fowl, whether she be mated or not."[1]

A male hybrid (mourning-dove × ?) was paired with a California mourning-dove which was supposed to be a female, but which turned out to be a male also. The hybrid attempted courting during an entire month, but his advances were repelled. He then began the course of incubation on the floor of the cage. I soon saw by his behavior,— i.e., his refusal to move and threatening to peck at me—that he was actually engaged in sitting, although he as yet had no mate and no eggs. This cock had never been mated, and I was surprised to find him sitting with all earnestness, and that too on the floor, instead of in the nest-box. He went on sitting in this way night and day for a week, when I placed a couple of eggs in his nest-box and placed him upon them. He at once took possession and continued his incubation faithfully night and day for a week, when I placed the two birds in another pen outside of the house in the hope of securing a mating. (Excerpts from C 7/48.)

The lack of a proper synchronization of the two cycles of activities in a pair of birds is illustrated by the following note. For a more detailed account, the reader is referred to the record of pair X-W 1 in Chapter VII. The female W 1 had been with another female and had started on her cycle. When paired later with the male (X) she was nearly ready to lay. She at once made advances, but he was intolerant and resisted. He first began sexual advances about the time the first egg was dropped, and continued these for four or five days, while the female had passed through her sexual period and was engaged in the work of incubation. Only at the end of his sexual period did the male first take part in incubation; this was three days late.

This male (X) was evidently not in the spirit of sitting, and the presence of the egg did not stimulate him to the act. The decisive stimulus for such an act is thus not external but internal—probably a feeling which comes over a male *periodically*, and which he will manifest perhaps only if the external stimulus is also present. He has no impulse to sit. He is ready for making a nest and to accept a mate, but he requres *time to*

[1] Fulton, *op. cit.*, p. 35.

generate the impulse to sit. This impulse naturally follows the period of sexual activity, and he was still in the latter period.

In a brief note under the heading of "Sexual periodicity," it is remarked that "this (sexual periodicity) is apparent especially in Australian pigeons—speckled–necks, geopelias, crested pigeons, etc. It is also marked in *Ectopistes*. The wood-pigeon and mourning-dove are good examples."[1]   (R 20, C 7/15.)

MANNER OF DISPLAY.

The crested pigeon (*Ocyphaps*) has a very striking mode of display. The tail is raised and spread, and at the same time the wings are lifted just enough to show the whole surface. The bird bows and coos and strikes its feet, and all these movements are rhythmically combined in a very effective way. I noticed one of the *Geopelia humeralis* also performing by quite the same combination of movements. One of the *Geopelia striata* has exhibited itself in the same way, and with a peculiar coo. It is remarkable that two genera so different as *Ocyphaps* and *Geopelia* should have the same peculiar manners.

Darwin had a red tumbler which had a coo like the "laughter," and "the habit, to a degree which I never saw equalled in any other pigeon, of often walking with its wings raised and arched in an elegant manner." (See Darwin, Animals and Plants, I, Chapter V, p. 198.) I have seen this in a white fantail and in a black tumbler (Japanese), and also in a white and crested male dove-cote pigeon. It is quite common behavior. It is done usually when a male sees a female—or a male—whose attention he wishes to attract.

In giving the "coo" to the female, the male *G. humeralis* bows and follows the female *with the tail raised to about 45 degrees or a little more, or nearly vertical, and spread*, so as to show it to the best advantage. He uses it in the same way when trying to drive another cock. So the same behavior serves at least two purposes—one to frighten and the other to display.

I saw this display first in the male bronze-wing. This male was on the ground; the female was on the step-ladder looking down, as if about to fly down to the male. He noted this and began to display as if to attract her. He faced her, held his head on a line with his body, raised the wings without spreading or only a very little; raised his tail a little and held it spread; and meanwhile he stood actually on tip-toes and lifted first one and then the other foot, and so raised one side of the body and then the other in a way to exhibit his iridescence in different lights. This performance is remarkable—so well adapted to the end it serves. As the male wags his whole body slowly from side to side, he moves his head and neck back and forth from side to side. The lifting of the foot with the side raised, and then the other foot and side, right and left alternately, is done with an exquisite lightness of foot, as if the bird were overflowing with delight at the approach or the prospect of approach of his mate. I have seen this done by three males within a week or two, and by one of them several times.

A male mourning-dove (11) walks slowly and grandly along the perch, displaying himself to the female. She does not at once respond. He is anxious to mate, and shows no disposition to be aggressive, or to force her in any way. He sits at a little distance, watches her, now and then wags his wing at her, preens his feathers, inviting her to him. When he walks back and forth he raises the feathers of his back and rump, swells up his neck, and makes himself generally attractive. In strutting the wings droop a little, the primaries being held a little lower and not quite so tightly closed as ordinarily. The feathers of the back project backwards, rising above the level of the wing-feathers. The male sometimes flies to a point nearer the female, and as he alights *strikes* his feet against the perch, spreads his tail, and raises his wings a little, so as to show them and the tail at the

---

[1] For sexual periodicity in man, see Loisel, Comptes Rendus, Oct. 29, 1900.

same time. This combination of movements is often seen in male domestic doves. I noticed that when the female came to his call in the box the male winked (half closing and opening his eye-lids) and smiled (slightly opening his beak), expanding the lower mandibles so that we could almost see into his mouth. The California mourning-dove male performed in all these ways toward me when I answered his call.[1]   (R 20, R 17, R 29, E 28, Em 7.)

STRIKING OR STAMPING.

A male *G. humeralis* is mated with a blond ring-dove. This pair is kept in a pen in the library. The male has just flown up from the floor to the perch beside the female. As he alighted he arched his neck proudly and gave the perch several quick strokes with his feet; I think both feet were used in alternation. The strokes could be heard very distinctly. The mourning-dove behaves in the same way when he alights beside his mate and sometimes he does this when both are on the floor. He runs up alongside his mate and gives these strokes as if in a burst of joy to be with her. The crested pigeon and the geopelias do this also.

A male mourning-dove flies from his perch to the nest-box and then back again near to his mate; each time, as he alights, he lifts each foot once, one after the other, and strikes it against the perch so as to make quite a loud thump, or rather two thumps in quick succession. It is done so quickly that it is difficult to see the movement, although the sound is quite marked. The crested pigeon strikes his feet in quick succession when spreading and cooing before his mate. I think the ground-dove of Florida does something of this kind, but I have only an indistinct recollection of it. *G. humeralis* does the same. *Leucosarcia* stamps once with each foot, one after the other, and holds its head down and its tail raised when it threatens to attack. This is sometimes repeated two or three times, the bird watching its antagonist to see if it shows signs of fear. If it does not take the hint, this may be followed up by a raising of a wing as if to strike. In presenting the wing, the bird moves sidewise up close to his antagonist before striking just like the crested pigeon. The stamping, the lowering of the head, and the raising of the tail seem here to be a modified form of the display of the crested pigeon.

A male *G. humeralis* began early in the morning to parade on the perch before the female; his performance settled into a repetition of jumping from one perch to the other, then stamping with his feet as he straightened up, and presenting always his beak to the female.

The stamping behavior of a mourning-dove (3) was observed. Each time that this male alighted on the perch he *struck his feet against the perch*; one foot being made to strike after the other. The two strokes were quickly made and were loud enough to be heard quite distinctly. At the moment of alighting he "swelled up" a little, held his tail somewhat expanded, and raised his head with an expression of pride in his appearance. The male mourning-dove No. 11 also *strikes* his feet against the perch as he alights, spreads his tail and raises his wings a little, so as to show them and the tail at the same time.

The behavior of the white-faced pigeons (*Leucosarcia*) in mounting reminds one strongly of the bronze-wing (*Phaps*) and of *Geopelia*; it is, in fact, essentially the same. The male white-face usually flew to a wide shelf (12 inches) 5 or 6 feet from the floor. He then struck his feet a few times in rapid succession against the shelf, so that one could hear the thumps plainly across the yard, even at many yards distance. When the male flew up and the female was already on the shelf the latter would straighten up and strike her feet on the shelf; then turn her back with the tail held down, the head up, and the tail spread just a little; then walk off, wheel around, and run up to him—stamping her feet. The male then would offer his beak, she would accept, and be fed with a few shakes,

---

[1] Further material on this and the following topic may be found in Chapter X.

then he would press his neck over hers, and, if she inclined, would mount. After waiting for a few moments on her back, he would dismount with a slap of his wings over his back. As he struck the shelf after dismounting he opened his mouth and gave a squawk somewhat like the bronze-wing, and would strut away from her. The pair would then bow, approach, strike their feet on the shelf, and repeat the mounting, often several times without a single union. (R 20, R 29, C 7/7, Em 7, Sh 8/13.)

CHARGING AND DRIVING.

The charging and driving of the female occurs when other males are present, as is noted in the subsequent topic on "Jealousy." The same activity occurs during the mating period in connection with courting, the quest for a nest-place, and in nest-building. It may also occur occasionally even during the incubation period. The purpose of the act is to force the female to the nest.

A male bronze-wing began to *charge* at the female, driving her quite vigorously. He evidently recognized her as his mate, but for some reason not clear he *charged* at her as if displeased at something. As soon as the birds were put together in the same cage the male began to *charge* the female, though they soon began their love-making and for two hours they kept up hugging, etc. The male began by charging, clucking all the time as he drove the female. It was more evident now that he wanted to start the female off *for a nest-place.* (Excerpts from fifth cycle, R 33.)

On January 13 and 14, 1906, the male bronze-wing was again seen driving the female. He would charge at her with his wings lowered a little and his back and feathers elevated, emitting a short cluck so rapidly as to leave nearly no interval between them. The notes and manner appear petulant. (From second cycle, R 28.)

The following description, covering the period from Jan. 29 to Feb. 2, 1906, refers to a pair of bronze-wings while seeking and building a nest (during the third cycle). On Jan. 29 the male is anxious to build a nest. The birds have been uneasy for 3 or 4 days, running along the side of the pen next to the window, as if they desired to get out. They have been in the nest-box several times, but have not yet decided to accept it and begin work. The male just now took to driving the female, charging at her, and giving her a spiteful peck, as if for some reason displeased with her. I then saw him do what I have not before noticed. He first charged at her; then lowered his head until his beak *touched and rested on the floor,* and with tail raised and slightly spread and wings a little raised, he gave plainly two little coos, in doing which he opened his beak at each note *without once lifting his beak.* The beak opens but little. The notes sounded like oo-oo, with a very short but distinct separation between the syllables—the first shorter, the second somewhat fuller and longer. This could be called the "driving coo," and is analogous to the cooing of a domestic pigeon. I later heard this oo-oo or cu-coo several times from two other males. It is, in fact, not a frequent behavior, but it occurs with all males in connection with charging and driving, whether for a nest or because the male is jealous of the presence of another bird.

On Jan. 30, the male is driving the female again. He gives the "short cluck" as he charges at her and walks about after her. The cluck is repeated slowly, except when he rushes at her, when it runs off hurriedly as a bubbling sound as before described. While following her up in this way, he sometimes resorts to the "driving coo." He has just given this twice. This time I feel sure that there are three notes and that the male is driving the female to find a nest as quickly as possible.

The notes sound sharper than yesterday, more as if beginning with C and could be represented thus *coo-coo-coo,* the second and third notes being shorter and closer together.

On Jan. 31 the male is again driving the female. He seems quite excited, holds his head high, the feathers of his forehead loosened so as to stand apart and out straight. The female tries all the while to get out of the cage in the direction of the window. The call described above certainly has three syllables, the second and third partially linked together.

On Feb. 2 the female has been sitting on the nest calling for straw and the male has been at work. The female left the nest and came to eat when I put food in their cage. The male at once became excited and began to drive her. He gave the cluck—the short petulant note described above, and followed it with the quicker rattling note as he charged and pecked at her. She seemed to understand his displeasure and returned to the nest without food. He then quieted and, after helping himself to the new food, returned to carrying straw. This shows what the driving clucks, the charging, and the coos mean. They mean, in this case, "go to the nest!"

The following case of charging occurred during the incubation period of the second cycle of the bronze-wing pair; it occurred about four hours after the second egg was laid. It was not connected with the initiation of a new cycle, as this did not begin until 5 days later. On Jan. 18 the male took his turn at nesting the egg; when he came off about 2 p. m., he charged at the female and even pecked her rather roughly twice. She soon went to the nest. (R 28.)

PERIODICITY OF THE DISPOSITION TO FIGHT.

The disposition to fight comes on in all pigeons at the time of mating and choosing a nesting-place. While fixing on a nesting-place, the male of a pair of homers (C. tabellaria) was pugnacious, fighting off other males whenever they came near. The disposition to fight and drive other doves away from the neighborhood of the nest becomes strong as soon as nest-building begins, and it is still stronger when the eggs are laid and after hatching. My young male passenger-pigeon (Ectopistes) while mated with a blond ring-dove bristled up, raised his wings, and scolded fiercely at the crested pigeons in the adjoining pen as soon as the first egg was laid. The wood-pigeons (C. palumbus) have been cooing and courting for at least 2 weeks, and the male drives off the 3 or 4 others of his own species from the nest. He seems to care little for the presence of birds of another species. A pair of white-faced pigeons (Leucosarcia picata) became savage, during the nesting period, towards the others of their own species, and I had to remove all of them; other species, however, were left with this pair. (R 4, R 11, Sh 8/13.)

In pigeons, as in many other birds, this disposition to resist enemies shows itself as soon as a place for a nest is found. While showing a passionate fondness for each other, both male and female become quarrelsome towards their neighbors. The white-winged pigeon (Melopelia leucoptera) of the West Indies and the southern border of the United States is one of the most interesting pigeons I have observed in this respect. At the approach of an intruder the birds show their displeasure in both tone and behavior. The tail is jerked up and down spitefully, the feathers of the back are raised, as a threatening dog "bristles up"; the neck is shortened by drawing the head somewhat below the level of the raised feathers, and the whole figure and actions are as fierce as the bird can make them. To the fierce look, the erect feathers, and the ill-tempered jerks of the tail there is added, moreover, a decidedly spiteful note of warning. If these manifestations are not sufficient, the birds jump toward the offender, and if that fails to cause retreat, the wings are raised and the matter is settled by vigorous blows.

The pugnacious mood is periodical, recurring with each reproductive cycle, and subsiding like a fever when its course is run. The birds behave as if from intelligent motive, but every need is anticipated blindly, for the young pair, without experience, example, or tradition, behave like the parents.

It seems to me that this mood or disposition,[1] although in some ways appearing to be independent of the disposition to cover the eggs, can best be understood as having developed in connection with the latter. It has primarily the same meaning—protection to the eggs—but the safety of the eggs and young depends upon the safety of the nest, and this accounts for the extension of this period to cover all three stages, building, sitting, and rearing.[2]

JEALOUSY.

The males of mourning-doves display as much jealousy as the common doves and the ring-doves, driving their wives at the approach of other doves. I have been amused at seeing a male guard his mate. In one instance I had placed a small cage with ring-doves within the larger mourning-dove pen. The male mourning-dove would drive his mate away from the cage to the farther end of the pen, then he would run back to the cage and peck at the ring-doves, and then back to his mate, billing her most affectionately. This behavior he kept up from day to day. Such behavior is a proof of his making a most decided distinction between his "mate"—a ring-dove—and other ring-doves.

In the case of a pair of homers I noted that just as soon as the nest-place was settled they both began to spend some time on the nest every day; copulations were then frequent (about a week before laying). A little later the male began to follow the female jealously and closely if other males were about. This male became so anxious to guard his mate that he would not, in many cases, stop to fight off a strange male, not even from his own stand; instead, however, he kept his eye constantly on his mate, paying no heed to another male, except to drive his mate out of reach of the latter.

The female bronze-wing, of a pair just beginning to build a nest, came to the floor at a time when several geopelias were strutting about in the adjoining cage; the male bronze-wing, for this or some other reason, tried to drive the female away. In doing this he drooped his wings a little, raised the feathers of the back, and charged at her, uttering a rapid note—a "cluck" rolled off rapidly. The charge was made several times and reminded me of similar charges that I have seen made by *Geopelia* males.

When a male dove sees another male in the act of soliciting or mounting his mate, he appears to understand instantly what the purpose is, and often makes frantic efforts to intercept the act, giving the danger signal in the loudest and most excited form and flying directly at the offending bird.[3]   (R 20, R 4, R 33, R 7.)

---

[1] This disposition to fight is further described near end of Chapter VI; see topic "Resistance to intruders".—ED.

[2] Woods Hole Biological Lectures, p. 326.

[3] If one male sees another male copulating, he immediately flies to knock him off. It is surprising the distance at which the bird will recognize this. If two birds make ready to copulate on the barn, a male which is on the roof of the house will immediately spy them and fly to prevent it. It is not that the intruder wishes definitely to take the place of the first male. His chief aim seems to be to knock the first male off. He then generally struts about and coos to the female, but only in a very general way. I have seen female birds act similarly, but the female's object of spite seems to be the other female, whom she attacks, pecking her head. Espinas says that even dogs do not seem to know when other dogs are copulating, and do not show jealousy. The birds certainly seem to understand what is going on. I think Espinas is wrong in regard to dogs. Dogs seem very peaceable in the matter, yet they seem to know what is going on. You commonly see half a dozen males after one female. Among pigeons, a bird sees and knows what is being done just as surely as you or I do. (Conv. 7/2/10, W. C.)

11

Reprinted from *Bull. Amer. Mus. Nat. Hist.*, **68**(7), 483–500 (1935)

The Courtship of Gould's Manakin
(Manacus vitellinus vitellinus) on Barro
Colorado Island, Canal Zone

FRANK M. CHAPMAN

MY STUDIES ON BARRO COLORADO

My studies of Gould's manakin were made on Barro Colorado Island, in the Canal Zone, between January 20 and March 24, 1932,[1] and December 9, 1934 and April 17, 1935. In 1932, observations were made on 43 days; in 1934–1935 on 62. They were usually begun at

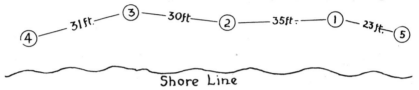

Fig. 4. Position of ''Courts'' in the Laboratory Group. 1932.

7:30 to 7:45 and continued until 9:30 to 10:00 A.M. Occasionally I was afield soon after daybreak, also at mid-day and in the late afternoon.

In 1932, manakin court groups were found on the small peninsula 300 yards east of our laboratory pier (Laboratory Group); on the slopes west of the upper part of Fuertes Estero (Fuertes Group); near

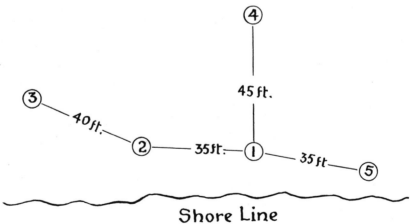

Fig. 5. Position of ''Courts'' in the Laboratory Group. 1935.

the end of the Miller Trail (Miller Group); at No. 23 Armour Trail (Armour Group); and (by Dr. Ray Carpenter) on a small peninsula

---

[1] An abridged, popular report of these observations was published in 'Natural History,' Nov.–Dec., 1932, pp. 470–480.

west of No. 6 on the Fairchild Trail (Fairchild Group).   Studies were made chiefly at the Laboratory and also at the Fuertes Group.   Both contained five courts.

In 1935, thanks to the stable conditions that prevail on Barro Colorado, the Laboratory, Fuertes, Armour, and Fairchild Groups were again active.   A group was found at No. 2, Donato Trail, which may have been occupied in preceding years.   The location of the Miller Group was not re-visited.   Studies were made chiefly at the Labora-

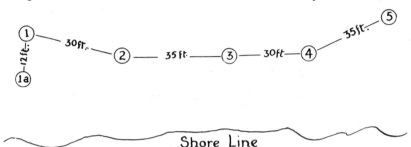

Fig. 6.   Position of "Courts" in the Fairchild Group.   1935.

tory Group (5 courts), Fairchild Group (7 courts), and Donato Group (5 courts).

Male Gould's manakins, when at their courts, can be approached to within a distance of eight to ten feet.   Beyond the cover afforded by the vegetation, no concealment, therefore, was required in the study of these birds.   It should be obvious, then, that whatever merit this study

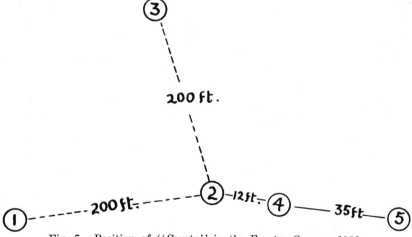

Fig. 7.   Position of "Courts" in the Fuertes Group.   1932.

of manakins may possess is due in no small part to the coöperation of the birds themselves.   Observations were recorded in a field book as they were made and were subsequently "posted" on sheets under the subject heading to which they belonged.

In addition to notebook, pencil, and field-glass, my equipment consisted of a leather-bound cushion, which provided a comfortable, dry, and comparatively insectless observation post, a "snapper," with which to imitate the sounds made by the birds, and mounted specimens of both sexes of Gould's and the yellow-thighed manakins, with feathers and Duco cement with which to modify them, the use of which is described beyond.   I employed also, with only indifferent results, both "still" and motion cameras.

The weather records of Barro Colorado as a meteorological station have been used in preparing this report.

Outline of Courtship Habits

In order that the more detailed descriptions of the courtship habits of Gould's manakin here presented may be considered in relation to the whole of which they form a part, I present an outline of the bird's activities and organization during the breeding season, as I understand them.

Like the ruff (*Machetes pugnax*), and certain, perhaps all, hummingbirds, some grouse, Paradise-birds, and other birds, Gould's manakin conducts its courtship at a *lek,* or regularly frequented locality. Here, during a mating period of not less than eight months, several males gather and are visited by females.   At such localities, always in the forest, each male clears a small space on the ground (here called the "court") which becomes the focal point of his existence.   Here his rights are rarely disputed, here he produces certain sounds and presents certain performances in order to attract the attention of the female and induce her to take part in his display as a means of stimulating her sexually and securing her consent to coition.

Beyond this connection, at a time when her eggs are doubtless ready for fertilization, it is believed that the adult female has no other association with the male.   In the sense of pairing, like the ruff, grouse, and hummingbirds, she has no mate.   Unaided, she builds her nest, incubates her two eggs, and rears her young.

It will be seen that this type of courtship organization, with the court as the daily frequented center of sex life, offers an exceptional opportunity for the prolonged and continuous study of both individual and group life. The male's claim to ownership of a clearly marked area demonstrates a phase of territorialism, while the visit by the female to the male in his court seems equally well to illustrate the theory of sexual selection.

EXTENT OF BREEDING SEASON

In 1932 I did not begin to study manakins until January 20, when their breeding season was well under way. In 1934 I reached Barro Colorado on December 7. On the 11th, manakins were first discovered on their courtship grounds, but it was not until the 25th that their courtship *whirrs* and *snaps* were noticed. Between that date and the 28th these sounds were heard at four different localities, indicating that, on Barro Colorado, Gould's manakin begins to court the first half of the last week in December, in other words, at what is usually the beginning of the dry season.

Of interest in this connection is the fact that in both 1926 and 1934 the yellow-thighed manakin (*Pipra mentalis minor*) was first seen courting on December 25.

In 1935 my observations ceased on April 17. While some individuals then seemed less responsive, sexually, than they had been earlier in the season, group activity suggested that the courtship season was far from ended.

Whether more than one brood is raised is unknown, but observations made on Barro Colorado show a high percentage of nest fatality. It is more than probable, therefore, that many females make at least two, perhaps more, attempts at nesting during the same season, and thereby prolong the nesting period.

The activities of courtship are exhibited both in the morning and afternoon. Early in the season they cease at about 9:30 A.M.; later they continue until about eleven o'clock. An essentially similar variation is shown in the afternoon.

The males may be found at or near their courts throughout the day. At intervals of two or three hours they may absent themselves for as many minutes in order to secure food. (See beyond.)

My own incomplete observations, supplemented by those of Drs. Gross and Van Tyne on Barro Colorado and by Mr. Harrower at Gatun, presented beyond, show that courtship and nesting season extends at least to the latter part of August. It covers, therefore, a period of not less than eight months and includes boh dry and wet seasons. When we discover that the intense, competitive sexual life of these birds requires that they be ever ready to respond to the wants of the female, this is indeed a surprisingly long period. It is true we have no data to show that the same male functions continuously during this time. But my daily observations of the same individual show that he is sexually active for nearly four months, and it is not impossible that he may continue to function to the end of the nesting season. On the other hand, it is not impossible that different individuals may have different nesting dates and hence that this eight-month period may represent the nesting season of the species rather than of the individual. But the stability of group and court organization during my nearly four months of continuous observation is not in favor of this theory.

Seasonal Variation in Time

In 1934–1935 my studies of Gould's manakin began December 9, about two weeks before the inauguration of the nesting season, but, unfortunately, in 1932 they were not begun until January 20. So far as season is concerned, therefore, comparison of the two years cannot properly be made before the last-named date. Moreover, the stuffed female used in these studies throughout the 1934–1935 season was not available in 1932 until February 16, thereby further restricting the comparable periods. Nevertheless, enough comparable data remain to show that there was a marked difference both in the dates and the character of the development of the two seasons, the phenomena of 1932 appearing earlier and being, on the whole, more pronounced than in 1935. Thus, as before stated, the date on which coition with a stuffed female was first attempted in 1932 was February 16; in 1935, March 11. Moreover, the ardor displayed by the male of February 16, 1932, warrants the belief that he would have accepted the stuffed bird at an earlier date. Other activities of court-life seemed to have been performed with greater enthusiasm, and the responses to the mounted bird were more evident in 1932.

This delay or indefiniteness in the advance of the season was also manifest in the blooming of certain trees. Thus the flowering of neither the guayacan (*Tabebuia guayacan*) nor jacaranda (*Jacaranda copaia*) reached its usual climax. Of the former we saw only scattered trees from time to time. At no period could more than half-a-dozen flowering trees be seen from the laboratory on the mainland across the lake, while in former seasons as many as 140 inflorescent trees have been counted at one time. Of the jacaranda only scattered blooms were noted; not one tree acquired the mass of flowers so characteristic of this species.

It was the emphatic opinion of those of us who were working on the island during the winter of 1934–1935 that the season was exceptionally wet and cold. On the whole the records of our rainfall and temperature seem to warrant the strength of our impression. I append the data.

Rainfall

	December	January	February
Average.—1927–1933	8.81	1.71	0.86
1931–1932	2.67	1.20	0.97
1934–1935	15.35	1.66	5.91

Temperature

	December	January	February	March
Mean.—1931–1932	?	80.6	80.3	81
1934–1935	80.2	79.6	79.4	?

Compared with the average and also with the season of 1931–1932, it will be observed that the rainfall of December 1934 and February 1935 is much higher, while that of January is essentially the same. It seems possible, however, that the effects of the very heavy rainfall of December may not have been apparent until January and that the birds were further affected by the unusual precipitation of February.

Add the slightly lower temperature of 1935, as compared with that of 1932, and we have, in my belief, sufficient cause to account for the differences observed in the habits of manakins during these two seasons.

Opposed to this theory, however, is the fact that manakins nest until August and thus, well into the wet season. Hence the variation above recorded may be due to other than climatic causes or to insufficient data.

The Manakin and Migration

While the manakins of Barro Colorado, as species, are permanent residents, and, as individuals, probably do not leave the island, the regular return to their mating grounds, at approximately the same time each year, to establish territories to which the female will be invited, is prompted by fundamentally the same motive that induces a migratory bird to return to its nesting grounds. Indeed, a migration in miniature may be observed when the male, inspired solely from within, leaves its perch to visit the heart of its territory for the purpose of sexual display and possible mating and, this purpose accomplished, returns to the place whence he started.

The Courtship Grounds and Court

The courtship grounds of Gould's manakin find an equivalent in the *lek* of the ruff. They are situated in the forest, the same locality being used year after year. The males return, or migrate, to them about two weeks before they begin to clear the spaces on the ground in which they display their courtship activities. During this preliminary period I have heard only their vocal notes, *chee-póoh* and *pée-yuk* and its variants, and have seen no evidence of competition for possession of the spot in which the "court" is to be situated.

What I have termed the "court" of Gould's manakin is a space on the forest floor cleared by the bird of all moveable material (see Fig. 11). The making of the court indicates the arrival of the nesting season and it at once becomes the focal point of its owner's life. To induce the female to come to court is now the chief object of the male's existence. His jurisdiction over his court is rarely questioned. If he can induce a female to visit his court, he not only supplies the suggestive background for his advances but also surroundings in which he need fear no competition.

Courts vary in size and shape but are usually irregularly elliptical in outline and average two and a half feet long by twenty inches in width. They are placed in forests with an undergrowth of small saplings, several of which grow at or near their borders.

The bird removes leaves and other material from its court with its bill. This act seems to be more or less sexual in character and often terminates the court display of snapping. Small leaves are carried to a height of about three feet and dropped at an equal distance from the court. Large ones are taken in a more direct line to the border of the court. One that I saw removed measured $10\frac{1}{2} \times 3\frac{3}{8}$ inches; it was therefore slightly more than two and a half times as long as the bird that carried it.

The court is not only cleaned but kept clean. Leaves that I placed in courts were soon removed. To test the bird's possible sense of color I introduced into a court, at the same time, scarlet petals of the passion flower (*Passiflora vitifolia*) and green leaves of about the same size. There was no regularity in the order of their removal. Sometimes one, sometimes the other was taken first, indicating that the bird was no more responsive to red than to green.

Courts are associated in groups of from four or five to seven and possibly more, and are placed at varying distances from each other. Two courts of the Fuertes Group were only twelve feet apart, but two others of the same group were each 200 feet away from them in different directions. In 1932 the five courts of the Laboratory Group were set at approximately equal distances of about thirty feet from each other. In 1935 some of these courts were near to or on the sites occupied in 1932, but the plan of the group as a whole was different.

The least distance at which we have found court-groups, or *leks*, separated is 300 yards. The intervening area was occupied by very dense growth and included a bit of water about twenty-five feet wide and a difference in elevation of approximately fifty feet. The birds of one court were not heard by me from the other, but it is probable that females occupying the intermediate territory were within hearing of both groups.

Observation indicates that the same localities are used by courtship groups year after year. For example, groups were found 300 yards east of our landing; at the head of Fairchild Cove; at Armour Trail No. 23; and on the slopes above Fuertes Estero in 1932 and 1935. No observations were made in 1933 or 1934, but it seems probable that the localities named were also occupied in those years.

In view of the habit that prompts the female to seek the male, rather than the reverse, it is obvious that a group of males is more likely to attract the female than is a single bird. It also seems clear that unless the members of such a group are organized under laws which all observe, the confusion resulting from ungoverned competition on the occasion of a female's visit would defeat the aims of both sexes. Hence, the company and the court, the continuity of court location, and the law and order of court life.

NOTES AND CALLS

The male Gould's manakin is the author of many sounds of which none, in a musical sense, can be called a song. Some are of vocal, others of mechanical origin. The former are probably uttered throughout the year but doubtless possess additional significance during the breeding season. The latter appear to be restricted to the period of courtship.

The principal vocal notes are *pée-you* and *chee-póoh*. Both are high-pitched, clearly uttered, and of sufficient volume to be heard in the forest, under favorable conditions, at a distance of about 200 feet.

Of *pée-you* there are many variants. It may be given as *pée-yur* or *pée-yuk*, with the final syllable slightly trilled, as a pensive *pee-a-a* or a plaintive *pee-e-e*. *Pée-you* or *pée-yuk* is a note of awareness, address and response, or of inquiry or protest. In the latter sense it is invariably used when a mounted male is placed in a bird's court. At such times it expresses excitement and it may be uttered twenty-five to forty times a minute.

Chee-póoh is a note of address and response, never of protest or alarm. It frequently follows the rolling *whirr* and, like a signal, is repeated from court to court. It is a well-defined call and subject to little variation.

In addition to these calls the bird utters a variety of single syllabled *chees* and *pees* and fragmentary rolls, all more or less conversational or responsive in character. *Chee,* for example, a low, fine note, is uttered as a rhythmic response to the snap in court of a near or distant neighbor. The calling bird shows no excitement, utters its *chee* after each snap, when it seems to express awareness of and sympathy with the snapping bird's activity.

The mechanical calls of Gould's manakin are given only by the male, and, apparently, only during the breeding season, and hence function as songs. They include a *snap* or *crack*, a *snip*, a snapping *whirr,* and a reedy *whirr,* and are all made with the wing feathers.

The *snap* or *crack* is given only as the bird jumps from one perch to another, distant usually from two to four feet. It has somewhat the character of the sound produced by the explosion of a percussion cap and is apparently made by the sudden, violent impact of the secondaries, one upon the other, which evidently can be secured only by the muscular effort accompanying the act of jumping. As a rule the bird snaps only in or near court. A single *snap* is given with each jump and the frequency of production is determined by the length of the jump, whether the route travelled is around or across the court, and the rapidity of the bird's movements.

It is difficult to believe that so loud and so hard a noise as the snap can be produced by an object as soft as a feather. Doubtless, for this reason, it has often been attributed to a snapping of the mandibles. But if this were true we should expect to find the bill of the male heavier or otherwise different from that of the non-snapping female. Furthermore, if the snap were made by the bill it would not, presumably, be necessary to jump when snapping. On the other hand, the somewhat large, slightly curved and stiffened shafts and broad webs of the secondaries in the male are doubtless of functional value in producing the *snap*, and represent a stage of development which in *Machaeropterus,* as before stated, has reached its known maximum.

The *snip* is a restrained or nascent snap. It is also, in my belief, produced by the secondaries but does not call for the effort required by the *snap* and, hence, can be given while its maker is perched. It accompanies the arched jump over the court and is heard when several males in the tree tops are pursuing each other or are in joint pursuit of a female. At such times the effect of their combined *snips* resembles the sound produced by an exploding pack of diminutive firecrackers.

In the whirring or rolling *snap* the perching bird throws itself slightly forward, raises its wings above its back until they are within less than an inch of one another, but do not touch, snaps its secondaries and rotates its primaries in such a way as to produce a combined *snap* and *whirr.* This is done with a single effort lasting not more than a second or two, during which the wings are held aloft and the outline of their outer margin can be clearly seen while that of the feathers is blurred.

This sound, while not so loud as the jumping snap, is a surprising sound, and also denotes sexual excitement. It may be uttered anywhere, in court or out, and usually announces the return of that condition which prompts court display. It is frequently followed by the vocal call *chee-póoh* and is answered in kind by birds throughout the group.

The reedy *whirr,* the fourth of the mechanical calls produced by the male of Gould's manakin, is a comparatively low, short, vibrant call without a trace of snap, and, in my opinion, is produced by its deeply incised outer primaries. It usually follows a round of court snapping as the bird flies upward to a perch near the court; but it can be produced in less resonant tone by the perching bird. At such times the wings are half-raised.

Of these mechanical calls the *snap* and the snapping *whirr* are the loudest and hence doubtless the most important. On a still morning, over water, I have heard them at a distance of 300 yards. When courting, the males are confined by habit to an area a few square yards in extent and it is therefore essential that they have some effective means of announcing their presence to the females who may be in need of their attention. Lacking a voice of much volume, like many other birds, they use their wings to produce sounds that will notify receptive females of their location and invite them to share the functions of the court.

The Observance of Territorial Boundaries[1]

The success of the *Manacus* system of courtship is based on a rigid observance of territorial rights. With the exception of the two unusual instances, described under "Interrelations of Males," no living male manakin was seen in another's court and experiment shows that territorial boundaries extend a well-defined distance beyond the border of the court itself. When the courts are comparatively near one another the intervening area is divided between their owners, the amount belonging to each being in apparent relation to the birds' sexual development. Both distance and development were measured by the use of mounted birds. In 1932 only a female was available. In 1935 both sexes were employed, but the male provoked a greater response.

From many experiments I select the following: Courts 1 and 1-a of the Fairchild Group were but twelve feet apart. Their owners had no relations with one another, each perching near its own court. On March 12, a stuffed female was placed in Court No. 1 and a few minutes later was removed to Court 1-a. She aroused a similar and only slight response from each court owner and was soon replaced by a stuffed male, first in No. 1, then in 1-a. This bird was attacked in both courts, No. 1 acting more vigorously than 1-a. The stuffed male was then placed at different distances between the two courts when it was found that No. 1 attacked it up to a distance of seven feet from his court, No. 1-a remaining in his territory; but beyond seven feet it was attacked by No. 1-a, while No. 1 did not go beyond the seven foot line. March 17 this test was repeated with essentially similar results.

This division of the space between courts was found to occur up to a distance of seventy feet. In the Donato Group, this distance separated Courts Nos. 1 and 2. There, on March 8, 1935, the owner of Court No. 1 responded to the introduction of a mounted male in his court chiefly by the assumption of a pose of nine minutes duration. The owner of No. 2, on the contrary, was far more aggressive and, after jumping and snipping, just outside court, flew at the mounted male, perched on his head and picked at it furiously. This action was at

[1] Since writing this section I have read Dr. Ernst Mayr's classification of territories in his paper on 'Bernard Altum and the Territory Theory' (Proc. Linn. Soc. N. Y., 1933–1934, p. 33) from which it is clear that, as a territorialist, *Manacus vitellinus* belongs in Section "II. Mating station, but not feeding ground. (b). Not connected with nest (Ruff, many Tetraonidae, Paradisaeidae)."

once repeated whenever the bird was driven off and permitted to return.

When the mounted male was moved gradually from Court No. 2 to Court No. 1 it was found that No. 2's interest extended up to forty-five feet while No. 1 did not appear until the mounted bird was within twenty feet of his court.

An unusual illustration of the recognition of territorial boundaries was supplied on February 26, 1935, by the owners of Courts No. 3 and 4 of the Fairchild Group. At 8:53 A.M. a mounted female was placed in Court No. 4. At 8:54 the male returned, *whirred*, hopped about with beard partly extended, but made no close approach to the mounted specimen. This slight demonstration attracted the attention of No. 3 who came from his court, distant thirty feet, and perched about midway between the two courts. Meanwhile a living female appeared in Court 3 (whether in the male's absence or after his return I am unable to say) and jumped with its owner who *snapped* loudly. This more pronounced demonstration attracted the attention of No. 4, who now left his court, with its stuffed female, and went about halfway toward Court No. 3. There he *whirred, snapped,* and twisted his head in a peculiar spasmodic manner, indeed showed more excitement than he had before, but did not venture beyond what he evidently considered his territorial boundary.

Later study of these two birds showed that No. 3 possessed much greater sexual ardor. He indeed was one of the few males who in 1935 attempted to mate with a mounted female and, correlated with his condition, a test with a mounted male showed that he claimed two-thirds of the distance between his court and that of No. 4.

Wider territory was claimed not only by the most sexually active bird, but, as might be expected, by the dominant bird. This was shown by tests made with a mounted female between Courts Nos. 2 and 4 of the Fuertes Group, which were twelve feet apart, and between Nos. 2 and 3 of the Laboratory Group, which were thirty feet apart. In each case the dominant's territory was about fifty per cent greater than that of the submissive's.

In these, and many similar tests, it was observed that in no instance did the neighboring birds both claim the mounted bird at the same time, nor did either one trespass on the territory of the other. This recognition of boundaries makes for law and order. No time or energy is lost in futile disputes or needless conflicts and the birds may devote themselves to winning the attention of the female, now the chief object of their lives.

The disruptive effects of trespass are clearly shown by the prompt and vigorous response to the introduction of the mounted male in court, as shown in the succeeding experiments.

Interrelations of Males

When one considers the intense sexual ardor of the males and the fact that the mating activities of a number of individuals are conducted in a comparatively limited area, the fact that conflicts between them rarely occur is a tribute to the effectiveness of their group organization.

In the earlier part of the courtship season, before the attachment to their courts is fully developed, two to six males have been observed and heard actively jumping, *whirring,* and *snipping* twenty to thirty feet from the ground in the upper part of small trees. Rarely a female, of which the males seemed to be in pursuit, was observed, and on at least one occasion a male appeared to give chase to another, but, normally, no actual contact between males has been observed. A male's court is his castle and with rare exceptions I have not, under normal conditions, seen his right of possession disputed.

When neighboring courts are comparatively near each other the owner males, during periods of sexual inactivity, often occupy perches within six or more inches from one another. At such times one bird, and always the same, may try to win the attention of the other by actions which, in a measure, suggest the courtship of the female yellow-thighed manakin by the male. With short, mincing hops, the "courting" bird self-consciously approaches the "courted" bird, occasionally half-flitting his wings and quickly bowing his head; this manoeuvre is preceded or followed by a rapid, nervous turn of the head from side to side through an arc of about 120 degrees. At times this motion becomes so pronounced that the bird turns its entire body halfway around and back again. Not a note is uttered, but the beard of the "courting" bird is extended and in periods of extreme activity he appears to be much excited. At all times his attitude toward the male he is addressing is one of supplication or submission and I have called him, therefore, the "submissive." Meanwhile, the bird addressed makes no response and, beyond an occasional change of position, makes no acknowledgment of the apparent homage that is being paid to him. This demonstration may occur at frequent intervals. I have called the bird addressed the "dominant." In three out of four cases his yellow areas were darker than those of the submissive bird.

In only one instance has any question of court ownership been noted. This occurred in 1932, in Court No. 1 of the Laboratory Group. For a period of over a month (January 20–March 1) this court was occupied by two males; at the end of that time apparently one of them (I did not ascertain which, if either) formed a new court twenty-three feet beyond No. 1. One of these males was the darker and in my notes they appear as "Dark" and "Pale." At times Pale acted as a submissive to Dark, courting it as described above. At other times it seemed to be a rival of Dark, snapping and performing in the court as though it were the owner. Always it retreated before Dark who was ever the aggressor in pursuit. Again, for long periods, the two birds perched within a foot or two of one another. But in these apparently peaceful intervals Pale was on guard and, at any movement of Dark, half-raised its wings in anticipation of attack. This situation continued until, as I have said, a new court was formed beyond No. 1, I assume by either Dark or Pale, but without seeing them together I was unable to say which of the two remained in No. 1. On March 24, 1932, when a mounted female was placed in Court No. 1 the owner did not appear, but the owner of the new court, No. 5, came to it, evidence supporting my belief that this court had been built by one of the two birds which had claimed No. 1.

It is interesting to record that three years later a court was still placed at position No. 5 and one within twelve feet of position No. 1, and that No. 1 was submissive to No. 5, their relations being more pronounced than in any other observed case of this type of homosexuality.

When, in 1935, a mounted male of Gould's manakin was placed in Court No. 1, not only No. 1 responded but it was joined by its dominant No. 5 who showed his relations to No. 1 by threatening the mounted bird as actively as did No. 1, the owner of the court. Although, as related under the section on "Territorial Rights," the presence in court of an alien male was, as a rule, vigorously resented, the assistance by No. 5 to No. 1 was accepted by the latter in good part, as in apparent conformity with his relations as a submissive to No. 5.

THE ACTIONS OF THE COURTING MALE

The wide variation in the actions of courting males doubtless depends primarily on their sexual development, as that is affected by the time of the year and the character of the season. But they are also

143

influenced by the nature of the stimulation to which they are subjected, and probably also by temperament and by age.

In 1932, sex-life seemed better organized and its manifestations more pronounced than in 1935; the female more often visited court and there appeared to be less pursuit of her out of court. In the belief that high rainfall and low temperature checked the development of the breeding phenomena of 1935, I use here my observations of 1932 as more nearly representing the normal habits of the species.

While waiting for the appearance of a female, the event of supreme importance in his life at this season, the male perches quietly near his court. Should he have a close neighbor, the two birds often sit within a few inches of one another when there may arise between them the dominant-submissive relation described above.

Should the waiting bird be alone, he squats in relaxed pose on his red toes and seems to be dozing. But his wide open eyes and the frequent movement of his head show that he is constantly on the alert.

To the snapping in court of another member of the group he may, without changing his position, reply with a sympathetic *chee*, repeated rhythmically after each snap. A snapping *whirr* arouses more interest and, raising his wings, he replies in kind. If this signal spreads through the group, indicating that the alarm is not a false one, he proceeds to his own court to perform, or, when court-life is not fully organized, he may take part in a disorderly pursuit of the female, through the tree tops, by several *whirring, snipping* males.

On occasion he appears to be aroused by an inborn emotion. Without apparent cause he suddenly calls *chee-póoh,* becomes more alert, partly erects his beard, and raises his wings for a *whirring snap.* Meanwhile he has moved toward his court, his excitement increases, his beard is thrust forward, and with a series of cracking *snaps* he makes the round of the perches about his court.

Six to ten *snaps,* delivered as rapidly as two to the second, may leave him on the ground in court with beard still extended and yellow areas fully expanded; then, with head on one side, or with bill pointed upward, he assumes a rigid, gaze-pose and holds it for from several seconds to a minute or more.

After this rest, inspired to fresh exertions, he may jump rapidly forward and backward across court and, no matter how short the distance, in some invisible way, he turns in the air to alight, facing the point of departure.

The display is usually ended by an upward jump to a nearby perch, which is accompanied by a low-toned, reedy *whirr*.   This sound, rarely heard at other times, is without trace of *snap* and is made, I assume, by the narrowed primaries.

To this exhibition, No. 3 of the 1932 Laboratory Group added what I have called the dirigible pose.   With bill touching the end of a slender, broken sapling the size of a pencil and about eighteen inches high, it fluttered its wings while holding a horizontal pose; then, bill still pressed to the sapling, it slid down to the court and, with bill now touching a root, wings still fluttering, seemed to be standing on its head.   In spite of this remarkable demonstration, which was not infrequently repeated and seemed to imply the possession of much virility, its author, No. 3 of the Laboratory Group, was submissive to No. 2.

When the male has been prompted to visit court by the presence of the female, this demonstration is evidently designed to induce her to visit him in court.   But in her absence it apparently is inspired by cumulative, ungratified sexual desire, and we may think of his snapping dash around or to and fro across court as a pursuit of an imaginary female, while the subsequent display is a further outlet for excess sexual emotion.

His ineffective attempts to mate with a stuffed female so stimulated No. 3 of the Fairchild Group that he often left the mounted bird to jump and *snap* about his court.

When the performer was not sufficiently stimulated to *snap*, the jump from perch to perch was arched and accompanied by only a *snip*, and at times it was wholly noiseless.

This demonstration is supposed to indicate that the male is ready to "dance" with a female should she come to him.   Whether the dance will lead to coition depends doubtless primarily on the condition of the female.   But the failure of the male of March 20, 1935, in the Fuertes Group to respond to a stuffed female, but to repeatedly mate with a living one, indicates that in some instances the male, as well as the female, requires the stimulation of the dance to induce coition.

Whether during his long period of court-life, which may last for eight months, the male at times is not in condition to mate, I am unable to say, but it is difficult to believe that he would submit to the restrictions of court-life if he were not at all times ready to perform its functions.   The males that mated with a stuffed female were *always* ready to mate with her.

145

THE ACTIONS OF THE FEMALE AND HER RELATIONS TO THE MALE

The female of Gould's manakin is obscurely colored, her single call-note, *pée-you* or *pée-yuk,* is not often uttered, she apparently visits courts rarely, in short, her mating habits are so difficult to observe that my notes on them are far from satisfactory.

It seems clear that when the female is in condition to have her egg or eggs fertilized she visits the *lek* to receive proper attention from a male.   Whether she visits it at other times I am unable to say.   I have seen birds in female plumage in the trees above the courts whose presence did not excite the males.   They may have been young males, they may have been unripe females.   If, however, the males are able to recognize the oestrous female, why are they so often deceived by a mounted female, as described beyond?   On the other hand, it seems improbable that a female would nest in a *lek* within forty feet of the nearest occupied court without being annoyed by the males, who were doubtless at all times conscious of her presence, unless, in some way, she could inform them that she did not wish to be courted.

Usually I was made aware of the presence of the female by the activity of the males.   Presumably the first male to see her produces his snapping *whirr* often followed by the vocal *chee-póoh*.   The significance of this demonstration seems to be understood and it is repeated by bird after bird throughout the *lek*.   One or more of the birds then goes to court and snaps.   It will be observed that under these conditions the male does not go to the female, but to his court, in an apparent attempt to induce the female to visit him there.   In court he will not have to meet the competition of other males; in court the stage is set for the performance which may lead to the consummation of his desires.   For whatever be the sexual condition of the female, she apparently must be courted before she will receive the male.

Producing, therefore, their loudest sounds from fixed localities, the males issue their invitation and, in accepting one of the several available, the female seems to select her partner.   But before giving her consent to coition she apparently is to be further stimulated by a performance which may be called the mating dance.   At its full development this is a rhythmic jumping across court, or its immediate surroundings, in which, facing one another, the two birds jump at the same moment, pass in midair, and repeat from the same or new perches.   While the birds act in unison and thus give to their actions the character of a concerted performance, this so-called "dance" is possibly a mock pursuit in which the female, instead of making a genuine effort to escape the male, keeps just beyond him and possibly thereby increases his desire to secure her. The male *snaps*, the female is silent; there is an air of excitement about this performance which often produces no observable result, may lead to coition in court, or may be terminated by what appears to be a mating flight in which the two birds disappear.

It will be seen, then, that the completed cycle of this joint demonstration requires (1) that the sexual condition of the female prompts ·her to seek the male of whose suggestive calls she may be within hearing. (2) Arrived at the *lek,* the more evident and urgent invitations of the males further arouse her desires until she is induced to visit the court of one of her ardent wooers; an apparent instance of sexual selection. (3) Not yet aroused to the point of sexual consent she evades the male's advances by jumping across court as he springs toward her, but if the excitement of the chase sufficiently stimulates her she may mate with the male in court one or more times, or coition may follow the end of a mating flight.

When court organization is neither complete nor binding, or, possibly, when the female is not sufficiently developed to respond to the invitations of the male to visit his court, the males leave their courts and with much *snipping* pursue the female in the tree tops. The result is confusing and apparently the ends of neither sex are served.

* * * * * * *

12

Reprinted from *Trans. Linn. Soc. N.Y.*, **4**, 57–69 (1937)

Studies in the Life History of the Song Sparrow I

MARGARET M. NICE

CHAPTER VI Territory Establishment

Melospiza melodia, in my experience, is a typically territorial bird, behaving very much as does Howard's classic example—the Reed Bunting (*Emberiza schoeniclus*). Territory is of fundamental importance to the Song Sparrow on Interpont—the basis of its individual and social life for more than half of the year. Special ceremonies are concerned in the establishment of territory; the matter of song is closely bound up with territory, while males show a strong and lasting attachment to their individual territories.

A. The Establishment of Territory

That territorial behavior is deeply ingrained in my birds is evidenced from two things: the elaborate ceremonies that are involved in its maintenance, and the part it plays in the change from juvenal to adult singing.

When a new male Song Sparrow arrives in spring, the neighboring males at once try to drive him off. If he is a transient, he flies, but if a candidate for a territory, he stands his ground—and then the "territory establishment" begins.

The complete procedure consists of five parts: assuming the role; staking out the claim; the chase; the fight; and finally the proclamation of ownership of each bird on his own bit of land.

In the first part the two birds show diametrically opposed behavior. The invader—puffed out into the shape of a ball, and often holding one wing straight up in the air and fluttering it—sings constantly but rather softly, the songs being given in rapid succession and often being incomplete. The defender, silent and with shoulders hunched in menacing attitude, closely follows every move of the other bird.

The newcomer continues to sing flying in this peculiar puffed out shape from bush to bush that he wants to claim. Soon the owner begins to chase the intruder, but the latter, if determined, always returns to the spot he wants to claim. The chasing continues and at last finishes with a fight on the ground. After this the new bird is either

58

routed or both males retire to their respective territories, and sing loud and long, answering each other.

In less serious encounters the chasing and fight are omitted, the first and last parts only being indulged in. When, however, affairs are in deadly earnest, as in the spring when a summer resident returns and finds a resident has adopted his old territory, there is little wing fluttering and puffing, merely the singing, chasing and fight.

With a thickly-settled Song Sparrow population, territory establishment ceremonies of all degrees of seriousness may be seen throughout the year except during the molt; in fall and winter they are not common and occur only on mild days. At these seasons a bird will go some distance to start a "territory establishment" with another male with whom there is no question of real conflict over boundaries. In such cases the roles of despot and underling are freely interchanged. Excluding the very mildest territory establishment manifestations that are indulged in only by a young bird in the fall on some occasions when another Song Sparrow alights on a branch above him, the less serious the encounter, the more prominent is the posturing, bluff taking the place of action. As Howard, *86*, p. *37*, says "violent wing-action and violent contortions of the body are associated with postponed reaction."

When a summer resident returns to find his old territory pre-empted by another bird, at first the new arrival takes the role of the invader and is pursued by the bird in possession, but it does not take long for an old bird to reverse matters; after a fight or two he becomes the defender and drives his rival. Burkitt, *29*, tells of an old Robin Redbreast (*Erithacus rubecula*) being driven from his territory by a young bird; but this has not happened to my knowledge with the Song Sparrows; with them the old bird usually drives off the interloper, although sometimes he will take a neighboring territory. But as this sometimes happens under no pressure from other males, we cannot be sure that the old male was really intimidated by the young one.

Territory establishment ceremonies have not been worked out in such detail with any other species so far as I know. Howard writes of "butterfly-like" and "moth-like" flights, and of rapidly vibrated wings, and Pickwell, *147,* describes the boundary quarrels of the Prairie Horned Lark (*Otocoris alpestris praticola*) which show much resemblance to those of my Song Sparrows, except that the fight takes place in the air. But neither of these authors clearly differentiates between the behavior of the two participants, perhaps because they worked with unbanded birds. The Micheners describe what they think may be "a ceremony marking territorial lines" with Mockingbirds (*Mimus polyglottos leucopterus*), where one of the owners of the territory "came to the fence and approached the unbanded bird facing it and bowing and bobbing. One would step forward and the other back and then they would reverse," *123,* p. 126. Closely similar behavior is reported by Laskey, *100a,* with *Mimus p. polyglottos.*

It is reasonable to expect strongly territorial species to have special instinctive reactions by which territory questions can be settled. In order to observe and understand these, however, one must have individuals plainly differentiated; one must study the birds from the first taking up of territory; one must study two or three pairs intensively at first and finally there must be a sizeable population, so that territory establishment behavior can be shown. In 1935, for instance, when there were very few Song Sparrows, I saw almost no activity of this nature, although I was especially on the look out for it.

B. Territory and the Development of Song

Volumes could be written on the matter of song and territory, but I will confine myself to a brief treatment of two features.

With *Melospiza melodia* song is the chief means of proclaiming territory; the taking up of territory in late winter and the beginning of zealous singing coincide; while the main season of Song Sparrow song on Interpont is in March before the arrival of the females.

Territory has a powerful influence on the development of the Song Sparrow's juvenal warble into the short separate songs of maturity. A young bird may be warbling along peacefully by himself, but the moment a territory rival appears, the singing becomes almost typically adult. In late February a young bird may warble in low situations on his territory, but when he sits high in a tree proclaiming ownership, his songs are adult in form. The young transient males that pass through in March warble freely, but I have never heard a

young summer resident male warble in the spring on Interpont; upon
the arrival at the nesting grounds the bird reacts as an adult. With
the young residents the warble is given up in late February and never
reappears, all of the late summer and fall warbling coming from
young birds.

C. Summary

1. The Song Sparrow has a special ceremony consisting of posture, song and fighting for the procuring and defending of territory.

2. The new bird takes a humble, subservient role, the owner a dominating, threatening attitude.

3. The complete ceremony consists of five parts: assuming the role; staking out the claim; the chase; the fight; the subsequent proclamation of ownership.

4. Song is *Melospiza melodia's* chief means of proclaiming territory.

5. The young male has a continuous song of warbling character; but in territorial situations this is changed to the adult form of song.

References

29. Burkitt, J. P. 1926. A Study of the Robin by Means of Marked Birds. British Birds, 20:91–101.
86. Howard, L. E. 1929. An Introduction to the Study of Brid Behaviour. Cambridge. 136 p.
100a. Laskey, A. R. 1935. Mockingbird Life History Studies. Auk, 52:370–382.
123. Michener, H. and J. R. 1935. Mockingbirds, their Territories and Individualities. Condor, 37:97–140.
147. Pickwell, G. B. 1931. The Prairie Horned Lark. Trans. Acad. Science of St. Louis, 27:1–153.

CHAPTER VII
Territory Throughout the Year

The actual breeding season of the Song Sparrow lasts from 3½ to 5 months, but the territory is inhabited by the summer resident male from 6½ to 8 months and by the resident throughout the year. It is not, however, defended during the molt, nor the cold of winter, and only to a limited extent in fall.

A. TERRITORY IN THE FALL

The Song Sparrows normally molt in August and September, an occasional bird not finishing till October. Because of my absences from Columbus at this season I do not have much data on the molt of the adults. Wharton, *199*, in Groton, Mass., says the molt of his local Song Sparrows begins during the second 10 days in August and lasts from 40 to 45 days, but from my scattered observations I should expect it to last longer. Magee, *118a*, states that the wing molt of Purple Finches (*Carpodacus p. purpureus*) takes 10 weeks on the average. In 1930, perhaps in some way due to the unprecedented drought, the birds started to molt the middle of July and were through molting more than two weeks before their usual time.

1. *Singing in the Fall*

With the adult males there is a recrudescence in fall, in a lessened degree, of spring behavior so far as territory is concerned. Young males that have settled unmolested during the molt of the owner, are now driven off with appropriate territory establishment procedure, although other Song Sparrows are tolerated. Singing is heard again from some of the adult residents, while others are practically silent. During normal years the singing from summer residents is of irregular occurrence, but in 1930 there was a wonderful amount from both classes of males.

With many of the birds entirely through the molt the 10th of September instead of the last of the month as usual, with fine weather in September and an extraordinarily mild early October, and with the migration not taking place until its usual time in mid-October, we enjoyed a most unusual treat of Song Sparrow music. The summer resident 1M in 1929 sang Sept. 28, 29 and Oct. 4, but in 1930 from Sept. 17 to Oct. 11. Song was recorded from another summer resident—10M—Sept. 10 to Oct. 11 in 1930; Sept. 28 and Oct. 4, 1931; Oct. 9, 1932, and Sept. 28, 1933.

4M's early morning singing has started on the following dates: Sept. 29, 1929; Sept. 10, 1930; Sept. 28, 1932 (we returned to Columbus the day before); Sept. 28, 1933; Sept. 30, 1934; and Sept. 29, 1935. Considerable warbling is heard from juvenals in the fall—from residents, summer residents, transients and winter residents.

2. *Taking Up of Territories*

Many young residents take up their territories in their first fall and keep them for the rest of their lives; others try to do the same but are driven out by the owner when he completes his molt; still others do not settle down until February. I do not know whether this difference depends on age or other factors.

Some young summer residents also choose their territories in their first fall and return to them the following spring.

185M was caught in our garden Aug. 3, 1933, in juvenal plumage and was noted warbling 50 meters to the south from Oct. 4 to 6; on Mar. 16 he returned to the very same spot. In 1931 a right-banded bird warbled constantly west of our garden on Sept. 28 and Oct. 15, but I was not able to trap him; on Feb. 27 a right-banded bird returned and took up his territory in this same spot (112M). On Oct. 1 I banded 134M and found him Oct. 6 warbling south of the third dike; on Apr. 1 he returned and took up his territory about 100 meters to the south of this place, which at this time was entirely filled by other males.

Burkitt's, *28,* young Redbreasts (*Erithacus rubecula*) took up territories in July and August; Miller, *125,* found that with California Shrikes (*Lanius ludovicianus gambeli*) fall is the main time for taking up of territories; the Micheners report that young Mockingbirds (*Mimus polyglottos leucopterus*) do so in August and September, *123,* while British Stonechats (*Saxicola torquata hibernans*) settle in pairs on their territories in October, *101a.* But all these species defend their territories throughout the year. It is interesting to find the Song Sparrow, which defends his territory only during the breeding season, settling on it so early in life.

B. Behavior in Winter

It may be largely a matter of habit that keeps the adult residents of both sexes in the vicinity of their territories throughout the winter, if sufficient food and cover are present. Similar behavior is shown by the winter residents, in a few cases for a number of years, as with W6, as told in Chapter IV.

At this season the male resident may range over an area approximately 150 by 225 meters, a district six to ten times as large as the breeding territory. In cold spells birds may come unusual distances for brief visits to my feeding station, several from 270 meters, while two traveled more than 500 meters (57M and 58M, see Maps 9 and 13).

In cold, snowy weather Song Sparrows are apt to form into small flocks, the organization of which is very loose. On Jan. 16, 1931, I watched 50M leave his regular flock in our graden and join another below the first dike, the birds here paying no special attention to him. After staying with them for five days, he returned to his former companions. These flocks on Interpont are *not made up of "family parties" nor of "neighborhood groups,"* since they are composed of both residents and winter residents, and family ties are broken with the young when the latter are a month old; while mates, even if both are resident and winter near together, apparently pay no more attention to each other in fall and winter than they do to strangers.

C. Behavior in Spring

In late January or early or mid-February, depending on the weather, the resident Song Sparrows begin to take up their territories —isolating themselves through hostility to other members of their species and making themselves conspicuous by song.

1. *Song and Temperature*

Song gradually comes to an end in November, and no matter what warm and pleasant weather may occur in December, only occasional snatches of song are heard. (There have been three warm spells in December of three days duration and one of six days during the period of this study; mean temperatures ranged from 7.2°-14.4° C. (45°-58° F.), or 7.2°-15° C. (13°-27° F.) above normal, the median temperature being 10° C. (50° F.).) But in January song usually begins again, there having been from 4 to 16 days per month on which a fair amount of song was recorded from 1930 through 1935. Table VI shows the mean temperatures at which the Song Sparrows started singing.

64

TABLE VI

Lowest Mean Temperatures That Started Singing

Date of Start of Singing	Mean Temperature of Day of Start and Two Previous Days Centigrade			Fahrenheit			Normal Temperature of Day of Start C.	F.
Jan. 7, 1930 - - -	3.8	9.4	12.2	39	49	54	—1.7	29
Jan. 8, 1935 - - -	8.3	10	12.2	47	50	54	—1.7	29
Jan. 13, 1930 - - -	— 2.2	5.6	8.9	28	42	48	—1.7	29
Jan. 13, 1932 - - -	0	8.9	13.3	32	48	56	—1.7	29
Jan. 19, 1933 - - -	7.2	6.6	8.9	45	44	48	—2.2	28
Jan. 21, 1934 - - -	2.2	4.4	8.3	36	40	47	—2.2	28
Jan. 24, 1931 - - -	— 2.2	1.1	6.1	28	34	43	—2.2	28
Feb. 2, 1930 - - -	— 3.3	0.6	4.4	26	33	40	—1.7	29
Feb. 2, 1932 - - -	— 6.6	—4.4	2.2	20	24	36	—1.7	29
Feb. 2, 1935 - - -	— 2.8	1.1	0	27	34	32	—1.7	29
Feb. 7, 1934 - - -	— 1.7	—3.3	—2.8	29	26	27	—1.7	29
Feb. 9, 1935 - - -	— 3.3	3.8	2.8	26	39	37	—1.1	30
Feb. 11, 1934 - - -	—16.8	—8.9	—2.2	2	16	28	—1.1	30
Feb. 14, 1936 - - -	— 6.6	1.1	0	20	34	32	—1.1	30
Feb. 24, 1936 - - -	— 5.5	2.2	8.9	22	36	48	0	32

There has been some singing on the 7th and 8th of January following two warm days, and from the 13th to 21st following one warm day. From Jan. 21 singing has started in earnest when the previous day was only 3.3° C. (6° F.) above normal; by Feb. 2 singing has been heard on the first warm day, and by the 7th may reappear after an interval of bleak weather at a temperature slightly below normal. Singing appeared Jan. 7 and 8 at temperatures 14° C. (25° F.) above normal; from the 13th to 21st at 10°-15° C. (19°-27° F.) above normal; on the 24th at 8° C. (15° F.) above; on Feb. 2 from 2°-7° C. (3°-12° F.) above and Feb. 7 and 11 at 1.2° C. (2° F.) *below*. In 1936 when there had been no previous singing, it started on Feb. 14 at 1.1° C. (2° F.) above normal, and restarted on the 24th at 9° C. (16° F.) above.

That singing appears at progressively lower temperatures is clearly shown in Chart VIII, for which Prof. Selig Hecht of Columbia University kindly drew the curve and gave me its formula.

Ts.=54.2° F. —0.7d. (12.3° C. —0.39d.).

Ts.=the temperature at which singing starts, d.=day, 0.7=the constant indicating the slope of the curve. Or in other words the threshold of singing was 54.2° F. (12.3° C.) on Jan. 7 and decreased about ¾ of a degree Fahrenheit (about 2/5 of a degree Centigrade) each day.

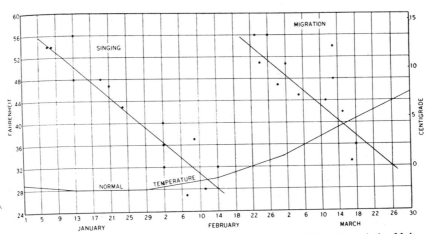

CHART VIII. *Threshold of Singing of the Residents and Migration of the Males. Dates of Start of Singing, 1930 to 1936, as shown in Table VI. Dates of Migration of Breeding Males, 1930 to 1935, as shown in Table III.*

It is of great interest that the curves for the threshold for the start of singing and for migrating should start at approximately the same temperature and have a similar slope, but the dates are a month and a half apart.

Singing and territory activity are well established the fourth week in January at a mean temperature of 6° C. (43° F.). This is also the average temperature at which the main migration of the males took place (Table III). It is of interest to note that 100 years ago De Candolle found that 6° C. or 43° F. was the threshold for growth with wheat and other plants. This "has formed the base used by Merriam (1894) in working out his life zones. This is also the base commonly used by meteorologists" (Shelford, *177*).

Temperatures at which the birds will *start* singing and those at which they *will* sing after once being well started are two very different things. If the Song Sparrows are once well started, they will sing to some extent at surprisingly low temperatures for a day or two. But a sudden drop in temperature, especially if accompanied by a bleak wind may stop singing temporarily, even as late as Mar. 6 (1932). There is also a difference between restarting and making the first start, as was shown in 1936. The birds that have been well

66

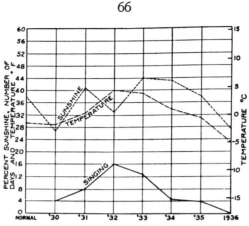

CHART IX. *Average Temperature, Percentage of Sunshine, and Number of Days on which Song Sparrows Sang in January from 1930 to 1936*

started, and then stopped by a bleak spell, begin more readily than did those in 1936 that got no chance to sing until Feb. 24, except for one day—Feb. 14. (During the last half of January, 1936, the highest mean temperature was 3.8° C. (39° F.); after that there was nothing but cold weather till Feb. 13 and 14, after which there was another cold spell lasting till the 23rd.)

Singing in January is not an automatic response to a certain temperature; it is influenced by the temperature of the previous days, and also by other weather conditions, being inhibited by strong wind, and

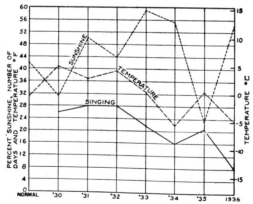

CHART X. *Average Temperature, Percentage of Sunshine, and Number of Days on which Song Sparrows Sang in February from 1930 to 1936*

157

sometimes apparently by cloudiness. It also depends on the individuality of the bird—some males starting to sing much earlier than others—, and upon whether or not he has already been singing.

The influence of light upon the breeding cycle has been much emphasized by Bissonnette, *22b,* Cole, *40b,* Rowan, *163,* and Witschi, *211a.* Let us see whether the percentage of sunshine appears to affect the singing of the Song Sparrows. In Chart IX the percentage of sunshine in Columbus in January from 1930 through 1936 is given, as well as the average temperature of these months and also the number of days on which singing occurred, while corresponding data for February are given in Chart X.

The amount of singing correlates very well with the average temperature of these two months throughout the seven years, but does *not* correlate with the percentage of sunshine.

An interesting case that bears on this point of the effect of temperature versus lengthening days is given by Laskey, *100a:* a certain banded Mockingbird (*Mimus polyglottos*) in Nashville, Tenn., began to sing on Feb. 26 in 1933 and on Mar. 4 in 1934, but on *Jan. 10* in 1935. "The temperature during January was unusually high and the excess for the month up to the 12th was 99 and reached 162 by the 19th." For 12 days he sang and courted his last year's mate, and then the temperature "dropped in one day from 59 to 14, followed by snow. Each bird retired to its own territory and they took no further interest in one another until Mar. 3." Excessive temperature had started courting activity at a time when the days had barely begun to lengthen.

2. Defense of the Territory

Hostile behavior towards territorial rivals begins at the time that singing is well established. Other Song Sparrows, that is, juvenal residents that have not yet started to sing, and winter residents, are tolerated. Perhaps this is a matter of personal acquaintance. 4M showed no hostility to two different young winter residents that stayed on his territory through February and one until Mar. 7. Warbling, being as it is, an expression of youth and of entire lack of intention to establish a territory, does not antagonize an adult male.

By March, however, all Song Sparrows are driven off, as are most other birds unless they are too large or too indifferent. House Sparrows (*Passer domesticus*) and Goldfinches (*Spinus tristis*) ignore the threats of the Song Sparrows that learn in turn to ignore these species. Field Sparrows (*Spizella pusilla*) are driven off with special vigor; nevertheless, two pairs used regularly to nest on Interpont in the midst of the Song Sparrows.

The Song Sparrow pair dominates most of the species that come into the territory; transients usually fly away, while the nesting birds merely avoid the threatened attack. The species driven off by both male and female Song Sparrows include: Juncos (*Junco hyemalis*), Tree Sparrows (*Spizella arborea*), Field Sparrows (*Spizella pusilla*), White-throated Sparrows (*Zonotrichia albicollis*), White-crowned Sparrows (*Zonotrichia leucophrys*), Fox Sparrows (*Passerella i. iliaca*), female Cardinal (*Richmondena c. cardinalis*), Red-eyed Towhee (*Pipilo e. erythrophthalmus*), Indigo Bunting (*Passerina cyanea*), Grey-cheeked Thrush (*Hylocichla minima aliciae*), Olive-backed Thrush (*Hylocichla ustulata swainsoni*), Hermit Thrush (*Hylocichla guttata faxoni*), Northern Yellow-throat (*Geothlypis trichas brachidactyla*), House Wren (*Troglodytes a. aedon*), Alder Flycatcher (*Empidonax t. trailli*), and Ruby-crowned Kinglet (*Corthylio c. calendula*). The approach of a Cowbird (*Molothrus a. ater*) is greeted with the anxiety note; if the enemy comes near the nest site it may be attacked by both of the Song Sparrows.

Territorial zeal is stated by Meise, *121*, to show a recrudescence at the beginning of each new nesting cycle, but this has not been my experience with the Song Sparrow. Territorial zeal typically diminishes as the season advances, unless a new territorial situation arises, such as the arrival of a new male, or as in the case when K2 nested outside of her mate's—1M—territory in the territory of her neighbor 4M.

D. SUMMARY

1. Some of the male Song Sparrows sing regularly in the fall.

2. In 1930 there was an exceptional amount of autumn singing, with all the birds through the molt two weeks or more early, and unusually warm weather in October.

3. 4M has shown a remarkable regularity in the beginning of his singing each fall from 1929 to 1935 with the exception of 1930.

4. Some young males, both residents and summer residents, take up their territories in their first fall.

5. Song Sparrows flock to a certain extent in cold, snowy weather. These flocks are not made up of family parties nor exclusively of neighborhood groups.

6. The resident males start their singing and take up their territories during warm weather in late January or early February.

7. Mean temperatures of 10° C. (50° F.) on Jan. 7 and 8, and of 9° C. (48° F.) on Jan. 13 will bring some singing, while singing will be well established in late January at temperatures from 8°-6° C. (47°-43° F.), and on Feb. 2 at 2° C. (36° F.), as shown in Table VI.

8. The threshold of singing was 54.2° F. on Jan. 7 and decreased 0.7 of a degree Fahrenheit each day, as will be seen in Chart VIII. This is similar to the threshold for migration, but occurs a month and a half earlier.

9. The number of days on which singing was recorded in January and February from 1930 through 1936 correlates well with the average temperature of these months, but not with the percentage of sunshine (Charts IX and X).

10. Song Sparrows try to drive from their territories most other species except those decidedly larger.

References

22b. Bissonette, T. H. 1931. Studies in the Sexual Cycle in Birds. IV. Experimental Modification of the Sexual Cycle of the European Starling *(Sturnus vulgaris)* by Changes in the Daily Period of Illumination and of Muscular Work. Jour. Exp. Zool., 58:281–314.

28. Burkitt, J. P. 1925. A Study of the Robin by Means of Marked Birds. British Birds, 18:250–257; 19:120–129.

40b. Cole, L. J. 1933. The Relation of Light Periodicity to the Reproductive Cycle, Migration and Distribution of the Mourning Dove *(Zenaidura macroura carolinensis).* Auk, 50:285–296.

101a. Lebeurier, E., and J. Rapine. 1936. Ornithologie de la Basse-Bretagne. (Suite.) Saxicola torquata hibernans (Hartert) 1910. Le Traquet patre britannique. L'Oiseau et la Rev. franc. d'Ornithologie, 6:86–103.

118a. Magee, M. J. 1936. The Wing Molt in Purple Finches. Bird-Banding, 7:73–76.

121. Meise, W. 1930. Revierbesitz im Vogelleben. Mitt. Vereins sächs. Ornithologen, 3:48–68.

123. Michener, H. and J. R. 1935. Mockingbirds, their Territories and Individualities. Condor, 37:97–140.

125. Miller, A. H. 1931. Systematic Revision and Natural History of the American Shrikes (Lanius). Univ. California Pub. Zool., 38:11–242.

163. Rowan, W. 1931. The Riddle of Migration. Baltimore. Williams and Wilkins. 151 p.

177. Shelford, V. E. 1930. Phenology and One of its Modern Descendants. Quarterly Review of Biology, 5:207–216.

199. Wharton, W. P. 1929. Notes on Banding at Groton, Massachusetts, 1922–1928. Bull. Northeastern Bird-Banding Ass., 5:98–105.

211a. Witschi, E. 1935. Seasonal Sex Characters in Birds and their Hormonal Control. Wilson Bull., 42:177–256.

13

Reprinted from *Auk*, **57**, 179–195, 205–211, 216–218 (Apr. 1940)

SOCIAL NESTING HABITS OF THE SMOOTH-BILLED ANI[1]

BY DAVID EDWARD DAVIS

INTRODUCTION

THE study of the social nesting habits of the Smooth-billed Ani, *Crotophaga ani,* a species which builds communal nests, was undertaken in order to increase the information about the subject of social parasitism and in order to help clarify the concept of territory. In this paper these two problems are considered in relation to the behavior of this bird.

The field work for this study was done at the Atkins Institution of the Arnold Arboretum of Harvard University at Central Soledad, Cienfuegos, Cuba. This Central is a large sugar estate, and the arboretum is an oasis in a desert of cane, since the only other places suitable for birds to live are roadsides, pastures, and a few spots unfit for cultivation. In addition to the field work at the arboretum, short excursions were made to the nearby Trinidad Mountains and also to Bahia Honda, in Havana Province.

The species studied, *C. ani,* belongs to a distinct subfamily, the Crotophaginae, of the Cuculidae, and occurs in the West Indies and South America. Included in this subfamily are *C. major,* inhabiting northern South America, *C. sulcirostris,* ranging throughout Central America, and *Guira,* a monotypic genus found in Brazil and Argentina (Peters, 1939). The habits of the members of the subfamily are discussed in innumerable collector's notes, which usually repeat the observations of the natives. *Guira,* the most primitive in generic characters, is poorly known but seems to be the least specialized in its behavior; these birds are reported to build their own nests, sometimes in colonies, and also to lay eggs in the nests of other birds. The life history of *C. sulcirostris* has been studied by Skutch (1935; 1937), and resembles that of *C. ani.* The habits of *C. major,* also social in its nesting behavior, have been observed by Young (1929). The behavior of *C. ani* is described by Gundlach (1874; 1895) who made very accurate but brief observations on the species in Cuba and by Young (1929) who briefly discussed the bird's habits in British Guiana. More recently Dr. Frank M. Chapman (1938), describing the history of the colony at Barro Colorado Island, has given the only

---

[1] Thesis submitted in partial fulfilment of the requirements for the degree of Doctor of Philosophy in the Division of Biology of Harvard University, June, 1939.

accurate account of the species. In addition, a mass of data concerning the genus, some of which is correct, was collected from travellers by Leverkuhn (1894).

I am indebted to many persons for assistance in this work. Mr. David Sturrock and Mr. F. G. Walsingham, of the Atkins Institution, made the complete facilities of the laboratory available, and Mr. William Leonard of Centrál Soledad granted innumerable favors. I am indebted to my friends there for making my stay most pleasant. This study was made during the tenure of an Atkins Traveling Fellowship and with the aid of a grant from the American Association for the Advancement of Science. I wish to acknowledge the painstaking criticism of Dr. Glover M. Allen. Without the encouragement of Dr. Thomas Barbour, this study never would have been made.

METHODS

A study of bird behavior in the wild is dependent upon the methods of field observation. To observe the bird's behavior I carried on studies from May 8 to September 13, 1937, and from April 8 to October 16, 1938. During these two breeding seasons I spent approximately 1300 hours in the field. The fact that the birds of a group sleep in the same tree made it possible to keep account of the different groups by coming to the garden at dawn. In trying to follow several groups intensively there are innumerable practical difficulties, and deficiencies in the records inevitably result. The identification of individual birds was made possible by the use of colored leg-bands. During the two years, thirty-three adults and fifteen young were trapped and banded. In addition, the fact that the yearlings are distinguishable by the bill for nine months after hatching aided greatly in identification. In order to study the nests at close range a blind twenty-four feet high was built and placed at several different nests.

The best method of presenting to the reader a picture of the bird's behavior is to use descriptive words, such as 'complaint,' 'perplexed,' 'mad,' and descriptive phrases. In this connection I wish to emphasize the fact that these terms are used merely for description, and that no knowledge of the state of the bird's mental processes nor of purpose is implied. Throughout this paper direct quotations from my field notes are preceded by the date and enclosed in quotation marks. The vernacular names 'ani' and 'judio' are used, and the terms 'colony,' 'flock,' and 'group' are used synonymously. For convenience each colony was named.

PART 1. SOCIAL HABITS

FLOCK BEHAVIOR

Crotophaga ani is a social species, living throughout the year in groups. Although occasionally a pair may live alone, usually the number in a flock is about seven; the largest group recorded by the writer contained twenty-four, and groups of fifteen are not uncommon. Since the number in the colony is continually changing, it is meaningless to derive an average size. In most flocks there is a surplus of males, although colonies composed of an equal number of each sex or having an excess of females are not uncommon. These flocks retain their individuality for a year or more. Although the yearlings remain with the flock throughout the year and may breed with the parent flock, in general there is a sharp decrease in the number composing a group (Table 1) just before the building of the first nest. In one case (Cypress) this decrease corresponded to the number of yearlings that were in the group.

The membership of each flock gradually changes, especially in the period just previous to nesting. Individuals of both sexes may leave the colony for no apparent reason and wander about searching for a colony to join. Table 1 shows the changes in number in the groups studied intensively during 1938 although the addition of new individuals is masked by the decrease due to mortality, and by departures.

TABLE 1

POPULATION CHANGE (1938)

Date Colony	4–16	5–8	6–5	7–3	7–31	8–28 (adults	9–25 only)	Dates nest started
Pond	20	18	15	14	14	13	13	6–10; 8–10
Gnhs	11	9	8	6	6	6	6	5–20; 7–10
Ceiba	11	11	9	5	4	3	3	7–20
Cypress	10	10	7	6	5	5	4	6–10; 9–16
Haemat	8	9	6–8	7	6	6	4	6–18
Phoenix	11	13	5±	7	6	6	6	6–25; 8–19

Some evidence suggests that a colony may divide into two parts. Certainly during the summer and probably during the entire year, new groups form. A typical history of a colony that developed but did not breed is as follows:

July 7. Four birds have taken up territory.
July 10. Still fighting for part of their area.
July 25. Five birds in the group.
July 26. Started to build a nest.
August 4. Still working on the nest; abortive type.
August 10. Fighting in the territory.
August 20. Group gone.

Less commonly these new groups may complete the nest and raise young. A colony may disintegrate if most of the individuals are removed. Although one nest held six incubated eggs, it was deserted when three out of four of the colony were shot. The remaining bird did not get a new mate.

The members of a flock pass the day together. Each colony sleeps as a group in a densely foliaged tree or in a bamboo clump. Sometimes the group may sleep in two parts and the sleeping tree may be changed occasionally, depending upon the number of suitable trees in the territory. In the tree the birds crowd as close together as possible in rows on a limb, pull the head down on the shoulders and fluff out the feathers. Shortly after dawn the birds come out and sit in a group on a branch, sunning themselves and trying to keep warm. Throughout the daytime the flock spreads out over the feeding area, using the 'judio' call to keep together. At evening before going to the sleeping spot the birds collect in a nearby tree and then fly in together.

The individuals of the flock respond to the actions of their fellow members in several ways. First, the birds are very solicitous for other individuals of the flock, crowding around with great excitement when one member is hurt. This behavior was observed when a dove fluttered off its nest and performed 'injury-feigning.' Second, the birds spend much time preening themselves and one another. When two birds are sitting adjacent to each other, one may suddenly stretch out its neck and raise its neck feathers in a movement resembling a reflex, which is stimulated by the touch of the bill of the other bird or arises spontaneously. The other bird then carefully preens its head and neck feathers, pulling out and swallowing all the loose ones. Then the situation may be reversed; or two birds may work on one simultaneously. Third, among the individuals it is probable that there is a flexible order of social dominance, although no criterion was found suitable for judging the rank of a bird. Sometimes two birds scuffle over an insect.

Albino birds occur and are reported by the 'guajiros' to live normally with their respective groups.

TERRITORIAL BEHAVIOR

Each flock of *C. ani* possesses a definite territory which it defends against strangers. The boundaries are well known to the members of the colony and are not passed. The individuals cannot be driven from their territory, even when pursued by a man. The size of the territory varies roughly in proportion to the number of birds in the colony, but is also related to the previous history of the group; a colony reduced in number will nevertheless retain the same territory for a long time. In large colonies (fifteen birds) the territory may occupy ten acres. The territory contains (1) a nesting and sleeping and (2) a feeding division. The size of each division is dependent upon ecological conditions. The nesting section contains several bushy trees for building the nest and sleeping at night, while the feeding area resembles savanna conditions. Except when engaged in nesting activities, the birds spend the day in the feeding division.

Throughout the year there are changes in the boundaries, as shown by the map (Text-fig. 1). A part of the territory not defended may be taken over by a neighboring group in a few days. For example, in July the W. of G. group was not present and the Ceiba group took over part of its territory. A group (Pond) slept about a quarter of a mile out in the cane fields for several nights. Since, for the first few days after the young are out of the nest, the colony sleeps in the feeding area, another colony may attempt to occupy the nesting division and sometimes succeeds in doing so. If in the dry season there are no moist places in which to find insects, the area is deserted, or contracted around the stream courses. The territory is vigorously defended against other individuals. Although any invasion is immediately repelled the colonies are on good terms with their known neighbors and settle boundary disputes quickly by 'agreement.'

Strangers of two distinct types try to invade the territory. One type comes in from a distance, flying high, sits in the top of a tree and calls *whew* several times. This behavior attracts the owners of the territory and the stranger leaves, still flying high, without attempting to join the colony. It seems likely that birds of this type are yearlings, not yet sexually mature. The second type of stranger tries to join the colony. Quietly it enters the territory and when the owners arrive it remains in the vicinity. One or more owners chase it throughout their area, sometimes flying high above the ground but more often circling near the ground. The stranger sleeps in the area but in a different tree and does not know the boundaries of the territory, but may fly into a neighboring territory, only to be chased again. Although strangers sometimes try to join the colony when it has a nest or young, they are less aggressive at this time. As soon as the young are hatched the strangers again aggressively attempt to join the colony.

The behavior used in driving away strangers is referred to as 'chasing' and is described in the following note. June 1, 1937: "Chas-

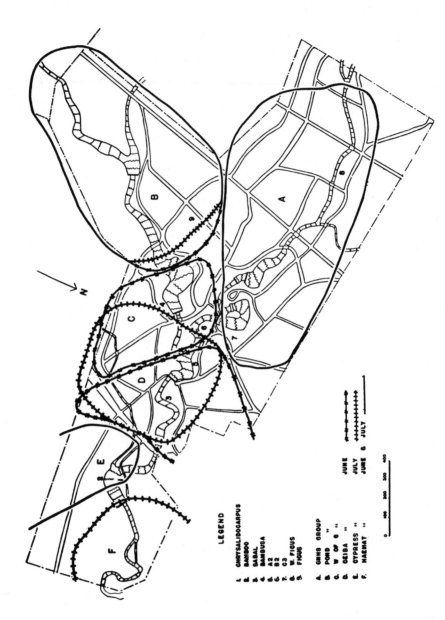

TEXT-FIG. 1.—Territories for June and July 1938.

ing is an interesting procedure. One bird follows another around and around and back and forth. Sometimes another joins in the pursuit. They fly for about twenty beats and then sail for a long distance. They are able to turn and wheel in the air quite well. . . . It (chasing) starts when two birds are in a tree and one starts 'conking.' The chased then starts to climb and the chaser follows, hopping up the limbs and twigs. Then the chased starts to fly and is followed." This chasing behavior may last for several days, until the stranger is driven out or succeeds in joining the colony, as shown in the following note, July 26, 1938: "Saw a chaser lose a fight with the chased but another bird came to the rescue. The birds chased all morning." Chasing occurs throughout the year but, since just before the breeding season the birds are moving around more, it is more frequent at this time. If the stranger comes while the nest is being built, the nesting may be interrupted and delayed.

Chasing is interpreted as part of the mating behavior by Young (1929). Although in appearance it resembles the mating flight of many birds, especially the ducks, the following facts make it necessary to discard this interpretation. (1) A bird may be on the nest while other birds chase. (2) Chasing may occur at any time of the year, or at any time during the nesting cycle, but does not occur in some colonies. (3) There is a definite mating behavior as described below. (4) There is a definite antagonism between the birds concerned and sometimes fierce fights ensue.

A behavior related to chasing may be called 'rushing.' Uttering a particular call (no. 3), the birds as a group fly from tree to tree. Since this behavior occurs after a period of chasing, it seems to be a method of defining the limits of the territory and searching for any strangers. Other than this 'rushing,' there is no special advertisement of the territory such as there is in passerine birds.

There is no fighting when birds from several different colonies are placed together in a cage distant from the birds' territories. The birds sleep side by side the first night.

The chasing, rushing, and fighting to defend the territory are very fierce. The impulse to defend is so strong that the birds fight a stranger even while a man is at the nest. Stuffed birds placed in the territory are attacked and destroyed, although at times the dummy is ignored, as shown by the following observations. June 24, 1938: "I put the dummy at the nest. After about 45 minutes four birds came in. Two sat on the nest, each for a few minutes. They ignored the dummy in spite of moving and shoving it at them. They then went away. III came in with a leaf but dropped it and went off." June 25: "The dummy is thoroughly damaged on the back and legs and is upside down. Three birds came back and III sat on the nest and complained a lot. They ignored the dummy although I moved it a lot. III brought in some leaves and sat ignoring the dummy." When attacking a strange bird, the defender spreads its wings and pecks viciously at the invader's head. Fights occur if, during chasing, the chaser overtakes the chased. As the two fall to the ground, screaming and fluttering, the whole colony forms a circle and watches the brief combat. A fight or display performance, described as follows, occurs occasionally. June 13, 1937: "Two birds are on the ground. One started to display in front of the other. It would strut around it and spread its tail and wings. Sometimes it would fall forward as it displayed and raise the tail high in the air. All the time the two were making little noises. At one time there seemed to be a sort of abortive attempt at a fight." The birds then separated and went into different territories. This behavior is probably a territorial dispute.

Possibly birds join a colony without fighting, as suggested by the following observations. July 7, 1938: "Saw AR and A-B sitting peacefully with two other birds in the sabal tree. Later saw two birds chase away a stranger that had 'whewed' from the top of the ceiba tree. AR is on excellent terms with the birds of the Ceiba group. It had its neck scratched and went up to the nest (location). I have watched the birds of this group fairly closely of late and they have been doing a lot of chasing, but I did not see them chase AR, although they could have very easily." AR was formerly a member of the Greenhouse group and was concerned with the nest on June 18. On July 3, the nest was destroyed and it is likely that AR left at that time and joined the Ceiba group.

Birds may leave the colony without being driven out. On July 26, 1938, ARR (female) was on good terms with the rest of the group. She left on July 27 and formed a colony with two males nearby. On July 30, a member of her former group chased her.

There is no difference in behavior among the sexes or immatures in fighting or chasing. Males may chase females and vice versa. When I arrived in April, 1938, birds hatched July 9, 1937, were defending their territory.

The birds will fight a mirror as would be expected, since, although they recognize the other members of the colony, they are not acquainted with themselves. This behavior occurred on October 4, 1938: "I put the mirror at the nest and two birds came on to the nest at once and started to *conk* and peck at the mirror. It puzzled them greatly especially when they went around behind it. After I took it away they explored the place looking for it and finally went on to the nest."

Fighting is frequently caused by a colony travelling as a group in search for a new territory. In the spring months when the birds are taking advantage of the regions made suitable for nesting by the rains, this travelling is often seen. For example, on April 20, 1938: "Six birds were seen in W. Ficus at 6.30 a. m. They seemed to have a purpose and eventually flew northwest in stages and then to the north out over the cane." In one case a pair which tried for several days to join the Pond group but was repulsed at the nest each time, later built a nest and laid five eggs. The birds arrive in an area and sit about in trees for a while to establish territory. If there is a claim for the particular area brute force decides the issue. The first stage is chasing; the second is rushing and the third is the acceptance of the boundaries or the withdrawal of the unsuccessful group. If there is no claim for the territory, it is at once defended against other groups.

By taking advantage of the fighting behavior, it was found possible to catch the birds, using a live decoy in a trap. A dead or mounted decoy is ineffective, probably because it is not moving and hence is not seen. The owners of a territory enter the trap and in a short time kill the decoy. The decoy does not resist. The birds continue to fight with the decoy until a person is within a few feet.

Several times an interesting conflict between the territorial behavior and the response to the alarm call was observed. Two colonies had nests close together, although their territories extended in opposite directions. When the nest of one colony was disturbed the birds of the other colony responded to the alarm call but never crossed the invisible boundary line of their territory.

CALL NOTES

The vocabulary of *C. ani* is varied and distinctive. Each call has definite uses and may serve to distinguish behavior patterns. The first six calls listed below are modifications of no. 1, but the other calls are not founded upon a common pattern. The calls have been given descriptive names which may assist to convey their characteristics. *Mimus polyglottos* mimics some of the calls, especially the 'judio' call and causes confusion not only for the investigator but for the anis themselves.

No. 1. 'Judio' is the flock call, used whenever a bird is flying for a distance or when the group is travelling to the feeding area, to notify the members of the movements of the group. This call resembles the Spanish word, thus giving rise to the Cuban name for the bird.

No. 2. The alarm call is used as a signal of warning or annoyance. When a person approaches the flock one member gives the call and the rest of the birds fly upward to a bush or tree and continue the alarm until the annoyance passes. The note is rapid and not nasal, but begins with an ñ sound. At each utterance the tail jerks. The young develop this call first, but require a month to reach perfection.

No. 3. The 'shout,' always repeated very rapidly, is used during the rushing behavior described above.

No. 4. The 'complaint' is a slow, high-pitched wail, similar to the alarm but higher in pitch, and is used only at the nest and mainly during the first few days of building. This call is a certain sign that work is in progress but does not distinguish between an abortive nest and a real one. The note is not given after the nest has been built but is used after pairing occurs. Hence it is a nest, not a pair, call. It should be noted that there is a special call for the nest but not for the territory as there is in many species. It was not possible to determine if this call is given by one sex only.

No. 5. The 'objecting' call is an undeveloped alarm which resembles a 'complaint' in tone, and is used at the nest when the birds are disturbed. First, the alarm call is used and then as the disturbance lessens the 'objecting' call is continued. Sometimes it is heard when a bird wants to exchange the duties of incubation with the bird on the nest.

No. 6. The 'get-up' call, repeated extremely rapidly, is used early in the morning before the birds leave the sleeping tree and sometimes late in the evening before the birds go to sleep. It is also used to attract the young birds out of the nest for the first few times.

No. 7. The 'chuck' resembles the sound *ah* and is used by the bird being chased but not by the chaser. Although the significance of this observation is not understood, several times in April this note was heard under circumstances which normally result in the alarm note.

No. 8. The 'chuckle' is used in the nesting tree when the birds are pairing.

No. 9. The guttural note occurs whenever the birds appear perplexed. This call is induced by a strange object such as the blind or a strange situation such as being caged.

No. 10. The 'whine' is a soft call used in mating. The pair sit in the nesting tree side by side and whine, so softly that the sound cannot be heard more than twenty feet.

No. 11. The 'conk' is used in fighting to defend the territory. After getting below the stranger, the attacker 'conks' a few times before attacking.

No. 12. The danger call is a 'quack,' signifying the approach of a hawk. The judios dive down into a bush or tree at this note. Immature birds give this call on seeing *Cathartes* and *Zenaidura*. Once, when the investigator was placing a decoy near the nest, this note was heard, perhaps indicating the highest pitch of excitement. Other species take advantage of this warning.

No. 13. The 'whew' is used by a stranger of the first type described above.

PAIRING AND NESTING BEHAVIOR

The nesting season begins shortly after the rains permit the vegetation to develop. At this time the diet changes, as shown by a study of stomach contents, from vegetable to animal matter (insects and lizards), and it seems likely that this change of diet is the stimulus for the development of the gonads.

The pre-incubation behavior takes place in several stages. (1) A pair or several birds spend part of the time away from the flock sitting quietly in a tree suitable for the nest. The manner of sitting differs from the type usually observed; the birds sit inside the tree, not on the top or outside. The birds sometimes hop about in pursuit of each other and after the first day or two occasionally break off a twig, but usually drop it. A bird may hold a leaf in its bill for a time or take one from another bird and then drop it. During this period occurs the whining note, which, except for the action of sitting side by side, is the only courtship behavior. The following note records typical behavior. May 21, 1938: "Four birds of the Cypress group were in the bamboo. One (ragged tail) whined for a long time and hopped around the bamboo, finally flattening itself out. The others showed no interest. Another . . . seized a twig but did not break it off and soon lost interest."

(2) Next comes the period of the complaint call, given persistently from a definite location. The observation on May 26, 1938, is typical of the behavior at the beginning of this period: "I went over to the Old Garden at 7.15 a. m., and found three birds sitting in the bamboo. One carried a leaf around after a while but did not know exactly

what to do with it and finally dropped it. Again it carried a leaf around. Moving the tail up and down seems to be characteristic of this time. The birds also complained a bit and whined some. A bird went up into the Ficus and pulled a twig and then was joined by another which also pulled. They sat for a long time and were still there when I left at 8.45." Although by this time the site for the nest has been determined, some birds may examine other locations. June 30, 1938: "At 7.00 a. m. there were two birds near the Sabal and three in it. They sat for a long time and then went over to the Bambusa and sat some more. Two birds left and two . . . more stayed and hopped around a long time. Then they went out and joined the others. Three went over to the Sabal and 'split-tail' went to the nest (location) in the Sabal. The whole group went over to the Bambusa. They cannot decide which to use."

(3) After the location of the nest is determined in the period of the complaint call, the birds begin the period of building. Twigs are broken off from a branch, never picked up from the ground. Usually one bird builds most actively but frequently as many as five carry in twigs at the same time. At this time the birds are so absorbed in nest-building that the approach of the investigator does not disturb them to any extent. The complaint call continues throughout this stage. Sometimes two birds may sit on a nest, side by side, as shown in this field note. June 7, 1938: "One bird in the Chrysalidocarpus complained and another came in from the Litchi. Both sat on the nest for three minutes and then went to the bamboo and sat."

(4) Following this stage for a few days little or no work may be done on the nest.

(5) In the last stage before the eggs are laid, the birds carry in leaves and occasionally a few sticks. An interesting action was observed in a colony which was building a nest after the first nest had been destroyed during incubation. A female sat on the nest and received sticks and leaves from the male, thus continuing perhaps the incubation which she had been performing on her former nest.

Copulation, the climax of pairing behavior, occurs usually in the first hours of the afternoon, but may occur at any time of day. The action shows no special behavior. When the birds are sitting in the nesting tree quietly, the male, without any preparatory performance, mounts the female. Because a stuffed bird is treated as a stranger and attacked, the birds do not copulate with it.

Many irregularities from the general outline above occur. The nesting behavior may stop at any stage. For example, on June 9, 1938: "One bird interested in the W. Ficus. It hopped around in the tree, sat for a while and complained all the time." No nest was built in this location. In some colonies the birds may break off twigs and then leave the territory, or a nest may be started but not completed. These incomplete nests are termed 'abortive' and, as discussed below, are comparable with the scrapes and trial nests of other species. After an abortion the colony usually starts another nest in another location. That a poor choice of nest site is not the cause of abandonment is shown by the history of the Pond group. This group built an abortive nest in a bamboo and then, after raising a brood elsewhere, came back and built on top of the abortive nest.

A type of irregularity occurring occasionally is desertion. After building a nest and laying eight eggs, one group deserted without obvious reason and built the nest in another place. The birds under observation never deserted the nest because of human interference.

Other aberrations of the breeding cycle occur. A bird was observed carrying sticks alternately to two different trees, indicating that the same bird may work on two nests. Although part of the group may start the nest, others in the colony may take no interest till later; the Bambusa group in 1937 consisted of fifteen birds, yet only nine eggs were laid, indicating that only two females had laid in the nest (evidence presented below shows that females lay between four and seven eggs, usually six). Only exceptionally do all the females of a group lay eggs at the same time. Often one bird, either male or female, is most interested in the nest, is most alarmed by the observer, and is most active in the building and incubation.

At the beginning of the laying period there often is a miscarriage of the building and laying sequence. Sometimes eggs are laid before the nest is completed and are covered in the bottom of the nest or dropped on the ground. One day a colony of fifteen birds dropped six eggs, some as far as twelve feet away from the nest. By placing the unbroken ones in another nest, these eggs were proved to be fertile. Although there is no evidence that the females fight on the nest, occasionally eggs may be knocked out of the nest.

If a nest containing eggs or young is destroyed the birds rebuild at once and with great speed in some cases, telescoping the stages into a few days. Table 2 shows that there is no relation between the stage of incubation or feeding at which the nest was destroyed and the length of time till a new nest is started.

TABLE 2

Rebuilding Dates in Relation to Nest Destruction

Colony 1937		Number birds at 2d laying	Date destroyed	Stage of nest	New egg date	Days
Old Garden		12	8–1	3 days incubation	8–12	11
C3	(2)	11	7–5	0 days incubation	7–13	8
C3	(3)	6	8–10	5 days incubation	8–16	6
Cocos	(1)	?	6–14	4* days incubation	6–25*	9*
	(2)	4	7–1	0 days incubation	7–14*	13*
	(3)	7	7–20	12 days incubation	8–3*	14*
Gray's		11	7–5	1 day hatched	7–10*	5*
1938					.	
Cypress		7	7–5	4 days incubation	7–10*	5*
Gnhs	(2)	7	7–3	1 day incubation	7–10	7

*indicates estimate of ±2 days.

Relationships Between Male and Female

Since the birds live in colonies and build communal nests, polygamy may occur. Copulation with more than one individual is the only criterion of polygamy, but since in this species copulation occurs only in very dense foliage, the number of copulations clearly observed is small. Of those copulations seen, none proves a departure from monogamy. In two colonies in which all individuals were identifiable (Ceiba and Cypress), the behavior indicated strict monogamy. Thus, in the Ceiba group all four birds were identifiable and only one pair was concerned with the nest and young while the other pair took no notice of the nest. There were only seven eggs, indicating that only one female had laid. This colony, therefore, consisted of two definite pairs. As another example, at the time the first brood was raised the Cypress group consisted of two definite pairs and two non-breeding birds. At the time of raising the second brood, the group consisted of one female (ABW) and three males (ARW, ARG, and ABB). All relations were harmonious and it was impossible to see any difference in the female's attitude toward the different males.

In many cases, however, polygamy is suggested by the composition of the colonies. In some colonies polygyny is indicated. A colony which consisted of two females and one male was collected and nine eggs were in the nest, indicating that both females had laid. Since the previous history of the group is not known, it is possible that there was another male at the time the eggs were laid. In its second nest, the Phoenix group, which consisted of seven birds, had 23 fertile eggs, and hence, unless some bird laid eight eggs, polygyny must have occurred. But on the other hand, the composition of other colonies suggests polyandry. The Bridge group, which had four eggs, consisted of two males and one female throughout its history. A colony that consisted of four males and one female was collected. In these two colonies there is no proof that more than one male copulated with the females, although at no time were sexually inactive birds collected in any colony. In conclusion, it seems that any of the three possible situations, monogamy, polyandry, or polygyny, may exist.

At the time of pairing there may be antagonism between the pairs. This behavior occurs in the early part of the breeding season, as shown by the following observations. May 23, 1938: "AW followed A into the C3 (a bamboo) and hopped about it for a long time. It was up in the bamboo and I could not clearly see what went on. There was a lot of whining, often scarcely audible. Then a third bird came in. I could not get a look at its legs and it may have been banded. It joined the others and there was some 'conking.' Presumably AW was mad but I did not see well enough to be sure. Then they went out to the west." June 10, 1938: "Three birds came into the Chrys nest at 5.30 p. m. and then 'ragged-tail' came in. A male attempted to copulate but the female was not very responsive. This occurred on a branch near the nest and they both went over to the nest where 'ragged-tail' was. They hopped around the bamboo and there were a few 'conks' and then they went out to feed." But in contrast, later in the season there is no antagonism. On July 9, 1938, the behavior of the birds ARG and ABW, which were certainly a pair, was recorded: "ARG was actively carrying twigs in to ABW who was sitting on the nest. Two birds came in and shoved ABW off the nest. ABW and ARG went out together, followed by the third bird and the fourth stayed on the nest. ARW came in and stayed on the nest and then copulated with the bird on the nest." July 11, 1938: "Four birds came in to the nest in the Bamboo and ARW and ABW sat on the nest together. Then ARW and a bird went off and ARG began to carry in twigs to ABW on the nest and did so industriously for about 15 minutes." These data indicate that at the time of pairing there is antagonism but that after the mates are determined there is harmony.

Both sexes incubate, although in some cases the male does more than his share. When a bird desires to incubate, it flies into the nest, usually with a leaf or sometimes with a stick. If the incubator does not wish to leave, it must be forced off or the newcomer is unable to incubate. The general behavior is recorded in the following observations. July 26, 1937: "A bird brought in a leaf and sat on the nest for a minute trying to get the other bird off; but then left. It soon came back and sat with head and breast over the other's back, called and then left. Again the performance was repeated. It came in with a twig but still the other stays on. Two more sticks are brought in in quick succession to no avail. The bird sat on the incubator and called but soon left. Another stick was brought in and the incubating bird took it. Then it (the first bird) brought in another stick. Now it is sitting on the incubating bird and calling. A third bird came in with a stick and both got off. The third bird got on the nest." July 30, 1937: "Several young have their eyes open. Four different birds brought in food at the same time. The incubator left. There are some birds sitting around all the time. Once two birds came in and both seemed to want to incubate. A third did not seem to know what to do with some food. All three stayed on the nest for a while. Finally two went off. Later two brought in food. The young did not eat (the food) offered by one and so the adult ate it." The delay in the departure of one bird is the basis for the belief that more than one bird incubates at one time. According to my experience two or more birds never incubate simultaneously.

RELATIONSHIPS BETWEEN ADULTS AND YOUNG

The young remain with the parent flock for a variable length of time and in some cases nest in the same tree in which they were hatched. For example, the young from two colonies, banded in 1937, were found in 1938 with the parent flock, sleeping in the same tree used in 1937. Nevertheless, in general, most of the young have wandered away by the breeding season. In one case the yearlings sat in the nest tree and seemed interested, but in a few days disappeared; since no chasing or fighting was observed, it is unlikely that they were driven out. The yearling cooperates with the parent flock to defend the territory.

Nearly all the adults in the colony feed the young, although some individuals are much more active than others. For example, in the Bambusa nest there were nine young, indicating that only two females had laid, but at least nine of the fifteen birds in the colony fed these young. In the Cypress group a female did not incubate, although, as shown by sections of the ovary, she had laid eggs. Nevertheless she did feed the young.

The adults call the young out of the nest by the 'get-up' call (no. 6). If the nest is threatened the adults give the alarm note and in addition click the bill; then the young, although they may be only four days old, climb out of the nest into the tree. The young return to the nest the first two nights after leaving it. One adult sleeps with the young for several days after they have left the nest but are still in the nest tree. After the young are able to fly, the adults take them out to the feeding area and the whole colony sleeps there. The young are fed by the parents for about a month after leaving the nest. If no nests are destroyed a colony in Cuba can raise three broods in one season.

The young are ineffective as a decoy to trap the adults.

RELATIONSHIPS AMONG THE YOUNG

The young of one brood stay with the parent flock while the next brood is incubated and hatched. Observations on the second brood are not satisfactory because the adults become so alarmed. The observer has spent three hours hoping that the birds would quiet down and resume normal behavior. This behavior is in great contrast to the behavior during the development of the first brood.

The young of the first brood may feed the young of the second brood, but they are less active than the adults. In Wares' nest in 1938 the young when 48 days old fed those of the second brood. Analysis of this behavior suggests that the young birds are mimicking the adults, as the individual does in all behavior, or that, as in the case of non-breeding birds, the stimulus of the young in the nest is a releaser which compels the bird to bring in food.

* * * * * * *

DISCUSSION

Territory.—The concept of territory, as developed by Howard (1920) and others (Friedmann, 1933; Nice, 1933; Mayr, 1935; Tinbergen, 1936; Evans, 1938), has become a most fruitful generalization for the study of bird behavior. According to Friedmann (1935): "The theory of the function of territory is that it so spaces the breeding pairs of birds as to insure enough food for the young close enough to the nest so that, in their search for food, the parents do not have to desert the young long enough for any harm to come to the latter. In the spring the male leaves the flock with which it has spent the winter, isolates itself on an exclusive breeding territory to which area it confines its activities, makes itself conspicuous by display and song, thereby attracting a mate and at the same time warning other males not to trespass." Each species studied in detail demonstrates a large or small difference from the general behavior pattern.

The characteristics of the territory behavior of *C. ani* show clearly that the defence of the piece of land is the important feature. Sex is not a factor: the whole colony defends the territory against a strange bird no matter of which sex it may be. It is possible that there could be a colony composed entirely of one sex and this possibility should be tested experimentally. Food is not a factor: the territory is not defended in order to maintain the food supply in this species, because the same territory may be owned by a colony no matter how many are in the group; for example, the number in a colony changed from eleven to six, but the same territory was defended. Since *Crotophaga* has no song in connection with territory it may be concluded that song is not a necessary attribute of territory but is, in passerine species, a secondary acquisition. *C. ani* has definite behavior patterns for the defence and establishment of the territory and resembles the wren-tits (Erickson, 1938) and the mockingbird (Michener and Michener, 1935) in maintaining a territory throughout the year. The characteristics, cited here, show that the behavior is related only to a piece of land.

The defence of a piece of land, it is reasonable to assume, developed from the defence of the nest. Although Friedmann (1929) considers that the defence of the area originated first, Tinbergen (1936, p. 7) implies that the defence of the nest is primal.   The nest is a tangible object which is always in the same place and can become the head-quarters.   For success in reproduction at least one member of the pair must defend the nest; other activities can take place almost anywhere. In various species the defence of the nest location is extended to other activities and in some species this enlargement is developed to such an extent that the same or a different territory may subserve any or all of the four main uses of territory, namely, mating, nesting, feeding of young, and winter habitat.   This viewpoint on the development of the defence of the piece of land renders unnecessary the teleological implications of Howard, so decisively attacked by Lack and Lack (1933).   This extension of the utility of the territory roughly parallels the development of the altricial nesting habit.   Since this development is continuous there are intergrades between all variations of territory and there can be no rigid categories.

The defence of territory in *Crotophaga* has no teleologic motivation (i. e., the birds are not consciously *fighting for* any thing) but is merely an extension of the defence of the nest site.   (The analogy of this extension to the development of an anatomical character, the crest on the bird's bill, is striking.   This large crest has no function but is merely an hypertrophy of the bill.)   It is true that territory has many functions (food, nesting site, etc.) but the teleologic interpretation of the behavior must be separated from the functional value of the territory.

The defence of the sex partner is a separate element, termed sexual fighting, and should not be confused with the defence of the nest and surrounding territory.   Mayr (1935) emphasizes the factor of sexual jealousy and states that "territory was originally developed only in connection with mating. . . ."   But the behavior of *C. ani* shows that sexual jealousy is not a factor in this species since many individuals live together harmoniously, and a stranger may join a group.

Definitions of territory have been attempted (Mayr, 1935; Tinbergen, 1936).   Although a definition is difficult, the nature of the concept may be clarified by a short characterization of its main features. Territory is a stage in an evolutionary process and is a characteristic of a species.   Dobzhansky (1937) points out the difficulties in trying to define a stage in an evolutionary process, in his case, a species. Despite these difficulties, territorialism may be described as the defence of an object (territory) which serves in reproduction.   Several salient points in this description should be noted.   The concept is con-

cerned with a behavior pattern in respect to a physical object. There
is no mention of purpose nor of individual nor of sex. The word
'serves' is intended to eliminate those cases in which the fighting is
in relation to sex partner. The concept is very inclusive and can
include other vertebrates as well and is functional, not teleologic in
its interpretation.

Breeding cycles.—Cycles, analogous to the estrous cycle of mammals,
have been postulated by Howard (1929), Huxley (1932), and others.
In addition Whitman (1919) refers to the 'synchronization' of the male
and female. The study of *C. ani* has obtained data of the type which
led to the development of these hypotheses. These data as outlined
below are behavioristic and extremely difficult to control and to test
experimentally.

One striking phenomenon is that a colony sometimes does not lay
eggs until a new member joins. Thus, Wares group in 1938 consisted
of seven birds (both sexes) until June 26 and did not show any indication of nesting, although other groups had already laid eggs by this
date. On June 26, a new bird (sex unknown) was present (eight
in the group), nesting behavior was first seen, and there were eggs
probably by July 10. There were eight eggs in the nest, indicating
that two females had laid. As another example of this phenomenon,
the Ceiba group originally consisted of about eight birds and showed
interest in several locations for a nest, but laid no eggs, although
other groups had eggs by this date. On July 7, a male joined the
group and incubation started on July 28. The group at that time
consisted of two males and two females, only one of which laid in
the nest.

A colony sometimes takes up a territory and builds an abortive nest
but does not lay eggs and soon leaves the area. This sequence of
events occurred in two colonies of five birds in 1938. A somewhat similar case occurred. The Haemat group built a nest and hatched six
young which were destroyed by a small boy. During the rest of the
season the group built three abortive nests and in one laid an egg
which was knocked out of the nest. During this time the group decreased in numbers from seven to four. Since a pair may raise a
brood, and since there is no correlation between the size of the colony
and the initial date of breeding or its success, it is known that a large
number of birds is not necessary for success in nesting.

Another characteristic of this species is that the number of days
between the destruction of a nest and the rebuilding varies greatly.

Table 7 shows that the interval between the date on which the nest was destroyed and the day the first egg in the new nest was laid, has no relation to the stage of incubation or feeding of the young and is not constant.   In addition Table 7 shows that the time between the day the first brood left the nest and the date on which the first egg of the second brood was laid varies between 20 and 41 days and has

TABLE 7

DAYS BETWEEN BROODS

Colony 1937	Adults at second nest	Brood left		Days	First egg second nest	Days
		first	second			
Bambusa	15	7–1	7–26	25		
B2	7	7–13	9–7	56	8–13	31
Wares	15	6–24	8–23	54		
1938						
Wares	6	8–14	10–5	52	9–2	20
Pond	13	9–4	10–18	44	9–25	21
Cypress	4	8–8	10–18	71	9–18	41
Gnhs	6	9–5	11–5	31	10–8	33

no relation to the size of the colony.   In this connection Nice (1937) has found that for the Song Sparrow (*Melospiza melodia*) the interval between the date a nest was destroyed and the day on which the young left the rebuilt nest was a constant (30 days) but that the interval between the dates the young left the nest in two successive broods varies between 30 and 41 days.   The renesting after destruction in *C. ani* must be controlled by psychic as well as by endocrine factors. The destruction of a nest is comparable to an abortion in mammals and in that class it is known that after an abortion a female will come into estrous period in a definite length of time for each species no matter (except for the last few days) at what stage of pregnancy the abortion occurs.   The Song Sparrow resembles mammals in that it produces young in a definite period of time.   But in *C. ani*, factors other than endocrine seem to control renesting.

Other evidence indicating cycles, is that some variation in egg-laying ability occurs as is shown by the fact that not all females lay in the nest.   In the Pond group there were thirteen birds in the colony and 29 eggs in the second nest (the first nest was inaccessible and was somehow destroyed), indicating that at least five females deposited eggs.   The third nest had twelve eggs, showing that three birds at the most (probably only two) had laid.   The size of the colony was reduced only by the loss of a male just after incubation had started

181

in the third nest. Since it is known that the same female can lay more than once in a season, it is difficult to understand why some females stopped laying for the third brood. One suggestion is that different females laid in the second and third nests; but this solution does not seem likely from the history of other groups.

In addition to variation in laying ability great variation of interest in the nest is shown among individual birds. Individuals may leave the colony at the time incubation starts. For example, ARR (female) left the Cypress colony in which there was a male apparently paired with her on July 27; incubation had started on July 25. She formed a colony with two males, built a nest, and started to incubate on August 18. As another example a male (A432804) left the Pond group (incubation started August 29) on September 1 and joined with another bird. When this pair was collected on September 25 the female had an egg in the oviduct and the male had functional testes.

Furthermore, individuals vary greatly in their activity at the nest. In the Cypress group one pair did most of the work although two females had laid eggs in the nest and a bird which had laid eggs did not incubate at all but did feed the young. In two other cases eggs were laid but the nest was deserted without cause.

An additional point in support of cyclic breeding is that the strangers are most aggressive in their attempts to join a colony if it is in the nest-building stage, although they try to join a colony at any time.

The data cited above show that in respect to breeding behavior there is a difference of some sort between individuals and also a difference in the same individual at different times. Now let us examine the sparse knowledge of endocrinology of birds for any evidence of cycles within the season of reproduction. In the male the testis is always functional except during the incubation according to Schooley and Riddle (1938) and as soon as the feeding of the young stops the male returns to complete function. In the female the relations between the several hormones are inadequately known. The relations of prolactin to brooding and the possible inhibition of FSH (Bates, Riddle, and Lahr, 1937) occur after the eggs are laid and give no clue to the factors influencing egg laying. LH is present (Leonard, 1937) but its function is unknown. A type of corpus luteum is present but it is a rapidly regressing structure. This summary indicates that there is no positive endocrine evidence of a cycle in breeding.

Several features concerning the endocrinology of *C. ani* may be presented at this point. The history of the ovary is of interest. After

ovulation the granulosa layer proliferates and forms a structure considered by Hett (1923) as a corpus luteum, which decreases in size. In
C. ani the corpus atreticum shows an interesting development. The
granulosa of eggs which have not undergone pre-ovulatory swelling,
proliferates and in *C. ani,* bursts, the yolk is apparently extruded from
the follicle, and lymphocytes enter and take up the fat globules. After
about ten days of atresia, this structure cannot be distinguished from
a corpus luteum. The function, if any, of these two structures is
unknown.

Another feature in reproductive physiology of *C. ani* is that throughout the season birds with small gonads occur. Birds of both sexes
were collected at the height of the breeding season. A testis, which
weighed one-sixth the normal, contained sperm, although these were
not normally distributed, and had none of the characteristics of
degeneracy. The ovaries contained eggs which are larger than in the
juvenile, but had not reached the pre-ovulatory swelling stage. One
explanation for these birds is that they are young hatched late in the
previous season (November) which are still too young to mature
sexually. Thus van Oordt (1938) has found that non-breeding Oyster-
catchers (*Haematopus ostralegus*) may be either yearlings or adults.
The testes of the yearlings are somewhat developed and have a few
sperm but those of the adults possess sperm and resemble those of
breeding adults. The ovaries of yearlings are less developed than
those of a non-breeding bird. This description agrees essentially with
the histological picture of the gonads of *C. ani.* In addition, Kendeigh and Baldwin (1937) find that about 15% of the young House
Wrens did not breed in the first year, and Lorenz (1931) finds that
young Jackdaws court and pair in the first autumn but do not breed
till the second year. Another explanation of the condition of the
gonads of these specimens is that they are "old virgins," as described
by Schooley and Riddle (1938). According to these authors, these
birds have ovaries which "have developed somewhat beyond the
juvenile state, but no ova have entered upon the final phase of growth
which is induced by a release of increased amounts of pituitary gonad
stimulating hormone." The basophiles of the pituitary are undiffer-
entiated and agranular and are smaller than pre-ovulatory basophiles.
The acidophiles are nearly normal and resemble those of nesting birds.
Although the pituitary resembles that of a senile animal, it is known
that these birds are no more than a year and a half old. This condition is essentially pathologic and would not be expected in a natural
population.

In reproductive physiology the problem of the factor which determines the cessation of laying in birds is important. Presumably either visual or tactile stimulation reacts through the pituitary to stop the laying process. But since in *C. ani* more than one bird lays eggs in the nest, there is a varying number of eggs, and hence neither sense can be effective. In this connection it is interesting that incubation starts slowly and is intermittent for a few days although most species start abruptly and sometimes before the eggs are laid or the clutch is completed. The fact that incubation starts simultaneously although the eggs may be laid over a long period of time (16 days in the Pond group) and presumably some of the females have completed their clutch suggests that the psychologic factors may be more important in arresting laying than any endocrine factors which may exist. The great variation (4–7) in the number of eggs laid per female may be related to the lack of a suitable stimulus to stop laying.

The hypothesis of Darling (1938) pertains to some of the apparently cyclic behavior seen in *C. ani*. Darling suggests from his work on the breeding of gulls that the birds must be stimulated by courtship to a high degree of emotional excitement before breeding can begin and that in colonial birds there is a great amount of reciprocal stimulation within the flock and hence a minimum number (threshold) of birds is necessary before the breeding can start. Therefore, in colonies where the number of birds is just above the threshold, the breeding is irregular and the number of unsuccessful nests is high. Darling's observations on which he based the above hypothesis can also be explained by the hypothesis of reproductive cycles. In the large colonies, since there are more birds, there is a greater chance of one bird meeting a bird in exactly the same phase with it and therefore in the large colonies there will be a greater success in breeding. Neither of these hypotheses is supported by experimental data as yet.

* * * * * * *

REFERENCES

ALLEE, W. C.
1938. The social life of animals. New York, 293 pp.

BALDWIN, S. PRENTISS, AND KENDEIGH, S. CHARLES
1938. Variations in the weight of birds. Auk, 55: 416–467.

BATES, R., RIDDLE, O., AND LAHR, E.
1937. The mechanism of the anti-gonad action of prolactin in adult pigeons. Amer. Journ. Physiol., 119: 610–614.

BYERLY, T. C., AND BURROWS, W. H.
1936. Effects of genetic constitution with respect to broodiness and prolactin content. Proc. Soc. Exp. Biol. Med., 34: 844–846.

CARPENTER, C. R.
1934. A field study of the behavior and social relations of howling monkeys. Comp. Psychol. Monog., 10: 1–168.

CHAPMAN, FRANK M.
1938. Life in an air castle. New York, 250 pp.

DANFORTH, STUART T.
1937. Ornithological investigation in Vieques Islands, Puerto Rico, during December, 1935. Journ. Agric., Univ. Puerto Rico, 21: 539–550.

DARLING, F. FRASER
1938. Bird flocks and the breeding cycle. Cambridge, 124 pp.

DOBZHANSKY, T.
1937. Genetics and the origin of species. New York, 364 pp.

ERICKSON, MARY M.
1938. Territory, annual cycle, and numbers in a population of Wren-tits (*Chamaea fasciata*). Univ. of California Pub. Zool., 42: 247–334.

EVANS, L. T.
1938. Cuban field studies on territoriality of the lizard, *Anolis sagrei*. Journ. Comp. Psychol., 25: 97–126.

FRIEDMANN, HERBERT
1928. Social parasitism in birds. Quart. Rev. Biol., 3: 554–569.
1929. The Cowbirds. Springfield, Illinois, 421 pp.
1933. Size and measurement of territory in birds. Bird-banding, 4: 41–45.

1935. Bird societies; in: A Handbook of Social Psychology, Worcester, Mass., Chapt. 5, pp. 142–184.

GUNDLACH, J.
  1874. Neue Beiträge zur Ornithologie Cubas. Journ. f. Ornith., 22: 113–166.
  1895. Ornitologia Cubana. Havana, 328 pp.

HERRICK, FRANCIS H.
  1910. Life and behavior of the Cuckoo. Journ. Exp. Zool., 9: 169–234.

HETT, J.
  1923. Das Corpus luteum der Dohle (*Coleus monedula*). Arch. Mikr. Anat., 97: 718–838.

HOWARD, H. E.
  1920. Territory in bird life. London, xiii, 308 pp.
  1929. An introduction to the study of bird behavior. Cambridge, Univ. Press, xi, 135 pp.

HUXLEY, JULIAN S.
  1932. Field studies and physiology: a correlation in the field of avian reproduction. Nature, 129: 166.

KENDEIGH, S. CHARLES
  1934. The rôle of environment in the life of birds. Ecol. Monogr., 4: 299–417.

KENDEIGH, S. CHARLES, AND BALDWIN, S. PRENTISS
  1937. Factors affecting yearly abundance of passerine birds. Ecol. Monogr., 7: 91–124.

LACK, D., AND L.
  1933. Territory reviewed. Brit. Birds, 27: 179–199.

LEONARD, S. L.
  1937. Luteinizing hormone in bird hypophyses. Proc. Soc. Exp. Biol. Med., 37: 566–568.

LEVERKUHN, PAUL
  1894. Über das Brutgeschaft der Crotophagiden. Journ. f. Ornith., 42: 44–80.

LORENZ, KONRAD
  1931. Beiträge zur Ethologie sozialer Corviden. Journ. f. Ornith., 79: 67–127.
  1937. The companion in the bird's world. Auk, 54: 245–273.

MAKATSCH, WOLFGANG
  1937. Der Brutparasitismus der Kukucksvogel. Leipzig, 152 pp.

MARSHALL, F. H. A.
  1936. Sexual periodicity and the causes which determine it. Phil. Trans. Roy. Soc., London, B, 226: 423–456.

MAYR, ERNST
  1935. Bernard Altum and the territory theory. Proc. Linn. Soc. New York, no. 45–46, pp. 24–38.

McILHENNY, E. A.
  1937. Life history of the Boat-tailed Grackle in Louisiana. Auk, 54: 274–295.

MICHENER, H., AND J. R.
  1935. Mockingbirds, their territories and individualities. Condor, 37: 97–140.

NICE, M. M.
  1933. The theory of territorialism and its development. In: Fifty years progress of American Ornithology, 1883–1933, pp. 89–100.
  1937. Studies in the life history of the Song Sparrow. I. Trans. Linn. Soc. New York, 4: 1–247.

NOBLE, G. K., AND VOGT, WILLIAM
 1938. An experimental study of sex recognition in birds. Auk, 52: 278–286.
NOBLE, G. K., WURM, M., AND SCHMIDT, A.
 1938. Social behavior of the Black-crowned Night Heron. Auk, 55: 7–40.
PETERS, JAMES L.
 1939. Birds of the World, vol. 4, in MS.
RIDDLE, O., BATES, R., AND LAHR, E.
 1935. Prolactin induces broodiness in fowl. Amer. Journ. Physiol., 111: 252–260.
RIDDLE, O., AND BRAUCHER, P. F.
 1934. Body size changes in doves and pigeons incident to the stages of the
  reproductive cycle. Amer. Journ. Physiol., 107: 343–348.
RITTER, W. E.
 1938. The California Woodpecker and I. Berkeley, 340 pp.
SCHOOLEY, JAMES P., AND RIDDLE, O.
 1938. The morphological basis of pituitary function. Amer. Journ. Anat., 62:
  313–349.
SKUTCH, ALEXANDER
 1935. Helpers at the nest. Auk, 52: 257–273.
 1937. In litt.
TINBERGEN, N.
 1935. The behavior of the Red-necked Phalarope (*Phalaropus lobatus*) in spring.
  Ardea, 24: 1–42.
 1936. The function of sexual fighting in birds; and the problem of the origin of
  "territory." Bird-banding, 7: 1–8.
VAN OORDT, G. J., AND BRUYNS, M. F. M.
 1938. Die Gonaden übersommernder Austernfischer (*Haematopus ostralegus* L.).
  Zeitschr. Morph. Ökol. der Tiere, 34: 161–172.
WHITMAN, C. O.
 1919. Behavior of pigeons. Carnegie Inst. Washington, publ. 257, 161 pp.
YOUNG, C. G.
 1929. A contribution to the ornithology of the coastland of British Guiana.
  Ibis, (12) 5: 1–38.

14

Reprinted from *Quart. Rev. Biol.*, **17**(2), 115–134 (1942)

THE PHYLOGENY OF SOCIAL NESTING HABITS IN THE CROTOPHAGINAE

By DAVID E. DAVIS

Biological Laboratories, Harvard University

INTRODUCTION

THE reproductive behavior of birds is based on the construction and maintenance of a nest. Normally a male and a female select a nest site and build a complicated structure for raising the young. Although nesting behavior is not limited to birds, nevertheless the habits reach their climax in variety and complexity in this class. The variations in nesting behaviors are innumerable and frequently demonstrate phylogenetic sequences.

The Crotophaginae, one of the six subfamilies of Cuckoos, has developed the unusual habit of building communal nests in which several females lay eggs. The phylogeny of this remarkable system of raising young is the subject of this paper. The family Cuculidae as a whole shows a tendency to develop peculiar breeding habits. Of the approximately 200 species of Cuculidae, about 80 are parasitic (Makatsch, 1937), representing two of the six subfamilies. This parasitism is found in several stages of development. Some species perhaps are just starting on the road to parasitism; for example the eggs of the Yellow-billed Cuckoo (*Coccyzus americanus*) have been found in the nests of at least 12 species of birds (Herrick, 1910). Other species are always parasitic. The least specialized of these lay eggs in the nests of numerous species. In other cases, as represented by the European Cuckoo (*Cuculus canorus*), the birds are individually host-specific (Friedmann, 1928a). Thus individuals of the species each parasitize particular host species, although in the same region other individuals may lay eggs in nests of other host species. The climax has been reached by certain species, such as *Eudynamis honorata*, which parasitize only one species in a region. The Crotophaginae, however, instead of developing parasitism as other cuckoos have done, have specialized in the direction of social nesting habits.

Before considering the background of the subject let us now proceed to review briefly the subfamily Crotophaginae. This group of birds consists of the monotypic genus *Guira* and the genus *Crotophaga*, the latter represented by three species: *major*, *sulcirostris*, and *ani*. The distribution of the subfamily extends from Rio Negro in Argentina north throughout the West Indies to Florida and throughout Central America to Texas. The birds have noisy, ostentatious habits and everywhere are abundant and conspicuous members of the avifauna.

At this point, in order to orient the reader, a brief description of the behavior of the species is inserted. All four species of the subfamily possess the same general behavior patterns and for the purposes of this paper may be described in general terms. The species prefer fairly open and moist habitats and are conspicuous in behavior, all using loud notes and making no effort to conceal themselves. The birds associate in flocks of about a dozen individuals, spending the days together and sleeping in the same tree. Each flock defends its territory by fighting. At the breeding season members of the flock cooperate in the building of a single nest in which several females may lay eggs. The males assist in the construction of the nest and in the care of the young. After the breeding season the young stay with the flock for a variable length of time and may help in the care of the next brood, and may even breed with the colony next season. During the non-breeding season the birds tend to congregate in moist areas where insects, their favorite food, are most abundant. The composition of the flock varies over a period of time; a bird may leave, or, after a period of fighting, join the group. In connection with the flock behavior are many calls and actions which are of value to the group as a unit.

These species, whose behavior has been briefly

115

described, show the development of a system of social nesting. The phylogeny of breeding habits within a group of animals is usually obscured by the extinction of some of the species which represent stages in the development. This situation makes it necessary to speculate and to reason from analogy. In some cases, however, the evolution of behavior may be traced through a small group of closely related species, as Friedmann (1929) has done in the Cowbirds. The Crotophaginae is another group of birds in which we are able to trace the steps in the evolution of a peculiar type of breeding habits, as this paper endeavors to show. The phylogeny of other groups of animals has been studied from the behavioristic viewpoint, for example, the ants in Wheeler's classic studies (1923, 1928). Emerson (1938) analyzes the nests of termites to trace the behavioristic and evolutionary sequences.

In discussing the evolution of any particular condition in animals it is customary to seek "causes" which determine the "effect." This principle of Cause and Effect, however, encounters numerous philosophical difficulties and in biology is further complicated by the simultaneous action of several factors. For the purposes of this paper, then, we shall not discuss any cause of the development of social nesting habits but merely state the conditions under which the breeding behavior evolved. From the evidence here presented it will be seen that social nesting habits are the logical outcome of these conditions and that, in this case, were any one of the several factors lacking, the habits probably could not have developed. These same habits, of course, probably could have resulted from a different evolutionary history. In addition it should be noted that the unusual social nesting, as found in the behavior of *Crotophaga ani*, is the result of a combination of situations, no one of which is in itself unusual, but all of which when combined, permit the social nesting. These various conditions may be regarded as conductive to the development of the habit. Throughout this paper examples of other species will be cited to show that none of the particular behavior patterns is unique but that all occur in one or more groups of birds.

Before considering in detail the behavior of the birds, it is desirable to discuss some features of their anatomy and then their geographical distribution. In their internal anatomy the various species show no significant differences. Of the external features the development of the keel on the bill in the members of the group is spectacular. *Guira* has a typical cuculine bill but the genus *Crotophaga* has developed a large thin keel on the bill. In *Crotophaga major* the keel extends one-half way to the tip of the bill and in *C. ani* the keel is high and thin, extending to the tip. In *C. sulcirostris* the keel is grooved. In coloration *Guira* is again obviously the most closely allied to to the other cuckoos. Its body feathers are white but the tail, wings, and head are streaked with brown and some black, giving the appearance of a brown bird; a notable feature is the crest of feathers on the head. The genus *Crotophaga* is uniformly black, with a different amount of blue irridescence in the plumage of each species. Both *major* and *ani* have a remnant of the crest on the head as shown by a few feathers which are slightly longer than the others. The evidence from the external features indicates that *Guira* is positively the more primitive genus and that within the genus *Crotophaga*, *C. ani* is probably the most evolved. The development of the bill and the black color and the reduction of the crest are evidence of evolution within the group.

Concerning the geographical distribution of the subfamily it may be said that the group is either very young or more probably has been unable to spread very far because of a lack of ability to adapt itself to colder climates. *Guira* is found in Brazil and Argentina. *C. major* is found from northern Argentina through Brazil to Panama. *C. sulcirostris* occurs on the west coast from northern Peru through Central America to Texas. *C. ani* is the most widespread, occurring from northern Argentina to Yucatan and Florida and also through the West Indies. This distribution suggests a gradual dispersal from the highlands of Brazil.

With this introduction as a background we shall now consider the phylogeny of behavior. A preliminary section brings together notes on comparative behavior within the subfamily and reference to problems of discrimination and to "releasers". The stages in the evolution of social nesting are next analyzed. The development occurs at three levels, represented by the species (1) *Guira guira*, (2) *Crotophaga major*, (3) *C. ani* and *C. sulcirostris*. The primary factors conducive to the development of social nesting habits are (a) modification of territorialism (b) aberrant breeding, (c) the type of habitat. These three levels of development will be analyzed in a separate section for each of the above factors. Finally there is a discussion of the phylogeny of behavior within the group and its general relation to social parasitism.

COMPARATIVE BEHAVIOR OF THE CROTOPHAGINAE

The author has had the opportunity to observe the same species in very distant parts of its geographical range. The study of a species in one particular locality may be misleading because of certain local mannerisms. Since studies of the same species of bird, carried out by the same observer in widely separated places, are rare, these data are considered worth reporting in some detail. The following notes are recorded in order to afford a comparison of the habits of *Crotophaga ani* in British Guiana and in Argentina with the habits of the species in Cuba (Davis, 1940a).

In British Guiana each flock strictly maintains a territory exactly as in Cuba. However, the vegetation of certain areas permits an apparent variation in behavior. For example, along the Abary River are occasional open areas of savanna, always surrounded by rain forest. Many of these areas are not large enough for more than one group of *C. ani* and hence in these cases no territorial fighting can appear between groups, for there are no neighbors with whom to fight. However, the occasional stranger is driven out by fierce fighting. Another modification of the territory is that *C. major* is permitted to share the territory. Certainly there is competition for food between the two species.

The colonies found in Guiana were smaller than many found in Cuba. For example, on July 13, 1939, I found a colony with five adults and a nest containing one young bird. Another colony containing six adults had three young, two weeks old. A month later I found a colony with four adults (two females, one male, one of unknown sex) and a nest containing four young ready to leave the nest. It should be noted that in these groups probably only one female laid in each nest, for in no case were there more than six young birds.

Observations on *C. ani* were made also near Saladas in the Province of Corrientes in Argentina. Here the species is not sufficiently abundant to show territorialism clearly. One group of birds (12), including some juveniles, inhabited a particular area, near a small marsh. Also another colony, consisting of three adults, had a nest with three young. Several times an individual or two of *Guira guira* came near and even perched in the same tree with an individual of *C. ani* without causing the latter to show defense of the territory. Again it is of importance that, as in British Guiana, the territory is not defended against a species so closely related, both taxonomically and ecologically. In summary it may be said that in the general activities, as well as in the particular habits mentioned above, the behavior of *C. ani* is the same in British Guiana, in Argentina, and in Cuba.

Correlation of anatomy and behavior

An interesting relationship between anatomy and behavior is shown in the Crotophaginae. It is considered that a display behavior develops before the particular structure used in the display; the wing marks of the ducks are the classic example. Lorenz (1935) describes the uses in herons of certain structures for elaborate display. Chapman (1935) calls attention to the fact that a manakin of Brazil (*Chiroxiphia*) has an elaborate dance but no special structure such as is found in other members of the family. A corollary of this principle is that the structure remains after the behavior·has disappeared. The development of the crest in the Crotophaginae may be an example of this corollary. The large crest of *Guira* is raised in alarm and also when the alarm rattle is given. *C. ani* has no such alarm note but has retained some feathers on the head which are somewhat longer than the nearby ones. This little crest is never used. This retention of a structure after the behavior is lost follows naturally from the fact that the structure is developed after the behavior.

Conspicuous buccal markings are found in the nestlings of *Guira* and also of *C. major*. In *C. ani* these markings are present but much less distinct.

Social behavior

The similarity of the call notes throughout the group is of interest. *C. ani* has the most varied vocabulary and is the most social. The alarm call of *C. major* greatly resembles the rattle alarm of *Guira*. But the most noteworthy is the resemblance of the flight flock call, the note used to notify other members of the group of the movements of one individual. In *Guira* the note is weak and indistinct. In *C. ani* the note is nearly identical in pitch but is very loud and harsh. The use is the same in both cases. The flight flock call of *C. sulcirostris* is of the same pattern. The notes used in mating are in the whole group a uniform series of soft whines.

The Crotophaginae agree with most other birds in having two calls to indicate danger. The 'alarm' call notifies the members of the group of a walking enemy or suspicious situation. The 'danger' call is used to warn of an approaching hawk or other flying predator. Many species of birds have this separation by call of the types of danger. Makkink (1936) found in the Avocet (*Recurvirostra avocetta*) that there was one call for the chief enemy, the gulls, and another for all other enemies.

Social behavior has nearly always developed coincident with an extension of the time the young remains with the adults (cf. ants, wasps, and humans). In the Crotophaginae, however, there is no evidence that social nesting or flock behavior is a result of this extension. Nevertheless it is of great interest that the young of one brood have been observed feeding the young of the second brood in *C. ani* (Davis, 1940a) and also in *C. sulcirostris* (Skutch, 1935). Furthermore, as Skutch points out, it is not unusual for unmated birds of many species to assist at the nest. Skutch (1940) found that the young of some wrens remain with the adults and even feed the young of the succeeding brood. According to Leach (1925), the young California Woodpeckers (*Balanosphyra formicivora*) also feed the young of the second brood. Thus we see that the cohesion of the family in several species of birds is the result of communal nesting, but not the cause.

The type of peck order or social hierarchy is of great importance in a social group. In the Crotophaginae no suitable behavior was found which could be used to rank the birds. While feeding, contests for juicy morsels occasionally occurred but showed no peck order. Lorenz (1938) points out that when there is a rigid peck order in colonial birds only one pair can raise young. In the Jackdaws (*Coleus monedula*) this situation is overcome by the fact that the despot pecks only the nearest in rank. In the Crotophaginae if a peck order occurs, a similar situation probably obtains.

Discrimination

Under the term, discrimination, many problems of animal behavior are grouped. The problem in general is to analyze a series of behaviors to determine which characteristics of the stimulating object act on which sense organs of the reacting animal to influence a subsequent behavior. In birds the stimulating characteristics are usually some morphologic or behavioristic element. The reacting organs are generally visual or auditory. In this paper is discussed the evidence concerning the discrimination of three of the many objects occurring in the bird's life, namely, the sex-partner, other individuals of the species and the boundaries of the territory.

The evidence concerning the discrimination of the sex-partner is meager. The problem is to find out what features of a bird are the clues which make possible the determination of the sex. In appearance the sexes are identical in all species of Crotophaginae except that the male is larger. Experiments with dummy birds were performed on *Crotophaga ani*. These dummies were museum skins arranged to resemble a live bird. In most cases the birds attacked the dummies as if they were live invaders of the territory. Sometimes after the dummy had been in one place a long time the owners of the territory ignored it. Decoy (live) birds, placed in a trap in order to catch the birds, were treated exactly like strangers and killed in a short time. Young birds, however, were ineffective as decoys because the adults did not respond in either a helpful or an antagonistic manner. The above evidence suggests that the Crotophaginae belong to the "labyrinth fish" type of mating reaction according to the classification of Lorenz (1935). In this category the display of the male develops into fighting except when the stranger shows female behavior. However, the assignment of the Crotophaginae to this category depends on insufficient evidence. In no case was "female behavior" observed to result in a cessation of fighting. This is, of course, not contradictory evidence but merely lack of evidence.

An important modifying condition is the relationships imposed on the individuals by the flock organization. It seems likely that an individual must be admitted to the flock before it can be eligible for pairing. If this is true then all the fighting observed was in reference to membership in the flock and none in reference to the sex-partner. If this is the case then we must admit that we have no evidence as to the method of discrimination of the sex-partner.

Another problem is to determine which sense modalities are used in discrimination of other members of the species or of other species. There is no doubt that each member of the flock knows individually every other member of the group. (Cf. Davis, 1940a, 1940b, 1941a.) The problem is to determine which sense modality is used in this discrimination. A natural experiment occurs in the presence of the species *Ptiloxena atroviolacea*, an entirely black Icterid about one-third smaller than *Crotophaga ani*. *C. ani* persecutes this species and drives it from the territory. The behavior towards *Ptiloxena* consists of the same series of patterns used against an intruder of the *Crotophaga* species. Another black species in the region, the rare *Corvus nasicus*, was seen only once and at that time was being persecuted by a group of *Crotophaga ani*. *Holoquiscalus niger*, another Icterid, was not observed to be persecuted—a situation probably due to the ecological separation of the habitats. As further evidence it may be noted that, according to the natives of Cuba, albino birds live with the flock in a normal manner; the writer never observed an albino bird. The above facts, showing that black birds of other species in addition to dummy birds are attacked, suggests that the discrimination is by means of vision and not by audition. That the auditory sense modality is used in certain species is shown by the work of Noble, Wurm, and Schmidt (1938) who found that herons (*Nycticorax nycticorax*) recognize their sex-partners by voice. Experiments with mirrors (Davis, 1940a) show that individuals of *Crotophaga ani* do not recognize themselves but fight against their own image. This evidence eliminates the olfactory sense modality, already improbable, from consideration.

Now, having outlined the evidence and suggested the conclusion that these birds utilize the visual sense modality for the discrimination of other individuals, it is necessary to consider what characteristics of the stranger stimulate the reacting bird. In general the problem is to determine whether behavior or physical appearance is the basis of discrimination. The evidence from the reactions to *Ptiloxena* and to the dummies suggests the conclusion that strange birds are recognized not by their behavior but by their appearance. Were behavior the means of discrimination only birds which behaved in a certain manner would be attacked. But birds are driven out which resemble *Crotophaga* in appearance although their behavior has no resemblance to that of an invading bird. The tolerance of *C. major* by *C. ani* in Guiana is an apparent exception which may be explained by assuming that the *ani* have learned

to recognize the *major* as individuals. It must be remembered that for the moment we are concerned with the recognition, not of the sex-partner, but of a stranger. Nevertheless, the data pertaining to the recognition of a sex-partner are useful in an attempt to analyze the situation in the Crotophaginae. Referring to other species, the Kingbirds (*Tyrannus tyrannus*) (Davis, 1941b) drive from their territory birds of various species whose behavior varies greatly and whose only common characteristic is an appearance different from that of the female. Tinbergen (1939) refers to errors of identification by the Snow Bunting (*Plectrophenax nivalis*). In these cases the behavior of the female is of the type which should indicate that she belongs in the territory. However, the male does not recognize the physical features of his mate and starts to fight until her features are recognized. Tinbergen also found that the male Bunting requires some time to learn to recognize his mate and until that time threatened her whenever she came into his territory. Lorenz (1935) states that a young bird must learn a certain number of characteristics of the parent before becoming able to recognize and thus not threaten it. The Jackdaw is a species which certainly does recognize the individuals of the flock by appearance, not by behavior. Also, Lorenz points out that the social life of birds differs from that of insects because there is a personal recognition of the individuals in the group. Noble and Vogt (1935) found that some species (*Agelaius phoeniceus* and *Geothlypis brachydactyla*) which differ in the plumage of the sexes, recognize the female by physical appearance. Noble (1936) carried the work further and found that the 'moustache' of the Flicker (*Colaptes auratus*) was necessary for sex recognition. In contrast, other species, which frequently are identical in the plumage of both sexes, recognize individuals by behavior (Noble and Vogt, 1935). Lorenz (1935) found that young herons recognize their parents by behavior. This evidence from other species suggests the possibility that even in the Crotophaginae the initial recognition of a stranger is by means of the behavior of the individual. This suggestion is supported by the work of Noble and Curtis (1939). In the Jewel Fish (*Hemichromus bimaculatus*) the sex of a stranger is recognized initially by behavior but after pairing the mates recognize each other by individual characteristics of appearance. In the Crotophaginae it is difficult to determine exactly what the behavior might be which would indicate that a strange bird is not a member of the group. However, the lack of knowledge of the territory may be noticable to the owners. Then on close inspection the owners may see that the bird is a stranger and drive it out. Also there is the possibility that a lack of certain positive behavior patterns identifies a bird as being a stranger.

In addition to the question of discrimination of individual birds there is the problem of the discrimination of the boundaries of territories. The birds remain so strictly in their own territory that there must be an exact knowledge of the very trees and shrubs which mark the edge. In Cuba it was most amusing to watch the members of a group follow the electric mowing machine which the birds had learned was a source of an easy meal of insects. Each group followed the machine to the exact boundary of the territory but did not trespass beyond, in spite of the strong attraction of food. The birds clearly recognized the individual trees or bushes which marked the boundary.

Members of the Crotophaginae are able to recognize individuals which seem identical to the human observer. Yet these same birds are unable to recognize that such a bird as *Ptiloxena* is of another species. These facts suggest a conflict in the interpretation. It seems likely that this apparent discrepancy may be resolved by a consideration of the probability that birds learn one or two special characteristics of each individual bird and unless these are present, consider the individual to be a stranger. Lack (1939) found that only a few characteristics of the English Robin (*Erithacus rubecula*) were required to cause the owner of the territory to attack a dummy. In humans the ready deception accomplished with disguises shows how very few marks we need to recognize our friends.

Releasers

The fact that a certain few characteristics are used to discriminate among individuals of the same species is further evidence that social behavior is largely controlled by specific combinations of special stimuli. These stimuli are anatomical structures or behavior patterns which are in themselves unusual or are combined in unusual behavior sequences. Such sequences are of so improbable a nature that they do not occur in the experience of the species except under the proper circum-

stances. Therefore, the automatic and inflexible response to the stimulus occurs only at the proper time. Such stimuli or series of stimuli have been called releasers (Von Uexkill and Lorenz) or sign-stimuli (Russell, 1934). The studies of the Crotophaginae have provided examples of several releasers which it is proposed to mention at this time. On one occasion a dove (*Zenaidura macroura*) fluttered off her nest, performing typical injury-feigning, right into the midst of a group of anis. The birds at once acted toward the dove as if it were a wounded individual of their own species. They crowded around the dove, fluttered and used a chuckle note. They did not use the 'conk', a note uttered while attacking a stranger. As another example the mutual preening has many of the characteristics of releasers. The moment that the bill of one bird touches the neck of its neighbor the neck is raised and the feathers fluffed out. This occurs in all species of the subfamily. This action may be no more than a simple reflex. The feeding of the young of the second brood by the young of the first brood is probably a social companion action. It is possibly released not by the young in the nest but by the action of the other adults.

The imperfections of releasers are probably as informative as the perfections. Witness the interesting conflict between the response to the alarm call and the strict observance of the boundaries of the territory. When the investigator disturbed a nest, the alarm call was used and the birds from nearby groups came near but they never passed the boundaries of the territory although it was obvious that the alarm call was a strong attraction. As another example of an imperfection of releasers, individuals of the Mockingbird (*Mimus polyglottos*) were able to mimic the alarm call and released the alarm behavior of the nearby anis. This observation shows that the discrimination was not very exact for even the writer could distinguish after short practice between the mocker and the true alarm note.

MODIFICATION OF TERRITORIALISM

The concept of territorialism has developed from attempts to interpret various types of fighting behavior of animals. This concept has had great heuristic value and in addition has clarified the interpretation of the behavior of many species. In other species (Cowbirds) the analysis of the manifestation of territorialism has suggested a possible course of phylogeny of behavior. It is now proposed to present the relation of territorialism to the phylogeny of behavior in the Crotophaginae and then take up the modifications of the concept as suggested by the behavior of the Crotophaginae. This chapter emphatically is not a critique of the theory of territorialism or of the many criticisms of the concept.

In all the species of the subfamily the colony owns a territory which consists of two distinct parts. One of these is a clump of trees for sleeping and the other an area of fairly open land for feeding. From this territory, in general, other members of the species are expelled. Among the species, however, certain differences exist. *Guira guira* defends its territory only slightly; numerous intrusions are permitted, although on some occasions birds are violently driven out. The colony lives in a large but nevertheless definite area, and, since it is defended, it can be called a true territory. The important point is that within this colonial territory one or two pairs may have their own small territories. One or more pairs may separate from the colony and build a nest, setting up around this nest a territory from which intruders are expelled. However, the defence of this small territory is not vigorous and in many cases other birds of the colony use the nest, thus producing a communal nest. This weak defence of the territory has thus permitted social nesting.

In *Crotophaga major* the development of social nesting has continued. The birds remain in pairs, all cooperating to build one nest in which several females lay eggs. Probably not all females lay eggs at the same time and also probably a single pair does occasionally build a nest and raise young. The colony defends the territory for the whole group, although weakly. Lack of defence is probably more apparent than real and depends on a mutual understanding. Such behavior occurs in other species. For example Ryves (1929) found that in Redshanks (*Totanus totanus*) "once an individual territory has been definitely secured by a pair of birds and approved by the other pairs of the same species, a distinct mutual understanding if not friendship developed among them."

In *Crotophaga ani* the communal nesting has reached its climax. Each colony defends its territory most aggressively and without exception attacks strangers. The marital relations are very flexible. Polygamy is the general rule for it is certain that in some cases polyandry occurred and in others polygyny prevailed. Yet in some groups monogamy was strict and there was no dissension among pairs. In one case a pair created a territory and defended it alone. Wetmore (1927) cites two instances in which he believed that a single pair built a nest.

Crotophaga sulcirostris is apparently at the same level of development as *C. ani*. The observations of Skutch (Bent, 1940) indicate that the birds have the same marital relations as does *C. ani*.

The phylogenetic development of social nesting coincides with the disappearance of territorial defence by the pair and the appearance of territorial defence by the colony. Other species have developed social breeding habits of several types, which can be roughly contrasted with the stages in the Crotophaginae. Several examples of these species will now be cited to illustrate for comparison the three levels in the development of crotophaginine social nesting.

At the phylogenetic level of *Guira* may be cited the numerous species which spend most of the year in flocks but separate out into pairs for breeding. A few examples will suffice. Tinbergen (1939) found that Snow Buntings separated from the flock to acquire territory. The Oyster-catcher (*Haematopus ostralegus*) (Huxley, 1925) has a nesting territory but a common feeding ground. The coveys of Bob-white (*Colinus virginianus*) (Errington, 1933) break up in April to pair for breeding. These examples show that a weakened territorial defence by a pair under certain conditions could permit the flock to remain together and lead to communal nesting.

Another bird shows the behavior at the level of *C. major*. In the species *Yukina bruneiceps* (Timaliidae), according to Yamashima (1938), "several pairs of birds join in building one nest." His data for four nests show that four or six birds, laying two or three types of eggs, composed each flock. *Corydon* (de Schaunesee, 1928) may have similar habits. In addition a comparison with the Jackdaw (Lorenz, 1931, 1938) and the Rook (*Corvus frugilegus*) (Yeates, 1934) is instructive. In these species, although there is a colonial territory, the strict maintenance of pair territories has prevented social nesting. Some Magpies (*Pica*) (Linsdale, 1937) live in similar colonies. The Weaver Finch, *Philetarius*, (Friedmann, 1930) nests in colonies and the colony defends a territory. Some colonial birds have no colonial territory, as the Oropendolas (*Zarhynchus wagleri*) (Chapman,

194

1928). Thus a colonial territory does not necessarily result in communal nesting unless, among other contributing factors, the pair territory is weakened.

At the level of the development of *C. ani* and *C. sulcirostris* the breeding of the California Woodpecker may be mentioned. The woodpeckers live in colonies and several pairs cooperate in excavating a nest hole and raising the young (Leach, 1925; Ritter, 1938). Two Australian genera, *Corcorax* and *Pomatorhinus* (Friedmann, 1925) are reputed to nest communally. Thus at each level in the phylogeny of the Crotophaginae, examples from other groups may be cited, each with a different modification or weakening of the territorialism of the pair.

Breakdown of territorialism in the Crotophaginae

Before analyzing the factors responsible for the breakdown of territorialism within the Crotophaginae, it is desirable to note the assumption that territorialism was originally present in the group. Territorialism is now considered to be a widespread phenomenon in the vertebrate classes. Breder (1936) has summarized the behavior of many fishes (Centrarchidae) and Noble (1938) and Noble and Curtis (1939) have analyzed the relation of sexual selection to territorialism. In those fishes which build nests the male guards the nest site. When the female arrives and copulation occurs, the eggs are deposited. The male in many species guards the eggs and then the young until they are able to fend for themselves. The defence in these species has the characteristics of avian territorial defence and can be differentiated from sexual fighting as such. In the Amphibia data indicate that the male commonly has a calling and mating station but as yet these data are insufficient to be certain that this is analogous to the territories of birds. For lizards the work of Evans (1936, 1937, 1938) and Evans and Clapp (1940) has shown that the species *Anolis carolinensis* maintains a territory in accordance with the characteristics of avian territories. For birds, the evidence is overwhelming that most species are territorial. The reviews of Nice (1933) and Lack and Lack (1933) indicate the extent of the behavior. For the Cuculidae in particular, Makatsch (1937), Friedmann, (1928b) and Davis (1940c) indicate that territorialism is nearly universal within the group. Chance (1922) found that the female European Cuckoo will lay in an unusual fosterer's

nest rather than leave her territory. However, there are records of two females using the same territory (Gosnell, 1932). Among the African Cuckoos, Friedmann (1928a) found that many parasitic species have territories which are dependent upon the availability of nests to parasitize. However, *Lampromorpha* has a faulty defence of territory and *Eudynamis* has lost the territorial instinct. Therefore, since territorialism occurs in most other cuckoos, we would seem justified in assuming that it was originally present in the Crotophaginae. The manner in which the behavior patterns of territorialism were weakened and modified may now be considered.

Several factors probably have contributed to the breakdown of territorialism for the pair. The first is that sexual fighting (fighting in relation to the sex-partner) is absent or extremely weak in the whole subfamily Crotophaginae. This situation may have permitted freer relations between the various members of the colony and a loosening of the sexual bond. As to the origin of the lack of sexual fighting it may be suggested that a contributing factor is the habit of living in colonies, thus producing within the group an individual familiarity which precluded sexual fighting. A second factor which contributed to the breakdown of territorialism is the lack of a song. In most territorial birds song is one of the most effective methods of maintaining a territory. None of the subfamily has any note with the characteristics of song. The possession of a song might have prevented the breakdown of territorialism.

Some observations on the defence of the territory deserve mention. According to the concept, the owners drive out birds of the same species. In the case of *C. ani*, other species, such as *Corvus nasicus*, and an entirely black member of the Icteridae (*Ptiloxena*) are driven out on all occasions. The defence of territory against the latter species must not be confused with the act of stealing eggs from the nests. The manner of stealing eggs differs in notes, in flight, and in flock behavior from the act of driving a *Ptiloxena* out of the territory. Although this species is smaller than *C. ani* nevertheless this defence reaction is probably due to an error of identification of the individual. But on the other hand, in British Guiana where *C. ani* is quite as aggressively territorial, *C. major* is permitted to remain in the territory and even sleep within a few feet of some individuals of *C. ani*. *C. major*, to the human observer, more closely resembles *C. ani* than does *Ptiloxena* yet apparently there is some feature which serves to distinguish the species. Errors of discrimination are common in other species. Song Sparrows (*Melospiza melodia*) (Nice, 1937), Kingbirds (Davis, 1941b), and the English Robin (Lack, 1939a) frequently drive other species out of the territory. In these cases the behavior is considered to be an exaggerated territorialism.

The observance of territorial boundaries differs among various species of birds. Thus, for example, there is variation in the reactions toward a predator. In *C. ani* when a predator (human, cat) comes to the nest tree of the colony the neighboring birds come near, attracted by the alarm call. But these neighbors do not enter the territory of the disturbed group. The attraction of the alarm call does not override the disposition to remain within their own territory. Some other species, as for example the Willet (*Catoptrophorus semipalmatus*) (Vogt, 1938), will go outside a territory and cooperate in the pursuit of an enemy. Fautin (1940) found that the Yellow-headed Blackbird (*Xanthocephalus xanthocephalus*), although strictly territorial, cooperated to drive out predators. L. Miller (1930) found that Horned Owls (*Bubo virginianus*) ignored humans when driving an invader out of the territory.

Biological value of territory

The function or biological value of the territory (as distinguished from the psychological cause of fighting) has been discussed at length in the literature. Food supply is the most obvious function. It has been shown (Davis, 1940a) that in *C. ani* the size of the territory of different colonies which contain the same number of individuals varies greatly in the same ecological habitat. Furthermore, in British Guiana, *C. major* may live within the territory of *C. ani* and certainly is a competitor for food. Food, then, is probably not an essential function of territory but an incidental result among these birds. As a matter of fact no function of territory in these species is clearly apparent other than the primary function, the provision of a mating station and nest site. It must be remembered that the function of territory does vary greatly among various species. As an example of the possibilities of variation in the function of territory, Skutch (1931) observed that in the Hummingbird, *Amazilia*, 50 per cent of the nesting failure was due to the birds stealing material from other hummingbirds' nests. Territorial defense would have prevented this robbery. The function of territory may even change during the year. Thus the Micheners (1935) found that in Mockingbirds (*Mimus polyglottos*) summer territory depended on the nest and was defended only by the male. On the other hand in the winter the male and/or the female may defend a territory for feeding purposes. In January the behavior changes from the winter to summer type of defense. This situation obtains in Shrikes (Laniidae) (A. H. Miller, 1931) and in the English Robin (Lack, 1939a).

These variations in the function of territory emphasize the fact that to be of any value the territorial concept must be sufficiently broad and flexible to describe and integrate the behavior of many species of animals. Meise (1936) shows the great variation in function among different species by classifying territories under various headings: (1) the relation between the size of the territory and area (*Lebensraum*), (2) duration of territory, (3) number of birds using it, (4) biological function. This exhaustive analysis provides a basis for the classification of the territorialism of any particular species. Friedmann (1933) points out that species vary in expressing territorialism in time and space. Both must be considered. As Nethersole-Thompson (1934) suggests, territory operates in relation to the peculiar needs of the species. Even within the same species territorialism may vary. For example, *Larus marinus* (*l. c.* p. 23) is territorial in some areas and colonial in others. Territorialism is a means to success in reproduction and not an end in itself. The study of the nesting habits of the Crotophaginae furnishes further evidence that territorialism is a variable behavior and can not be rigidly circumscribed by a definition if the concept is to have any usefulness. The patent development within one group of closely related species of a colonial territory from a pair territory, emphasizes the usefulness of an inclusive term.

Several attempts have been made to define strictly territorialism. These definitions are not in terms of function. Every definition presents only one point of view and thus automatically excludes many species. The latest attempt is the definition of Lack (1939a) in which he states that territory is an isolated area defended by one individual of a species or by a breeding pair against intruders of the same species and in which the owner makes itself conspicuous. This definition

196

suffers from the same drawbacks as any definition of territorialism (cf. Davis, 1940a: 206). It is too narrow and excludes too many species from the concept. Colonial birds, such as the Jackdaws, Crotophaginae, Gulls (Kirkman, 1937), and the California Woodpecker are eliminated. Further, it excludes polygynous species such as *Cassidix* (McIlhenny, 1937), Yellow-headed Blackbird (Fauntin, 1940), *Euplectes* (Lack, 1935), and lizards (Evans, 1938). This definition carries with it the tacit exception that an intruder of the same species which is a potential mate is not driven out and further it does not take into account several birds such as the Chickadee (*Penthestes atricapillus*) (Odum, 1941) which defend a territory but do not make themselves conspicuous by song or position. For these reasons it seems preferable to align the concept of territorialism with such concepts as 'species' and not try to define it strictly.

It is desirable to call attention to two principal types of fighting. Fighting in reference to the piece of land (territorialism) is the subject of this discussion. The fighting by the Crotophaginae in reference to the sex-partner has not been discussed here for lack of evidence. This lack is probably due to the masking of sexual fighting by territorial fighting. The behavior of the Crotophaginae shows clearly that the defense of territory is a separate element from sexual fighting. Fighting over the sex-partner does not enter into the motivation. The motivation is the defense of the nest and the piece of land. The whole colony defends the territory no matter what the sex of the defenders or intruders may be. It is becoming clear that the separation of the elements of territorial defense and sexual fighting should have been recognized earlier. The behavior of other species shows clearly that there are several causes of fighting among birds. For success in reproduction there are three principal reasons for defending the nest site: (1) Predators must be driven off; (2) A sex-partner must be secured and defended against other members of the species which attempt to obtain the partner; (3) A spot suitable for pairing, for placing the eggs, and for raising the young must be secured. It is proposed to restrict the term territorialism to the last-mentioned type of fighting. The most important separation is the distinction between fighting over the piece of land and fighting in relation to the sex-partner. As examples in other species, Vogt (1938) finds that in the Willet the defense of territory gradually

supersedes the defense of the female as the breeding period advances. Eventually sexual fighting is no longer necessary because territorial defense excludes all males. Defense of territory and of the female are the two causes of strife among Willets and often mask each other. Tinbergen (1939) describes the sexual fighting in the Snow Bunting. After the female arrives fighting increases and the male will fight outside the territory to defend the female. The birds drive away other birds of the same species, but, while the female fights for the male only, the male defends the female and also the territory. In this species territory defense is a behavior of the male only. These considerations demonstrate the variation in the causes of fighting. Territory is a very important cause and has been the subject of this discussion of the Crotophaginae.

It is now important to consider the origin of territorialism. Since the nest is the only factor in the bird's economy which is fixed, it seems likely that the territorial defense originated from the defense of the nest site or mating station. This viewpoint is in agreement with data on other birds and vertebrates. Noble (1940) states that in primitive birds (Night Herons) territories retain their piscine function of being primarily a place where sexual bonds are formed. In fishes (Centrarchidae) the territory is the place where copulation occurs. Secondarily, the territory has become the place where eggs are laid and develop. Hubbs and Bailey (1938) found that the Smallmouthed Black Bass (*Micropterus dolomieu*) defends the nest and that copulation occurs only on the nest. Huxley (1914) found that in Great Crested Grebes (*Podiceps cristata*) coition occurs only on the nest. Venables and Lack (1934), in the same species, observed that territorial behavior is clearly correlated with the position of the nest, and again (1936) that territory is associated with the nesting platform. This situation resembles the behavior of fish greatly. Lack (1939b) states that the function of territory in the Blackcock (*Lyrurus tetrix*) is to reduce the amount of interference between males at copulation. Schuz (1936) studying the White Stork states that "one often has the impression that it is first the nest that is defended and second the mate." Friedmann (1929) found that the Cowbird (*Agelaioides badius*) chooses the nest and then the territory. Also (1928a) he finds that in African Cuckoos territory is dependent upon the availability of nests to parasitize. Nethersole-Thompson (1933) includes interspecific fighting (for example, the fighting between woodpeckers and starlings for a nest-hole) as territorial. The behavior of these species indicates that the place of copulation (nest-site) is the original territory. But among the data contradicting the view is the fact that many highly evolved species choose territory before the nest. In some cases the female builds a nest outside the territory. Lorenz (1938) believes that the territory of the herons has shrunk to small size. Tinbergen (1935) interpreted his data on Phalaropes (*Phalaropus lobatus*) to indicate that the nest site is not an essential part of territory. Selous (1933) expresses the viewpoint that the territory is symbolical of the female and that the male defends the territory as he would the female. However, this contradictory evidence can be reconciled with the notion that the mating spot is the original territory. A division of nesting activities between the members of the pair can permit great variation in the relations between the mating station (territory) and the nest.

The evidence presented above suggests that the defence of the spot of ground on which sexual bonds can be formed and the nest built is the origin of territory. This viewpoint contains several assumptions. It is assumed that the defense of the piece of land can be extended forward and backward temporally and also can be extended spatially. The forward extension of a behavior pattern is commonplace in the lives of animals. Many species for example, incubate before the eggs are laid (*Ectopistes*, Whitman, 1919). Lack (1940) suggests that feeding the mate during courtship is an extension forward of feeding during incubation. Thus the extension in time of the defense of the nest site and the mating station encounters no difficulties. This viewpoint of the origin also assumes that the bird is able to recognize a potential competitor for the nest site. This recognition permits the explanation of the fact that many species stop active defense of the territory after the nest is built and recognize the territories of other birds by mutual agreement (Nice, 1937). The neighboring birds are known not to be competitors for the piece of land, whereas strange birds are known to be either competitors for the land and/or for the mate.

ABERRANT BREEDING HABITS

As was mentioned in the introduction, the Cuculidae show throughout the entire family a tendency to have unusual breeding habits. In the Crotophaginae these aberrations have developed into communal nesting. This proclivity is presumably based upon some peculiarity of the endocrine system although at present we know practically nothing about what the exact situation could be; nevertheless several conditions may be mentioned.

Presumably birds, like mammals, may be divided roughly according to the spontaneous or non-spontaneous method of ovulation. It is of course clear that the exact meaning of 'spontaneous' breaks down when courtship is considered. In mammals the term refers to species such as the rat, in which ovulation occurs at regular intervals without external stimulation. Other species (non-spontaneous), such as the rabbit, under natural conditions ovulate only after the stimulation of coition. In birds the external stimulation of coition has been superseded by the more intricate courtship performance. In many cases ovulation can be produced by a very slight external stimulus such, for example, as stroking the neck of pigeons. It becomes very difficult then to determine exactly what species are spontaneous and what are nonspontaneous. Marshall (1936) divides birds roughly into two groups in respect to ovulation; those which ovulate spontaneously and seldom mate for life or have elaborate courtship, and the non-spontaneous ovulators which frequently possess elaborate courtship performances. These generalizations do not hold for Parrots, which mate for life and ovulate spontaneously. The Rhea, which is known to have polygynous nesting habits, probably ovulates spontaneously. Harper (1904) suggests that polygynous birds are spontaneous and that monogamous birds are nonspontaneous. Craig (1913) points out the spontaneous (Phasianidae) and the non-spontaneous (Columbidae) types of ovulation. Allen (1925) although not using the terms spontaneous and non-spontaneous, points out the difference in the type of egg-laying as exemplified by the hen and the pigeon. He further suggests that sight of a nest is the stimulus which causes certain wild ducks to deposit eggs. It must be remembered that there are three distinct phases in the production of an egg. The growth of the egg and follicle is the first stage and apparently is under the control of the follicular stimulating hormone (FSH). The actual ovulation is the second stage and is controlled by the luteinizing hormone (LH). The last stage is the passage down the oviduct and the actual deposition of the egg in the nest. Just what factors influence this last event is not clear at present. The term spontaneous must be used to refer only to ovulation and not to laying.

There is considerable evidence to indicate that the Crotophaginae as well as the rest of the Cuculiformes are spontaneous ovulators. Farley (1924) reports that a pet which lived in his house, dropped eggs frequently on the floor. The birds drop eggs frequently on the ground even far away from the nest. Although Friedmann (1929) found that Cowbirds frequently drop eggs on the ground, in this case these aberrations are probably the result of social parasitism. The habit of spontaneous ovulation in the Crotophaginae is closely related to the lack of courtship performance. Since the function of courtship behavior is, primarily at least, the stimulation of ovulation, spontaneous courtship and lack of courtship are compatible. In birds such as the Crotophaginae which ovulate spontaneously there is no need or use for courtship. On the other hand, in those species which ovulate non-spontaneously, courtship display does have a use in identifying the sexual ripeness of the individual. Thus Noble (1938) found that in many fishes the display was a means of identifying not the sex but the sexual ripeness of the individual. A further development of this condition is the use of display to stimulate and synchronize the sexes, as Carpenter (1933) found in the pigeons. Huxley (1916, 1921) has suggested that the function of courtship is not selection of a mate but stimulation to copulation. In animals with courtship performances the external stimulus releases the internal endocrine mechanism responsible for ovulation. Craig (1911) found that the pigeon, which normally ovulates after courtship and is considered to be non-spontaneous, will ovulate even if there is no male present, but under these conditions only late in the season. Virgin females may not lay at all without a male. On the other hand, in animals which ovulate spontaneously, the stimulus to ovulation is brought about by an automatic release of the endocrines responsible for it independent of external stimuli. Courtship thus appears to be an external regulator superimposed upon the internal processes.

As a further correlate of spontaneous ovulation and the lack of courtship may be mentioned the lack of pair formation. Courtship, especially mutual post-ovulatory performances, serves to bind the members of the pair together in some species and to prevent the dissolution of the couple. The courtship performances, either by one of the pair or by both birds, serve to maintain the pair. Thus, in the Crotophaginae, the lack of bonds between the members of the pair and spontaneous method of ovulation are contributing factors permitting communal nesting. It should be emphasized that it is not intended to imply that spontaneous ovulation must necessarily be followed by communal nesting but that spontaneous ovulation permits the lack of courtship and of the bond between members of the pair, thus making possible and contributing to the development of social nesting.

Parenthetically it should be noted that many birds with most elaborate courtship (game birds, birds of paradise, mankins, hummingbirds) do not pair. In these groups the elaborate courtship is necessary to induce copulation.

Number of eggs laid

The two types of ovulation, spontaneous and non-spontaneous, exemplify one phase of the general problem of internal and external control of a neuro-endocrine mechanism. As background for a discussion of another phase of this problem, it is desirable to discuss the number of eggs laid by each female. In both genera of the Crotophaginae the number laid by a female is usually six, although sometimes it is five or seven (cf. Davis, 1940a, 1940b, 1941a). Skutch (Bent, 1940) believes that *C. sulcirostris* lays four eggs, a figure probably too low. Since several females lay in the nest there may be great variation in the number of eggs deposited in one nest. Thus the number of eggs laid by one female cannot be determined by the number in the nest. The number laid can be determined, however, by microscopic examination of the ovaries. The fact that a constant number is laid would suggest that the size of the clutch is fixed by an internal endocrine mechanism. The evidence as reported by Farley (1924), whose pet bird laid eggs for a long period of time, is not described in sufficient detail to be of value in this connection.

That there can be a difference in the method (external or internal) controlling the production of eggs is shown by the consideration of the factors which determine the number of eggs laid. Some birds lay until there is a certain number of eggs in the nest. The Flicker as reported by Phillips (1887) is the classic example. If eggs are removed from the nest this bird continues to lay until, in one case, as many as 73 eggs have been deposited, although it is possible (in litt.) that several females laid in this one nest. Other species lay only a certain number no matter how the number of eggs in the nest may be manipulated (Davis, unpub., *Larus argentatus*). Craig (1913) recognized these two types of birds and pointed out the Flicker and the pigeon as examples. In the first type (which the author suggests be called indeterminate) the external conditions determine the number of eggs laid. In the second type (determinate) the number is fixed by the internal hypophyseal-ovarian mechanism. The observations of Hann (1937) that in Ovenbirds (*Seiurus aurocapillus*) brooding began after laying the penultimate egg even though 3, 4, 5, or 6 eggs were laid in the clutch, suggests that the brooding impulse may regulate the number of eggs laid. The fact that the Flicker has elaborate courtship (Noble, 1936) suggests that in this species ovulation is non-spontaneous. At present, knowledge is insufficient to state certainly that indeterminate egg-laying is found in birds which ovulate non-spontaneously and that determinate egg-laying occurs in those which ovulate spontaneously.

It is of interest that another bird in which communal nesting has developed, the California Woodpecker, belongs to the Picidae which are prolific egg-layers. The classical example of the Flicker which laid 71 eggs in 73 days shows to what extent the birds can lay eggs. Eggs of various woodpeckers are frequently found in the nesting holes of other birds, and the group as a whole is famous for laying a varying number of eggs. The ducks are also prolific egg-layers and frequently lay in other birds' nests. Allen (1925) has indicated the importance of this habit in the eventual culmination of parasitism in the South American duck *Heteronetta*. Friedmann (1932) lists many species of ducks which have laid eggs in the nests of other species.

The above discussion is concerned with one aspect of the general problem of the interrelations between the nervous and the endocrine systems, namely, the internal or external regulation of the hypophyseal-ovarian mechanism. Another problem concerning the breeding and the relations of the gonads to the nervous system is the fact that, in *C. ani* and others of the subfamily, incubation by several females starts simultaneously. The individual bird, however, may not have completed its clutch or may have finished laying several days before. It should be noted, however, that in most cases incubation does not start until a few days after the egg-laying has stopped. It seems that continued egg-laying by one bird inhibits the incubation by the other birds. Possibly imitation may be the reason the birds, which have completed ovulation some time previously, start to incubate. It is known in general that the behavior of a bird towards the contents of the nest is dependent upon the contents and can be changed radically by changing the contents of the nest. For example, the substitution of eggs caused *C. ani* to brood for 24 days although the normal length of incubation is about 13 days. A Robin (*Turdus migratorius*) will feed young which are substituted for newly-laid eggs (Davis, unpub.). Contrasting with these species are others which have a fixed duration for each phase of the breeding cycle. Thus the Noddy Tern (*Anous stolidus*) (Watson, 1908) did not change behavior when chicks were substituted for eggs. Apparently this is a species difference in the control of behavior by external and internal stimuli.

Unusual aspects of reproduction

A further factor of unknown significance to the breeding habits is the fact that the Crotophaginae have a continuous molt. In general birds molt just before and/or after the breeding season. It is known that this molt is connected with the endocrine system in several respects. Just what connection, if any, there may be between the continuous molt and the endocrine situation in the Crotophaginae is not clear. It should be noted that other birds such as the Ptarmigan (Salomonsen, 1938) have a continuous molt. The observations of the Micheners are of interest in connection with the loss of territorialism as correlated with the molt. These investigators found (1935) that the Mockingbirds maintain territory throughout the year but that there is a slackening of territorial defense during the molt. The English Robin maintains a territory except during the molt (Lack, 1939a).

A word of caution is necessary. Endocrinology is not a panacea for the explanation of all sexual behavior. The task is to find out what relations exist. For example, Lack (1939a) finds that the English Robins establish territories and select mates in the fall when the gonads are regressed. This indicates that the gonadal hormones at least are not responsible for these particular behavior patterns or that estrin inhibits territorialism in the female when this hormone appears in the spring (cf. plumage in chickens, Domm, 1939). A further complication is the common occurrence of song in the fall. Brewster (1898) observed several species of swallows attempting copulation and carrying mud on August 22, after the migratory flock had formed. These examples indicate the magnitude of the task.

Aberrant breeding habits are frequently connected with an abnormal sex ratio in the population (Mayr, 1939). In fact when there is an unusual sex ratio, unusual nesting habits may be suspected. In the Crotophaginae the sex ratio is greatly in favor of the males. The question now arises as to whether the breeding habits are the result of an altered sex ratio or the converse. Sarasin (1924) believes that polyandry is the cause of social parasitism in the Cuckoos. As to the actual reasons for the abnormal sex ratio we know nothing positively although many hypotheses have been suggested.

As a manifestation of abnormal breeding habits a consideration of the frequency of feeding the young is of interest. In *Guira* the young are fed at long intervals, seldom less than one-half hour (Davis, 1940b), but in *C. ani*, although each individual bird comes to the nest only at long intervals, since several birds are carrying in food the young are fed fairly frequently. Compared with other birds, Herrick (1910) found that young American Cuckoos (*Coccyzus*) were fed every 25 minutes and sometimes as often as every four minutes. Passerine birds sometimes are fed at much shorter intervals. Baldwin and Kendeigh (1927) found that a female House Wren (*Troglodytes aedon*) fed the young 18 times in 34 minutes. Bigglestone (1913) found that Yellow Warblers (*Dendroica aestiva*) fed the young over a period of ten days more than 16 times per hour. In the Crotophaginae in addition to the factor of the infrequent feeding is the fact that not all the females incubate even though they may have laid eggs. These facts all indicate a most casual care of the young birds. Neither is the behavior of the adult birds at the nest highly developed. There is no ceremony at the nest at the exchange of incubators or at feeding. The adults do not carry away the egg shells or excreta from the nest. In some

Cuckoos (Herrick, 1910) the adults await the defecation and then remove or eat the fecal sac. No such behavior exists in the Crotophaginae. The excreta are not even contained in a sac.

The analysis of the factors responsible for the construction of abortive nests and the performance of symbolic activities is pertinent to the problem of aberrant breeding habits. It has been shown for *C. ani* that, at the first part of the season, many nests are started and then neglected. The usual explanation of this type of behavior is that the bird's reactions are not yet of sufficient intensity; a statement which explains nothing. Tinbergen (1939) describes these behaviors as "unfinished actions so typical of maturing instinctive behavior." That there can be a lack of adequate stimulus for the completion of a nest is shown by the work of Ali (1930) on a Weaver Finch (*Ploceus philippensis*). He found that the number of females per male in this polygynous species depends on the number of nests the male can build. At the end of the season many cock nests are built by males whose 'zeal' is insufficient to finish the nest. But on the other hand, it must be remembered that in many cases the abortion of a nest may be caused by the lack of a suitable location. Bennett (1938) found that female Blue-winged Teal (*Querquedula discors*) inspected many nesting sites before selecting a suitable location. Abortive nests were frequently started in unsuitable locations—as generally occurs in the duck family.

The question of mutual stimulation among members of the flock must be considered in relation to the egg-laying and the abortion of nests. This hypothesis has been developed by Darling (1938) to explain the behavior of groups of gulls. His evidence indicated that reciprocal stimulation among the members of the flock was conducive to greater success in nesting. In the flocks of Crotophaginae, mutual stimulation may be important in the stimulation of egg-laying and incubation, although the data do not show that large colonies were more successful than small ones. Other species show stimulation by the group. Lack and Emlen (1939) found that all birds in a colony of Tricolor Redwings (*Agelaius tricolor*) are in the same state of breeding. But nearby coloneis may differ greatly in their respective stages of breeding. Hoogerwerk (1937) found that the White Ibises, which nest in groups on a platform, were in the same stage of nesting on each platform. Apparently courtship (in the broad sense) can be effective beyond the actual mate.

TYPE OF HABITAT

The importance of the influence of the habitat on the development of social nesting habits is difficult to assess. *Guira* inhabits areas of open park-land savanna and sleeps in clumps of dense trees. Also, the birds tend to nest in thickly foliaged trees. Since there are not many groups of such trees the tendency is for the birds to come together in flocks. The original habitat of *Guira* was probably the Chaco and the campos of Brazil. These areas are characterized by open stretches of grass or marsh with scattered clumps of trees. The birds, able to feed out in the open, are thus forced to come together to sleep and nest. Although these islands of trees sometimes are very large, nevertheless there is the tendency for the birds to gather together. It seems likely that the flock habit developed in part under the influence of the habitat. It is of interest that the California Woodpecker, another species with comparable habits, also occurs in areas where the vegetation is distributed in islands. Ritter (1938) states that this woodpecker inhabits forest islands and that the boundaries of the island limit the area of one group of birds.

Crotophaga major and *C. ani* also inhabit more or less open areas, although they return to trees for sleeping and nesting. *Major* is frequently found along the forest edge but never in the thick forest. It lives along the borders of streams, probably because of the nature of the food. Both *Guira* and *C. ani* inhabit mesic habitats. The latter is strictly a moist habitat bird; in Cuba, in the dry season, the birds are found only along the streams, and in Argentina the birds stay throughout the year along streams. The habit of feeding in more open areas and of sleeping and nesting in a clump of trees, presumably developed by the ancestral birds and still to be observed in the primitive *Guira*, has been retained by the genus *Crotophaga*. These considerations suggest the conclusion that the type of habitat in which the species originated and the tendency to divide the territory into a nesting and a feeding habitat have been factors conducive to the development of social nesting.

The habit of sleeping in large flocks occurs in other species, as, for example, *Molothrus* and *Muscivora* as reported by Davis (1940b). In these species, however, there has been no factor conducive to the breakdown of territorialism and no social behavior has developed. Friedmann (1935) points out that colonial nesting persisted in, or developed in, those species which use separate feeding and nesting ground. When the same ecological habitat is used for both feeding and nesting, colonial nesting does not usually develop. Many species have the territory divided into two distinct parts. A. H. Miller (1931) found that shrikes had a definite headquarters in their territory where the birds slept. Mayr (1926) found that a finch (*Serinus*) used different associations for various purposes such as living, dancing, and feeding. Steinbacher (1939) found that Gallinules (*Gallinula chloropus*) defended a resting place, pairing territory, and a nesting territory. Pettingill (1936) found that the male Woodcock (*Philohela minor*) inhabited the woods in the daytime and the fields at night. These selected examples show that the ecological separation of the territory into several parts is a common occurrence among birds.

PHYLOGENY OF BEHAVIOR

The phylogenetic development of the flock habit is clearly shown by the behavior of the four species of the Crotophaginae. As mentioned in the introduction, *Guira* is the most primitive member of the subfamily both anatomically and behavioristically. *C. major* is the next in the phylogenetic scale. *C. ani* and *C. sulcirostris* resemble one another and represent the climax of development. In *Guira* the flock is a loosely coordinated group of birds, feeding in the same vicinity, sleeping together and, although frequently nesting in pairs, regularly nesting communally. In *Crotophaga major* the flock is more united and always nests communally, although pairs which live singly may nest singly. The flock behavior reached the climax in *C. ani* and *C. sulcirostris*. In these species the group lives together in a closely-knit organization and nests communally. It is of interest to contrast this behavior with that of other flocking species, such as the Oyster Catcher (Huxley, 1925). In this species pairs leave the flock to go to the breeding territory for a while and may collect in flocks until the eggs are laid. But the strict maintenance of territorial boundaries prevents the encroachment by the flock upon the territory of a pair. Hence social nesting is impossible.

The breakdown and redevelopment of the territorial behavior is also a phylogenetic trend. In *Guira* the defense of the territory by the pair is weak and that by the colony is also weak. In *C. major* the defense of a colonial territory is positive but nevertheless not vigorous. In *C. ani* the defense of the territory by the whole colony is fierce and effective and pair territory has vanished.

This summary of the phylogeny of territorialism in the Crotophaginae suggests the consideration of the relation of this subfamily to the development of parasitism in the other Cuculidae. Herrick (1910) suggests that a maladjustment of the nest-building and egg-laying results in parasitism—a statement which is an outline of the problem but not a solution. Allen (1925) emphasizes the fact that no one theory should be expected to account for the development of the several types of parasitism. He further outlines the distinction between egg-parasitism and nest-parasitism. This scheme has been verified by subsequent research and is followed in this paper. The cuckoos and ducks (egg-parasitism) developed parasitism by dropping eggs occasionally in the nests of other birds, although also building their own nests as, for example, in the Redhead (*Nyroca americana*). On the other hand, the Cowbirds (nest-parasitism) developed parasitism by stealing the nests of other birds or using old nests. The latter trait is common in the Mourning Dove (Nice, 1922) and habitual in the Solitary Sandpiper (*Tringa solitaria*). In these species certain behaviors (regurgitative feeding, precocial habits) precluded the development of parasitism. Davis (1940a) strongly supports these ideas. Friedmann (1928b) accepts this suggestion as a partial solution for Cowbirds. In addition, he emphasizes, there must be a loss of territorialism.

The evidence from the phylogeny of the Crotophaginae indicates that the Cowbirds and the Cuckoos have in common the modification of territorialism but that the other patterns of behavior in the species precluded the identical development of the habit. In the Cuculidae the courtship is weak or absent and the ovulation is spontaneous. Territorialism is retained in a much modified form but the spontaneous ovulation permits the laying of eggs in other birds' nests. From occasional laying in other nests some species developed the habit of always laying parasitically. That the tendency to lay eggs promiscuously occurs in nonparasitic members of the family is shown by the fact that Pemberton (1925) found the egg of a *Geococcyx* in the nest of a raven; he wrongly interprets this occurrence as deliberate parasitism. In the Cowbirds, in contrast to the Cuckoos, the courtship is retained and elaborated and ovulation is probably non-spontaneous. In these species the usurpation of a nest is necessary when the birds do not build their own nests. From the usurpation of a nest it is an easy step to laying in the nests of other species. These Cowbirds have developed parasitism in a very different manner and even the result, although superficially very similar, is different in many details.

The place of the Crotophaginae in the phylogeny of the parasitic cuckoos is worth consideration. Makatsch (1934) believes that the Crotophaginae represent a stage in the development of parasitism; that after a communal nesting habit, the next stage is true parasitism of other species. The evidence presented in this paper shows that the Crotophaginae represent not a stage in the development but an offshoot. Daguerre (1924) suggests that communal nesting developed as parasitism of one pair on the other. This would assume that there occurred the development of two phases within the species; one parasitic and the other non-parasitic. The fact that in some cases all birds take part in incubation and feeding precludes this interpretation.

The consideration of the phylogeny of behavior and the relation to parasitism in general suggest the desirability of a few inferences as to the future of the group Crotophaginae. At the present time the group seems to be at the climax of possible development. The species are universally successful as is shown by the fact that they are among the most abundant birds in their range. Low temperature seems to be the only barrier to their distribution. Their social habits help the individuals to avoid enemies. Nevertheless one purely hypothetical course of evolution is open to the group. It has been found that the birds are extremely indolent in their attentions to the nest. Individuals which laid eggs did not incubate. Birds which incubated did not feed the young. It is possible that a parasitic phase could develop within the species. Those individuals which lost the instinct of building the nest could survive by laying in the nest built by others. Thus a parasitic phase might develop. If this phase should evolve some morphological characteristic we should then have the evolution of a new species—perhaps the evolution of two new species. In this connection it should be remembered that many of the parasitic birds (Cowbirds, Viduinae) parasitize closely related species. It is, however, not intended to suggest that all these actually did develop parasitism in exactly this manner.

SUMMARY

The Crotophaginae, one of the six subfamilies of Cuckoos, demonstrates the phylogeny of communal nesting. The birds live in flocks and build one nest in which several females lay eggs. The subfamily consists of four species, distributed throughout South and Central America. The development of social nesting occurs at three levels, represented by the species: I. *Guira guira*, II. *Crotophaga major*, III. *C. ani* and *C. sulcirostris*. The primary factors conducive to the development of social nesting habits are (a) modification of territorialism, (b) aberrant breeding habits, and (c) the type of habitat.

During the investigation the behavior of *Crotophaga ani* was found to be essentially identical in three widely separated localities; Cuba, British Guiana, and Argentina. The call notes of the members of the subfamily are similar as well as other behavior patterns. Some evidence concerning the problems of discrimination of the sex-partner, of the individuals of the species, and of the boundaries of the territory was obtained. The sex-partner and other individuals of the flock are recognized by individual appearance. Strange birds probably are first discriminated by their behavior and then by appearance. The boundaries of the territories are known exactly by the trees and shrubs. Some of the behavior patterns exhibited by the species may be classed as releasers or sign-stimuli. Thus the birds gave the same response to an injury-feigning dove as to an injured member of their own species.

Each flock of all four species maintains a territory. However, the details differ in each case. *Guira* defends the territory only slightly and frequently one pair builds a nest and defends a small territory within the territory of the flock. The flock of *C. major* is composed of pairs which unite to defend a territory. Thus the colony is really a group of pairs which cooperate to build one nest. In *C. ani* the communal nesting has reached its climax. Polygamy or promiscuity is the general rule and the whole colony defends the territory.

This phylogenetic development of social nesting coincides with the disappearance of territorial defense by the pair and the appearance of territorial defense by the colony. Several factors have contributed to the breakdown of territorialism for the pair. Sexual fighting is weak or absent. There is no song in any of the species. Courtship performances are lacking or simple. In these species the biological value of the territory is probably merely the provision of a mating station and nest site. The obvious development, within one group of birds, of a colonial territory from a pair territory, suggests the usefulness of an inclusive, flexible characterization of the term territorialism instead of a strict definition. The behavior of the Crotophaginae further shows that sexual fighting (in reference to the sex-partner) must be clearly separated from territorial fighting (in reference to a piece of land). Territorialism probably originated from the defense of a spot of land on which the sexual bonds can be formed and the eggs laid.

The aberrant breeding habits of the various species permitted the development of social nesting. Several minor factors are of interest. The species have a continuous molt. The sex ratio is in favor of the males. The birds are indolent in the care of the nest, feeding the young at long intervals, and incubating sporadically. Many abortive nests are started and many eggs dropped on the ground or laid in the nest before completion of construction. The important factor is that the Crotophaginae, judging by indirect evidence, ovulate spontaneously and thus do not require courtship performances and pair formation for successful laying. This spontaneous ovulation raises the general problem of internal and external control of a neuro-endocrine mechanism. As one example the data indicate that members of the subfamily lay six eggs normally, therefore the hypophyseal-ovarian mechanism is set to lay a fixed number of eggs no matter how many are in the nest.

The third factor conducive to the development of social nesting habits is the type of habitat. The original habitat was probably a region containing scattered clumps of trees, such as the Chaco or the campos of Brazil. The birds are thus forced to come together to sleep and nest in the trees. This primitive habit is retained in the more evolved species of the subfamily which use different parts of the territory for feeding and nesting.

This subfamily of Cuckoos has developed the social nesting habits, in contrast to the parasitism evolved by the other Cuckoos. Social nesting, however, is probably not a stage in the development of parasitism but a side branch in the evolution of a peculiar nesting habit.

LIST OF LITERATURE

ALLEN, G. M. 1925. Birds and Their Attributes. Marshall Jones Co. Pp. 338.

ALI, S. A. 1930. The nesting habits of the Baya (*Ploceus philippinus*). *J. Bombay Nat. Hist. Soc.*, 34: 947–964.

BALDWIN, S. P., and KENDEIGH, S. C. 1927. Attentiveness and inattentiveness in the nesting behavior of the House Wren. *Auk*, 44: 206–216.

BENNETT, LOGAN J. 1938. The Blue-winged Teal. Collegiate Press, *Ames, Iowa*. Pp. 144.

BENT, A. C. 1940. Life histories of North American Cuckoos, Goatsuckers, Hummingbirds and their allies. *U. S. Nat. Mus. Bull.* 176. Pp. 506.

BIGGLESTONE, H. C. 1913. A study of the nesting behavior of the Yellow Warbler (*Dendroica a. aestiva*). *Wilson Bull.*, 25: 48–67.

BREDER, C. M. 1936. The reproductive habits of the North American Sunfishes (Centrarchidae). *Zoologica*, 21: 1–48.

BREWSTER, WILLIAM. 1898. Revival of the sexual passion of birds in autumn. *Auk*, 15: 194–195.

CARPENTER, C. R. 1933. Psychobiological studies of social behavior in aves. *J. Comp. Psychol.*, 16: 25–98.

CHANCE, EDGAR. 1922. The Cuckoo's Secret. *London*. Pp. 239.

CHAPMAN, F. M. 1928. Nesting habits of Wagler's Oropendola (*Zarhynchus wagleri*) on Barro Colorado Island. *Bull. Am. Mus. Nat. Hist.*, 58: 123–166.

——. 1935. The courtship of Gould's Manakin (*Manacus v. vitellinus*) on Barro Colorado Island, Canal Zone. *Bull. Am. Mus. Nat. Hist.*, 68: 471–525.

CRAIG, WALLACE. 1911. Oviposition induced by the male in pigeons. *J. Morph.*, 22: 299–305.

——. 1913. The stimulation and the inhibition of ovulation in birds and mammals. *J. Anim. Behavior*, 3: 215–221.

DARLING, F. FRASER. 1938. Bird Flocks and the Breeding Cycle. Cambridge Univ. Press. Pp. 124.

DAGUERRE, J. B. 1924. Apuntos sobre algunas aves de la Provincia de Buenos Aires. *El Hornero*, 3: 248–252.

DAVIS, DAVID E. 1940a. Social nesting habits of the Smooth-billed Ani. *Auk*, 57: 179–218.

——. 1940b. Social nesting habits of *Guira guira*. *Auk*, 57: 472–484.

——. 1940c. A suggestion concerning territorialism in *Tapera naevia*. *Wilson Bull.*, 52: 208.

——. 1941a. Social nesting habits of *Crotophaga major*. *Auk.*, 58: 179.

——. 1941b. The belligerency of the Kingbird. *Wilson Bull.*, 53: 157.

DOMM, L. V. 1939. Modification in sex and secondary sexual characters in birds. *In* Sex and Internal Secretions. Editor, Edgar Allen. Williams and Wilkins Co. Pp. 227–327.

EMERSON, A. E. 1938. Termite nests—a study of the phylogeny of behavior. *Ecol. Monog.*, 8: 247–284.

ERRINGTON, P. L. 1933. The nesting and the life equation of the Wisconsin Bob-White. *Wilson Bull.*, 45: 122–132.

EVANS, L. T. 1936. Territorial behavior of normal and castrated females of *Anolis carolinensis*. *J. Genetic Psychol.*, 49: 49–60.

——. 1937. Differential effects of the ovarian hormones on the territorial reaction time of female *Anolis carolinensis*. *Physiol. Zool.*, 10: 456–463.

——. 1938. Cuban field studies on the territoriality of the lizard *Anolis sagrei*. *J. Comp. Psychol.*, 25: 97–125.

——, and CLAPP, M. L. 1940. The relation of thyroid extract to territorial behavior and to anoxemia in *Anolis carolinensis*. *J. Comp. Psychol.*, 29: 277–283.

FARLEY, J. A. 1924. Argentine birds. *Auk*, 41: 169–170.

FAUTIN, REED W. 1940. The establishment and maintenance of territories by the Yellow-headed Blackbird in Utah. *The Great Basin Naturalist*, 1: 75–90.

FRIEDMANN, H. 1928a. The origin of host specificity in the parasitic habit in the Cuculidae. *Auk*, 45: 33–38.

——. 1928b. Social parasitism in birds. QUART. REV. BIOL., 3: 554–569.

——. 1929. The Cowbirds. C. C Thomas. Pp. 421.

——. 1930. The sociable Weaver Bird of South Africa. *Nat. Hist.*, 30: 205–212.

——. 1932. The parasitic habit in the ducks; a theoretical consideration. *Proc. U. S. Nat. Mus.*, 80(18): 1–7.

——. 1933. The size and measurement of territory in birds. *Bird-Banding*, 4: 41–45.

——. 1935. Bird Societies. *In* A Handbook of Social Psychology. Editor, C. Murchison. *Worcester, Mass.* Pp. 142–184.

GOSNELL, H. T. 1932. Two cuckoos laying in the same nest without rivalry. *Brit. Birds*, 26: 226.

HANN, H. W. 1937. Life history of the Ovenbird in Southern Michigan. WILSON BULL., 49: 146–237.

HARPER, E. H. 1904. The fertilization and early development of the pigeon's egg. *Am. J. Anat.*, 3: 349–356.

HERRICK, FRANCIS. 1910. Life and behavior of the Cuckoo. *J. Exp. Zool.*, 9: 169–234.

HOOGERWERK, A. 1937. Uit het leven der witte ibissen (*Threskiornis aethiopicus melanocephalus*) *Limosa*, 10:137–146.

HUBBS, C. L., AND BAILEY, R. M. 1938. The Small-mouthed Bass. *Cranbrook Inst. Sci. Bull.* No. 10: 1–89.

HUXLEY, J. S. 1914. The court-ship habits of the Great Crested Grebe (*Podiceps cristatus*); with an addition to the theory of sexual selection. *Proc. Zool. Soc., London.* Pp. 491–562.

——. 1916. Bird watching and biological science. *Auk*, 33: 142–161, 256–270.

——. 1921. The accessory nature of many structures and habits associated with courtship. *Nature*, 108: 565–566.

——. 1925. Studies on the courtship and sexual life of birds. V. Oyster Catcher (*Haematopus ostralegus* L.). *Ibis*. Pp. 868–897.

KIRKMAN, F. B. 1937. Bird Behavior. T. Nelson and Sons, *London*. Pp. 232.

LACK, DAVID. 1935. Territory and polygamy in a Bishop-bird, *Euplectes h. hordacea* (L.). *Ibis*. Pp. 817–836.

——. 1939a. The behavior of the Robin—life history, territory. *Proc. Zool. Soc., London*. A109. 169–219.

——. 1939b. The display of the Blackcock. *Brit. Birds*, 32: 290–303.

——. 1940. Courtship feeding in birds. *Auk*, 57: 169–178.

——, and EMLEN, J. T. 1939. Observations on breeding behavior in Tricolored Redwings. *Condor*, 41: 225–230.

——, and LACK, L. 1933. Territory reviewed. *Brit. Birds*, 27: 179–199.

LEACH, FRANK A. 1925. Communism in the California Woodpecker. *Condor*, 27: 12–19.

LINSDALE, J. M. 1937. Natural history of Magpies. *Pacific Coast Avifauna*, 25: 1–234.

LORENZ, K. 1931. Beitrage zur Ethologie sozialer Corviden. *J. f. Ornith.*, 79: 67–127.

——. 1935. Der Kumpan in der Umwelt des Vogels. *J. f. Ornith.*, 83: 137–213, 289–413.

——. 1938. A contribution to the comparative sociology of colonial-nesting birds. *Proc. 8th Intern. Ornith. Cong.* Pp. 207–218.

MAKATSCH, W. 1934. Ueber die phylogenetische Entwicklung des Brutparasitismus. *Beitr. Fort. der Vogel*, 10: 61–65.

——. 1937. Der Brutparasitismus der Kuckucksvogel. Quelle und Meyer, *Leipzig*. Pp. 152.

MAKKINK, G. F. 1936. An attempt at an ethogram of the European Avocet (*Recurvirostra avocetta* L.) with ethological and psychological remarks. *Ardea*, 25: 1–62.

MARSHALL, F. H. A. 1936. Sexual periodicity and the causes which determine it. *Phil. Trans. Roy. Soc., London, B*, 226: 423–456.

MAYR, E. 1926. Die Ausbreitung des Girlitz (*Serinus canaria serinus* L.). *J. f. Ornith.*, 74: 572–671.

——. 1939. The sex ratio in wild birds. *Am. Nat.*, 73: 156–179.

MCILHENNY, E. A. 1937. Life history of the Boat-tailed Grackle in Louisiana. *Auk*, 54: 274–295.

MEISE, W. 1936. Neue Ergebnisse der Revierforschung. *Mitt. Ver. sachs. Ornith.*, 5: 1–23.

MICHENER, H., and MICHENER, J. R. 1935. Mockingbirds, their territories and individualities. *Condor*, 37: 97–140.

MILLER, A. H. 1931. Systematic revision and natural history of the American Shrikes (*Lanius*). *Univ. Cal. Pub. Zool.*, 38: 11–242.

MILLER, L. 1930. The territorial concept in the Horned Owl. *Condor*, 32: 290–291.

NETHERSOLE-THOMPSON, D. 1934. Some aspects of the territory theory. *Oolog. Record*, 14: 15–23, 79–93.

NICE, M. M. 1922–23. A study of the nesting of Mourning Doves. *Auk*, 39: 457–474, 40: 37–58.

——. 1933. The theory of territorialism and its development. *In* Fifty Years' Progress in American Ornithology. Pp. 89–100.

——. 1937. Studies in the life history of the Song Sparrow I. *Trans. Linn. Soc., New York*, 4: 1–247.

NOBLE, G. K. 1936. Courtship and sexual selection of the Flicker. *Auk*, 53: 269–282.

——. 1938. Sexual selection among fishes. *Biol. Rev.*, 13: 133–158.

——. 1940. The experimental animal from the naturalist's point of view. *Am. Nat.*, 73: 113–126.

——, and CURTIS, B. 1939. The social behavior of the Jewel Fish, *Hemichromus bimaculatus* Gill. *Bull. Am. Mus. Nat. Hist.*, 76: 1–46.

——, and VOGT, W. 1935. An experimental study of sex recognition in birds. *Auk*, 52: 278–286.

——, WURM, M., and SCHMIDT, A. 1938. Social behavior of the Black-crowned Night Heron. *Auk*, 55: 7–40.

ODUM, EUGENE P. 1941. (In litt.)

PEMBERTON, J. R. 1925. Parasitism in the Roadrunner. *Condor*, 27: 35.

PETTINGILL, O. S. 1936. The American Woodcock. *Mem. Boston Soc. Nat. Hist.*, 9: 167–391.

PHILLIPS, C. A. 1887. Egg-laying extraordinary in *Colaptes auratus*. *Auk*, 4: 346.

RITTER, W. E. 1938. The California Woodpecker and I. Univ. of Cal. Press. Pp. 340.

RUSSELL, E. S. 1934. The Behavior of Animals. E. Arnold, *London*. Pp. 184.

RYVES, B. H. 1929. Three Redshanks at one nest. *Brit. Birds*, 23: 103.

SALOMONSEN, FINN. 1939. Molts and sequences of plumage in the Rock Ptarmigan (*Lagopus mutus* (Montin)). *Videnskabelige Meddelelser fra Dansk Naturhistorish Forening*. 103: 1–491.

SARASIN, PAUL. 1924. Der Brutparasitismus des Kuckucks und die Zahlverhaltnis der Geschlechter. Wagner'sche Universitats-Buchhandlung, *Innsbruch*. Pp. 6–18.

DE SCHAUNESSEE, A. 1928. A collection of birds from Siam. *Proc. Acad. Nat. Sci., Phila.*, 80: 571.

SCHUZ, ERNST. 1936. The White Stork as a subject of research. *Bird-Banding*, 7: 99–107.

SELOUS, EDMUND. 1933. Evolution of habit in birds. Constable and Co., *London*. Pp. 296.

SKUTCH, A. F. 1931. The life history of Rieffer's Hummingbird (*Amazilia t. tzacatl*) in Panama and Honduras. *Auk*, 48: 481–500.

——. 1935. Helpers at the nest. *Auk*, 52: 257–273.

——. 1940. Social and sleeping habits of Central American Wrens. *Auk*, 57: 293–312.

STEINBACHER, GEORG. 1939. Zur Brutbiologie des Grunfussigen Teichhuhns (*Gallinula chloropus*, L.). *J. f. Ornith.*, 87: 115–135.

TINBERGEN, N. 1935. The behavior of the Red-backed Phalarope (*Phalaropus lobatus*, L.) in spring. *Ardea*, 24: 1–42.

——. 1939. The behavior of the Snow Bunting in spring. *Trans. Linn. Soc., New York*, 5: 1–94.

VENABLES, L. S. V., and LACK, D. 1934. Territory in the Great Crested Grebe. *Brit. Birds*, 28: 191–198.

——, and ——. 1936. Further notes on territory in the Great Crested Grebe. *Brit. Birds*, 30: 60–69.

VOGT, WILLIAM. 1938. Preliminary notes on the behavior and ecology of the Eastern Willet. *Proc. Linn. Soc., New York*, 49: 8–42.

WATSON, J. B. 1908. The behavior of Noddy and Sooty Terns. *Pub. Carnegie Inst.*, 103: 189–255.

WETMORE, A. 1927. The birds of Puerto Rico and the Virgin Islands. *New York Acad. Sci.*, 9: 245–571.

WHEELER, W. M. 1923. Social Life Among the Insects. Harcourt, Brace and Co. Pp. 375.

——. 1928. The Social Insects. Harcourt, Brace and Co. Pp. 378.

WHITMAN, C. O. 1919. The behavior of pigeons. *Carnegie Inst. Washington, Pub.* 257. Pp. 162.

YAMASHIMA, MARQUIS. 1938. A sociable breeding habit among Timaliine birds. *Proc. 9th Intern. Ornith. Cong.* Pp. 453–456.

YEATES, G. K. 1934. The Life of the Rook. Phillip Allan, *London*. Pp. 95.

15

Reprinted from *Amer. Mid. Nat.*, **24**(2), 273–290 (1940)

A Study of Prairie Deer-Mouse Populations in Southern Michigan

W. Frank Blair

Contents

Introduction

The present study was undertaken with the purpose of obtaining statistically significant quantities of information about the populations and the movements of the prairie deer-mouse. Data on several other phases of the life history of the species also have been obtained incidentally.

A field study of the prairie deer-mouse *(Peromyscus maniculatus bairdii)* by the use of a live-trapping technique was conducted during the field seasons of 1938 and 1939. This is one of several similar studies initiated under the direction of L. R. Dice. Strunk (unpublished thesis) studied the home ranges of the meadow vole *(Microtus p. pennsylvanicus)*, and originated the technique of setting live traps in a grid pattern on measured plots. Burt (unpublished manuscript) in working with the wood-mouse *(Peromyscus leucopus noveboracensis)* developed the technique that largely has been used in the present study. An important contribution was his method of trapping a given area during regular periods (usually one week) alternating with similar periods in which no trapping was done.

Workers at several other institutions also have used live traps in studying small mammal populations. Chitty (1937) worked in England with the field mouse *(Apodemus sylvaticus)* and bank vole *(Clethrionomys glareolus)*. Dalke and Sime (1938) studied the home range of the cottontail *(Sylvilagus*

273

floridanus mallurus). Hamilton (1937a, 1937b) has worked mostly with the meadow vole. Allen (1939) used live traps to study the cottontail *(Sylvilagus f. mearnsii)* in southern Michigan.

All of the field work was conducted on the Edwin S. George Reserve of the University of Michigan, in Livingston County, Michigan. The investigations in the 1938 season extended from March 29 to September 27, but the author was in the field continuously only from July 4 to September 27. In 1939, the period of field work extended from April 23 to September 23, and the author was mostly in the field on alternate weeks after May 23.

The facilities of the George Reserve were made available by F. M. Gaige, Director of the Museum of Zoology. B. J. Bujak and P. F. Blair, Jr., served as field assistants during a part of the 1938 season. I am indebted to the former for certain data pertinent to my study obtained by his use of nest boxes in the fall of 1938 and 1939. I am indebted to W. H. Burt for suggestions while planning the field investigations. Irving J. Cantrall kindly made available the identifications of plants which were made by Fred Hermann, of the United States Bureau of Plant Industry. The part of the field work concerned with the determination of the breeding rate of the deer-mouse was supported by a grant from the Michigan Academy of Science, Arts, and Letters.

Methods

In the 1938 season, plots 1, 3, (Fig. 1 and 2) and 4, comprising about 7.6, 9.0, and 3.8 acres, respectively, were trapped. In 1939, plot 4 again was trapped, and plot 1 was changed somewhat and enlarged to about 9.0 acres. The shift in plot 1 was made necessary by a pond that formed in one end due to an artificial change in the drainage. In addition to the regular trapping areas, in 1938, some 40 to 50 acres elsewhere were trapped for short periods and at various intervals, in an attempt to retake mice that moved from the regular plots.

In calculating the size of the plots, one-half the distance between two trap lines has been added on those sides where the blue-grass association extended beyond the limits of the trapping areas. Thus the calculated size of each plot is somewhat larger than the area on which the traps actually were set. This method has been used to minimize the error resulting from some mice having ranges that extend beyond the limits of the trapping area.

Small live traps of the single-catch type were used, and only one trap was placed at a spot. Most of the traps were of wooden construction, but some were made of copper. The wooden traps were superior to the metal for everything except the meadow vole, which often gnawed them severely. The traps were set in parallel lines. In the lines the traps were placed 60 feet apart. On two of the plots and on about one-third of another, the lines were 60 feet apart; on the remaining area, they were only 45 feet apart. Two principal considerations guided the spacing of the traps on the plots. It was necessary to space the traps closely enough to catch all of the resident animals and to give reliable data on the size of the individual home ranges. On the other hand, it also was necessary to place the traps sufficiently far apart that, with the personnel available, it was possible to trap relatively large areas. The spacings used proved a very satisfactory compromise, and also worked quite well for the wood-mouse, meadow vole, jumping mouse, and short-tailed shrew. Each trap on the regular plots was left in the same spot during the season's trapping.

The traps were baited with a mixture of millet, hemp, wheat, and canary seed. Each trap was provided with a small wad of cotton that could be utilized by the mouse for building a nest. The traps were set in the evening, usually between 5 and 7 o'clock. They were visited in the morning, usually between 7 and 9 o'clock. The traps were left unset during the day to prevent spermophiles from getting in them and dying from the heat. In the summer of each season the plots were trapped during alternate weeks. This method of spacing the trapping periods gives the residents of each plot a week to recover from the disturbances and possible deleterious effects of the previous week of trapping. During the spring, the trapping periods varied in length, because the weather was so cold much of the time that the mice would have died in the traps.

Each mouse was numbered by clipping one toe off a forefoot and by punching holes in the ears. In a few cases the hole in the center of an ear later would be torn out to the edge, apparently by snagging or through fighting. In spite of this, however, the numbers of such mice usually could be recognized. Each mouse was released at the point where it was taken, after its sex, approximate age, breeding condition, and the number of the trap in which it was caught had been recorded.

In the field notes the deer-mice were separated into 4 age groups. Those still completely in juvenile pelage were classified as immature. Those in part juvenile and part adult pelage were called subadult. Those in adult pelage, but definitely young animals, were classified as young adult. All others were considered adult. For purposes of summarization, the immature and subadult classes have been combined in the present report.

The statistical computations are by M. A. Taylor. Standard errors rather than probable errors are used. In the comparison of two means the border line of significant difference between the two is considered to be 2.7 or more times the standard error of their difference (see Dice and Leraas, 1936).

Ecological Associations

The prairie deer-mouse normally occurs on the George Reserve only on grassy areas, on which the principal dominants are two species of blue-grass (*Poa compressa* and *P. pratensis*). The former species is by far the most important herb on the well drained soils of the uplands, and in many places it forms almost pure stands. The latter species occurs on poorly drained soil in the swales and along the moist margins of ditches and ponds, where it generally is more luxuriant than *compressa* of the uplands. The amount of admixture of other herbs with the blue-grass on various parts of the Reserve apparently depends on the past history of the respective areas. Most of plot 1, which has a relatively pure, dense stand of blue-grass, is on a rather steep slope, and so far as known was never cultivated. However, it probably was utilized as grazing land. Plots 1 and 3, according to local residents, were cultivated before 1900, and today these plots have a greater variety and abundance of herbaceous species than plot 1. Other than blue-grass, the most conspicuous plant on the upland parts of plot 1 is common milkweed *(Asclepias syriaca)*. Some of the other herbs characteristic of plot 3 and 4 are found there, but they are not abundant. A panic grass *(Panicum scribnerianum)* is present in some abundance on a level area of well drained soil, in plot 1.

Plot 1 differs from the other regular trapping areas in having a considerable area of poorly drained soil with a luxuriant growth of herbs and some shrubs. In this situation, characteristic plants are: shrubby cinquefoil *(Potentilla fruticosa)*, bur marigold *(Bidens discoidea)*, boneset *(Eupatorium perfoliatum)*, swamp thistle *(Cirsium muticum)*, Canada thistle *(Cirsium arvense)*, timothy *(Phleum pratense)*, golden-rod *(Solidago graminifolia)*, black-eyed susan *(Rudbeckia hirta)*, and yarrow *(Achillea millefolium)*.

On plots 3 and 4, and less abundantly on the better drained parts of plot 1, the principal herbs in addition to the blue-grass are: common milkweed, yarrow, common mullein *(Verbascum thapsus)*, silvery cinquefoil *(Potentilla argentea)*, panic grass *(Panicum scribnerianum* and *P. tennesseense)*, wild bergamot *(Monarda fistulosa)*, pearly everlasting *(Anaphalis margaritacea)*, pussy's toes *(Antennaria neglecta* and *A. neodioica)*, galingale *(Cyperus filiculmis)*, daisy fleabane *(Erigeron ramosus)*, bush clover *(Lespedeza hirta)*, dropseed *(Sporobolus neglectus)*, and pinweed *(Lechea minor)*. Both plots 3 and 4 have small patches of sweet clover *(Melilotus alba)*. On one end of plot 3, there is some shrubby smooth sumac *(Rhus glabra)*. On this same plot, there is one large black oak *(Quercus velutina)*, and one black cherry *(Prunus serotina)*. There also is a small gully with a dense growth of shrubby dogwood *(Cornus* sp.*)*.

All of the grassy areas studied are considered, for the purpose of the present study, to be parts of the blue-grass association. On the George Reserve, the areas of blue-grass association alternate with oak—hickory woodlots. Each of the trapping areas, consequently, was bordered on two sides by the oak—hickory association. Plot 4 was bounded on a third side by a cornfield that adjoined the Reserve. On the other sides, the blue-grass association extended beyond the limits of the study areas.

Most of the small mammals of the blue-grass association are shown in Tables 1, 2, and 3. Of these, the wood-mouse *(Peromyscus leucopus noveboracensis)*, and the chipmunk *(Tamias striatus lysteri)*, are principally forest inhabitants, and the latter occurs in the blue-grass association only as a casual invader. The short-tailed shrew *(Blarina brevicauda talpoides)*, meadow vole *(Microtus p. pennsylvanicus)*, and jumping mouse *(Zapus h. hudsonius)*, are most abundant in the moistest parts of the association. Species that are known to range over the association, but were not taken in the traps, include: the common mole *(Scalopus aquaticus machrinus)*, striped skunk *(Mephitis mephitis nigra)*, red fox *(Vulpes fulva)*, and deer *(Odocoileus virginianus borealis)*.

TABLE 1.—Catches of small mammals on plot 1, E. S. George Reserve, Livingston County, Michigan

1938	Peromyscus maniculatus bairdii	Peromyscus leucopus noveboracensis	Microtus p. pennsylvanicus	Blarina brevicauda talpoides	Zapus h. hudsonius	Synaptomys c. cooperi	Tamias striatus lysteri	Mus musculus	Citellus t. tridecemlineatus	Mustela frenata noveboracensis	Total
March 29-April 1	3	..	..	..	..	..	..	..	..	..	3
April 19-22	4	1	..	..	..	..	..	..	..	..	5
May 11-14	8	..	..	..	..	..	..	..	..	..	8
May 21-23	8	2	3	..	..	..	..	..	..	..	13
June 15-20	15	8	10	..	..	..	..	..	..	..	33
July 11-17	12	21	41	3	..	..	1	..	..	..	78
July 25-31	13	21	31	9	..	1	..	..	..	..	75
August 9-16	14	19	37	12	..	3	..	..	..	..	85
August 24-30	12	11	33	8	..	5	..	..	..	..	69
September 7-13	9	6	38	13	1	4	..	..	..	..	71
September 21-27	12	11	28	17	..	5	..	1	..	..	74
1939											
April 23-26	2	3	..	..	..	..	..	..	..	..	5
May 23-29	3	10	..	..	1	..	..	..	..	..	14
June 8-15	2	16	1	2	16	..	..	..	..	..	37
June 24-30	..	15	..	1	29	..	..	..	..	1	45
*July 9-17					20						
July 23-29	2	9	1	..	25	..	..	..	2	..	39
August 6-12	1	10	1	4	23	1	..	..	1	..	41
August 20-26	..	8	..	7	13	1	..	..	1	..	30
September 3-9	..	8	..	4	8	..	..	..	1	..	21
September 17-23	..	4	..	6	15	1	..	..	..	..	26

* Traps set only on part occupied by jumping mice.

TABLE 2.—Catches of small mammals on plot 3, E. S. George Reserve, Livingston County, Michigan.

1938	Peromyscus maniculatus bairdii	Peromyscus leucopus noveboracensis	Microtus p. pennsylvanicus	Blarina brevicauda talpoides	Citellus t. tridecemlineatus	Tamias striatus lysteri	Zapus h. hudsonius	Synaptomys c. cooperi	Mus musculus	Total
April 20-22	8	2	..	..	..	..	..	..	..	10
May 11-14	10	3	..	..	..	..	..	..	..	13
May 21-23	10	4	..	..	2	..	..	..	..	16
June 15-20	15	14	2	..	8	..	..	..	..	39
July 4-10	16	26	2	..	8	1	..	..	..	53
July 18-24	18	27	6	1	4	..	..	..	..	56
August 1-8	12	19	2	..	3	..	1	..	..	37
August 17-23	20	16	14	..	1	..	..	2	..	53
August 31-Sept. 6	20	10	6	..	..	1	..	1	..	38
September 14-20	20	13	14	..	..	..	..	2	1	50

TABLE 3.—Catches of small mammals on plot 4, E. S. George Reserve, Livingston County, Michigan

	Peromyscus maniculatus bairdii	Peromyscus leucopus noveboracensis	Microtus p. pennsylvanicus	Blarina brevicauda talpoides	Zapus h. hudsonius	Tamias striatus lysteri	Total
1938							
July 10-16	32	6	..	1	..	2	41
July 24-30	35	8	..	1	..	..	44
August 9-15	29	6	..	1	..	..	36
August 24-30	14	6	..	..	..	..	20
September 7-13	19	5	2	..	..	..	26
September 21-27	22	5	2	2	..	..	31
1939							
April 23-26	14	3	..	..	..	..	17
May 23-29	8	8	..	..	..	..	16
June 8-15	10	11	..	.	..	..	21
June 24-30	9	9	.	..	..	..	18
July 23-29	11	14	..	..	1	..	26
August 6-12	10	10	.	..	..	..	20
August 20-26	10	8	..	..	..	..	18
September 3-9	8	7	..	..	..	..	15
September 17-23	7	2	..	..	..	..	9

Populations

The deer-mouse was found to be the most abundant small mammal of the blue-grass association. The meadow vole, jumping mouse, and short-tailed shrew also were numerous at times in some parts of the association. Data that were obtained on the meadow vole, jumping mouse, and short-tailed shrew have been previously reported (Blair, 1940, a, b, and c, respectively).

POPULATIONS OF SMALL MAMMALS

A comparison of the total catches of small mammals on the three study areas (Tables 1, 2, and 3) shows that most of the species of small mammals were common to all of them. The relative abundance of the different species, however, differed considerably on each plot. On plot 1 in 1938, the meadow vole was the most abundant mammal, and the deer-mouse was the next most plentiful resident species. The wood-mouse, which ranged from contiguous woods into the edges of the area, however, outnumbered the deer-mice during several of the trapping periods. On plot 3 in 1938, the order of abundance was reversed, and the deer-mouse exceeded the meadow vole in numbers. The wood-mice during some periods outnumbered on this plot both the deer-mice and the meadow voles. Spermophiles were fairly numerous in this area, while none was taken on plot 1. On the other hand, short-tailed shrews were much scarcer than on plot 1. On plot 4 in 1938, the prairie deer-mouse was by far the most plentiful mammal; the numbers of wood-mice were proportionally smaller than on the other plots; and other species of mammals were relatively scarce.

In 1939, the proportions of the various species differed from those in the previous year. On plot 1, the deer-mouse comprised a minor part of the population. The jumping mouse was the most abundant species, in contrast with its scarcity the preceding year. The meadow vole was extremely scarce. whereas it had been the most plentiful form in 1938. On plot 4, in 1939. the deer-mouse still was the most abundant small mammal, although the numbers of wood-mice were greater than during the previous year and approximately equalled those of the deer-mice. Other species were even scarcer than in 1938.

Plots 3 and 4 were nearest alike in the proportions of the several species of small mammals making up their populations. In these two plots most of the environmental features were fairly similar. Most of the area of both plots was relatively dry blue-grass association, which favors the deer-mouse. Both plots were bordered by woods from which wood-mice ranged into the blue-grass association. The greater numbers of wood-mice on plot 3 than in plot 4 apparently were due to the much longer woods—grass border. The marked dissimilarity of plot 1 evidently was caused by the presence there of a considerable amount of relatively moist blue-grass association and the contiguity of extensive areas of marshland. These moist situations, virtually lacking on the other plots, account for the relatively great abundance in plot 1 of the meadow vole, jumping mouse, and short-tailed shrew.

These data indicate that, at least in the two years of the study, the prairie deer-mouse was the most abundant small mammal on the relatively dry parts of the blue-grass association on the George Reserve. On and near the moist parts of the blue-grass association, the deer-mouse was less abundant than the meadow vole in 1938, and scarcer than either the jumping mouse or the short-tailed shrew in 1939.

POPULATIONS OF DEER-MICE

The catches of deer-mice on the trapping plots are shown in Tables 1, 2. and 3, and the calculated numbers per acre on each plot are shown in Table 4. It is evident that the concentrations of deer-mice were not the same on the different trapping areas. Plot 4, which was smaller than either of the others. consistently had the greatest gross catch and the greatest average number of deer-mice per acre in both years. Plot 3 usually had more mice than did plot 1, although the difference was not nearly so great as between 1 and 4.

A definite seasonal cycle of abundance was noticeable on each plot in both years, except on plot 1 in 1939 when the population was very small and eventually vanished. In the early spring, the population was small and made up of adults that survived the preceding winter. These animals began to breed in the spring, and those of their offspring that survived increased the population until a maximum was reached about the last of July. Then the concentration of deer-mice tended to decrease as many of the mice ceased breeding during the middle of the summer. It increased again in late summer and autumn, after the autumn breeding season began. During the winter, the population decreased to a spring minimum. It is interesting that on plot 4 in 1939 the largest population of the year was found in April. By late May, however, all but 5 of the 13 adults had disappeared, presumably through movement into other areas. Thereafter, the seasonal cycle followed the general pattern, although it was less pronounced than in 1938.

During no two trapping periods were exactly the same individuals present on any of the plots. The individual deer-mice taken on plot 3 during each trapping period in 1938 are shown in Table 5. There were certain resident animals that remained in the area throughout most of the trapping season. Examples on plot 3 are males No. 1847, 1848, and 1857, and females No. 1825, 1826, 1842, 1890, and 1900. Each of the other plots also had certain residents that remained throughout most or all of the trapping season. On plot 4, a female that was first taken about mid-July, 1938, remained there during the remainder of the trapping season, and also was on the same plot through the 1939 season. This particular individual apparently, therefore, remained in the same general locality for at least 15 months, if we assume that it remained there during the winter when no trapping was done. Some residents disappeared at irregular times from the plots. Several of these later were taken on other plots, indicating a movement from one plot to another. Others never again were taken, indicating that they either fell victim to predators or that they moved beyond the study areas. Some previously uncaught adult animals moved onto the plots, and a few moved through but failed to remain there. A few individuals left the plots, then returned after rather long absences. Examples on plot 3 are male No. 1815, which returned after an absence of about 4 months, and female No. 1816, which returned after an absence of about two months (Table 5). The young deer-mice taken either matured and remained on the plots, or they disappeared after one or two trapping periods. Some of those that disappeared later were taken in other plots, but others never were retaken. These data indicate that during the seasons studied the deer-mouse populations on a given area are continually changing. Some individuals remain for long periods of time; others disappear, and still others move in. Young animals mature to increase the adult population, or move away, or possibly are victims of predators, disease, or accident.

Because the field work has extended over only two seasons, the data on annual fluctuations in abundance are limited. There was a very marked decrease in deer-mouse abundance from 1938 to 1939. On plot 4, the average number per acre dropped from 9.2, the 1938 maximum, to 2.9 per acre at a comparable time in 1939. On plot 1, the concentration of deer-mice decreased from a maximum of 2.0 per acre in mid-June of 1938, to a maximum of only 0.3 per acre in late May, 1939. No deer-mice were living on the plot in the last part of the 1939 season. On plot 3, which was under observation by Bujak in 1939, the deer-mice were scarcer than during the previous year. Possibly the dry spring and consequent failure of much of the blue-grass to set seed accounted for the small populations of deer-mice in 1939.

Winter Carryover

The average life of a deer-mouse in nature on the George Reserve is less than one year, although individuals of another subspecies of the deer-mouse, kept in the laboratory, have lived for 7 and 8 years (Dice, 1933:147). When trapping was begun on plot 1 in April, 1939, not a deer-mouse remained of those I had marked there in 1938. One of the two mice recorded in late April, 1939, had been born in the fall of 1938 and was marked then by Bujak; the other was unmarked, and, therefore, presumably was born after trapping was discontinued in 1938. On plot 4, only two of the 13 adult deer-mice present in late April, 1939, had been residents of the plot during the previous summer, having been born in early summer of 1938. Of the other 11 mice, 5 are definitely known to have been born on the plot in the fall of 1938, and the others presumably were born then, as they were unmarked. Thus, most of the residents of the plot in 1938 had disappeared by the beginning of the 1939 breeding season. The population in the spring of 1939, therefore, was mostly made up of mice that had been born during the previous fall. Consequently there was an almost complete annual turnover in population.

The concentration of deer-mice varied on the different plots, but even on the most densely populated area the greatest average number per acre was only 9.2 individuals. There is a definite seasonal cycle of abundance. The relatively small spring population increased to a maximum in about late July; there was some decrease in numbers as sexual activity slowed during mid-summer; the population increased after the fall breeding got under way; then there was a decrease during the winter. There is evidence of marked annual fluctuations in abundance, for the deer-mice were much scarcer in 1939 than in 1938. The deer-mice surviving the winter were mostly those that had been born in the previous fall, indicating an almost complete annual turnover in population.

TABLE 4.—Average number of prairie deer-mice per acre on the trapping plots in 1938 and 1939.

1938	Average No. Per Acre Plots			1939	Average No. Per Acre Plots	
	1	3	4		1	4
March 29-April 1	0.4	Not trapped	Not trapped	April 23-26	0.2	3.7
April 19-22	0.5	0.9		May 23-29	0.3	2.1
May 11-14	1.0	1.1		June 8-15	0.2	2.6
May 21-23	1.0	1.1		June 24-30	....	2.4
June 15-20	2.0	1.7		July 23-29	0.2	2.9
July (first period)	1.6	1.8	8.4	August 6-12	0.1	2.6
July (second period)	1.7	2.0	9.2	August 20-26	....	2.6
August (first period)	1.8	1.3	7.6	September 3-9	....	2.1
August (second period)	1.6	2.2	3.7	September 17-23	....	1.8
September (first period)	1.2	2.2	5.0			
September (second period)	1.6	2.2	5.8			

TABLE 5.—Trapping periods in which each individual deer-mouse was caught on plot 3 during 1938 season. o indicates transient; + indicates immature; × indicates adult and young adult.

Field No.	April 20-22	May 11-14	May 21-23	June 15-20	July 4-10	July 18-24	August 1-8	August 17-23	Aug. 31-Sept. 6	Sept. 14-20
♀										
1812	×	×		×						
1816	×					×				
1825	×	×	×	×	×	×	×	×		
1826	×	×	×	×	×	×			×	×·
1830		×(dead)								
1842		+	×	×	×	×	×	×	×	×
1845		×	×	×	×	×	×	×		
1844			+			× →(moved 1150 feet away)				
1890	(from 865 feet away) →×				×	×	×	×	×	×
1899				×	×	×		×		
1900				+	×	×	×	×	×	×
1910				+	+	×	×	×	×	
2079						o				
2048						+				
2270								+	+	×
2190						(from 1320 feet away) →×				×
2337						(from 330 feet away) →o				
1896						(from 410 feet away) →×				×
2349									+	+·
2408										+
♂										
1814	×									
1815	×	×								
1721	→o (from 660 feet away)									
1824	×	× (dead)								
1828		× (dead)								
1831		×								
1832		×	×	×						
1847			×	×	×	×	×	×	×	×
1857			×	×	×	×	×	×	×	×
1848		+			×	×	×	×	×	×
1898				×						
1870			×	×	×	×				
1914				+	+					
1949					×					
1946					+	+	+	× →(moved 825'away)		
1889				(from 825 feet away) →×		×	×			
2049						+				
2116							×	×	×	×
2139						(from 3795 feet away) →×		×		
1946(a)							×			
2267						(from 1730 feet away) →×				
2225								+·		
2263								+	+	+
2276								+	+·	×
1897						(from 410 feet away) →×				
2359									×	×
2361									+·	+
2412										+
2414										+

Home Ranges

One of the principal objects of the field study was to determine the average size of the home range of the prairie deer-mouse. The method of calculating home range size already has been described (Blair, 1940a). With this method, one-half the distance between two traps is allowed beyond the outermost traps in which each mouse is caught. This obviously does not show the exact shape and size of the individual home ranges, but the figure thus derived is believed to approximate the actual area of the home range.

The sizes of the home ranges have been calculated on a monthly basis. Each monthly home range is based on the records from the two alternate weeks of trapping during the month. The ranges have been calculated on this monthly basis, because of the occasional movements of the mice. A deer-mouse sometimes would have a definite home range in some particular locality, but later it would move to a new home range on the same or on another plot. Therefore, in order not to confuse movements with home range it was necessary to restrict the period for which the home range would be calculated. It was possible, at best, to catch any one mouse 14 times during a month. Some individuals were taken every night, but most were caught less frequently. In many cases the calculated home range included some traps in which the individual mouse was not taken during the month. In large ranges the number of traps within the calculated range exceeded the possible number of catches per month, so the fact that the mice did not get into every trap within their respective ranges does not seem particularly significant. Nor does it appear to detract from the validity of this method of calculating the area of the home range.

Not all of the deer-mice were caught a sufficiently large number of times during a month to give adequate information about the size of their respective home ranges. Some apparently transient individuals were caught only once. Also, some residents that disappeared shortly after the beginning of a trapping period might have been caught only once or twice, and immigrants that took up residence near the end of a trapping period also might be caught only a few times during that trapping period. In order to exclude such incomplete records from the calculations of home range size no home range based on less than three catches per month is used. Another problem has been the elimination of home ranges that were only partially within the plots near edges where the blue-grass association extended beyond the trapping areas. Many of these are eliminated by excluding from the calculations ranges based on less than three catches per month. In addition, all mice that were taken in not more than two adjacent traps near an edge bordered by blue-grass association also have been excluded from the calculations.

A total of 153 monthly home ranges of prairie deer-mice have been calculated according to this method. Seventy adult males that were caught an average of 7.0 times each per month had an average monthly home range comprising 0.63 ± .04 acre, while 64 adult females that were caught an average of 7.1 times each per month had an average monthly home range of

0.51 ± .04 acre. There is no significant difference in the size of the male and female ranges, the difference in the means being 0.12 ± .06 acre. The largest female range comprised 2.29 acres on plot 3 in August, 1938, but only one other female range exceeded one acre. The smallest female range comprised 0.12 acre on plot 1 in August, 1938. The largest male range was 1.67 acre on plot 3 in September, 1938, the smallest 0.25 acre on plot 3 in July, 1938.

Ten home ranges of immature males, based on 5.4 catches per individual per month, averaged 0.61 ± .11 acre. The largest comprised 1.37 acre on plot 3 in July, 1938, and the smallest comprised 0.08 acre on the same plot in September, 1938. Nine home ranges of immature females, also based on 5.4 catches per individual per month, averaged 0.64 ± .09 acre. The largest comprised 0.99 acre on plot 3 in September, 1938, and the smallest 0.17 acre on plot 4 in July, 1938.

The home range of each deer-mouse tends to remain in the same vicinity over a rather long period of time, although the shape and size as computed by my method vary more or less from month to month. Each adult, therefore, generally remains on the same part of the plot as long as it stays on that plot. Examples of the length of residence of individual deer-mice on a plot are shown in Table 5. The longest that one individual apparently resided in the same home range was about one year. This female on plot 4 occupied during the 1939 season almost identically the same home range that it had during the summer and fall of 1938. If we may assume that the female stayed in the same area during the intervening winter, it is apparent that she occupied approximately the same range for about a year. The entire area that this female ranged over during the year, as indicated by the trapping records in the two seasons, comprised only 0.53 acre, which was only slightly larger than her largest monthly home range of 0.49 acre in July, 1938. This is an extreme case, of course, and most of the deer-mice remained in their respective home ranges for shorter periods of time.

The fluctuations from month to month in the shape and size of the calculated individual home ranges possibly are due to limitations of the trapping technique which allows at best only 14 catches during a month. I am inclined to believe, however, that the home range of a deer-mouse is not a definite bit of terrain with fixed limits. Instead, the home range is undoubtedly a constantly changing unit that varies with the changing needs of the individual mouse. The size and shape of the home range over which an animal habitually ranges probably are conditioned by many factors, particularly: (1) location of the home site or home sites, (2) available food, (3) distribution of preferred cover, (4) location of temporary refuges, (5) relations to other individuals of the same sex and species, (6) relations to individuals of the opposite sex of the same species, (7) relations to individuals of other species, (8) weather, (9) micro-climates.

Most of these factors change from time to time. The available food and cover, for instance, change with seasonal changes in the vegetation. At times of great abundance of food a mouse would have to range only a short distance from its nest to feed, but at times of scarcity it might have to range over a relatively large area. The relations of a sexually active individual to others of the same species undoubtedly are different from those of a non-breeding animal. With these changes in the determining factors the size and shape of the home range must also continually change.

The home range of each deer-mouse usually overlaps that of one or more other individuals of the same or of opposite sex. This overlap occurs at times of low concentration of deer-mice as well as at times of high. When the species is abundant in an area the amount of overlapping naturally is greater than when the species is scarce. The greatest overlapping of home ranges was found on one end of plot 4 in 1938. There in July, 1938, on an area comprising about 0.77 acre, the home ranges of 8 resident adult females and of 4 resident males broadly overlapped. In the following month 5 adult females and 4 males had similarly overlapping ranges in the same area. In September, 5 adult female and 4 male ranges overlapped in the area. One of the females was suckling young in July and in August, and all of the females apparently were pregnant in September. The home ranges of various breeding females likewise broadly overlapped in other parts of the same plot and on the other plots. This indicates that, in the prairie deer-mouse of southern Michigan, there is no antagonism and consequent territorial behavior among breeding females similar to that described by Burt (1940) in the wood-mouse. A striking example of the overlapping of the ranges of two breeding female deer-mice on plot 3 is shown in Figure 2. Females No. 1826 and No. 1842 occupied ranges that were largely coincident. Both were sexually active at this time as evidenced by the fact that the latter bore a litter on September 19 and another on October 16, and the former gave birth to young on October 6.

Even though a great many of the home ranges overlapped, there usually were some vacant spaces on the plots in addition to the vacant zone bordering the woods into which the deer-mice seldom ventured. The smallest amounts of vacant space were found on plot 4 in 1938, when deer-mice were abundant there. There were virtually no parts of the plot, except along the border of the woods, that were not included within the home range of one or more male or female deer-mice. On plot 4 in 1939 and on plot 1 and 3 in 1938, with relatively small populations of deer-mice, there were, in each month, rather large areas from which one or the other sex was absent and smaller areas that were not in the home ranges of either sex (see Figs. 1 and 2).

In general, the data on home ranges of the prairie deer-mouse indicate that in southern Michigan the average home range of both immature and adult males and females comprises about one-half acre. The home range of each individual tends to remain in the same vicinity over a rather long period of time. There is some month to month fluctuation in the size and shape of the calculated home ranges, apparently due to differences in the daily movements of the individual mice produced in turn by their changing vital needs. The home range of each deer-mouse usually overlaps that of one or more individuals of the same or opposite sex. This is true of breeding as well as non-breeding mice. Notwithstanding the overlapping, there usually are some unoccupied spaces in the plots.

Refuges

During the 1938 season several of the deer-mice and wood mice were followed after being released from the traps until they entered holes or found other places of refuge. Most of the mice showed by their actions that they knew the terrain well, and they went directly to a hole or other refuge. On several occasions an individual used the same route when going from a trap to a refuge. In a very few cases a mouse acted as if lost when released from

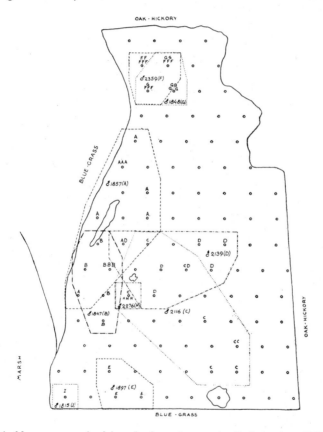

Fig. 1. Home ranges of adult male deer-mice on plot 3, September, 1938. Letters indicate traps in which each individual was caught and the number of times it was caught in each. Broken lines show the limits of the calculated home ranges.

the trap. In some cases a mouse would appear lost for a few moments, and would run a few feet away from the trap. Then, apparently becoming oriented, it would reverse direction and go directly to a hole. From my observations of the released mice it seems certain that each individual generally knows well the terrain within its home range, and that each tends to follow certain definite trails. In many cases the mice were lost to sight in heavy blue-grass or other relatively dense vegetation, so the holes they entered could not be located. However, enough mice were followed to definite holes or other refuges to give a fairly reliable indication of the types of refuges that they sought. Some of the holes to which the mice went probably were their "permanent" homes, but others probably were only temporary refuges. From the evidence at hand, therefore, it may only be said that all of the holes to which the mice went served as refuges at a time of danger, and that presumably some of the holes were the actual homes of the mice.

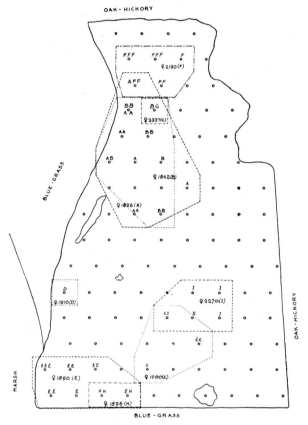

Fig. 2. Home ranges of adult female deer-mice on plot 3, September, 1938. Letters indicate traps in which each individual was caught and the number of times it was caught in each. Broken lines show the limits of the calculated home ranges.

Fifty-seven deer-mice were followed to some refuge or other a total of 113 times. The mice went 45 times to small holes that they apparently had made themselves. They went 42 times to spermophile holes. All of the spermophile holes apparently had been previously abandoned by their makers. The deer-mice went 18 times to old tunnels of the common mole. In 4 cases the mice went to logs in the blue-grass association. In three cases they went to old, abandoned skunk dens. One deer-mouse went to an abandoned fox den. Thus 40% of the times the deer-mice went to holes that they apparently had made. while 57% of the times they went to the abandoned holes and tunnels of other mammals. Only three per cent of the times did they go to logs.

Usually the refuge to which a deer-mouse fled was located within the limits of its home range determined from the trapping records. A total of 123 records of the distances traveled by deer-mice before they reached a refuge or were lost to sight in dense vegetation are available. In 114 (93%) of these cases the deer-mouse did not leave its home range. In 7 (5%) of the cases the mouse went less than 30 feet beyond the limits of the plotted home range, and in only 2 (2%) of the cases did a mouse go more than 30 feet beyond the limits of its known home range. The fact that in 93% of the cases the mice did not leave their calculated home ranges and in another 5% of the cases went less than 30 feet beyond the calculated range supports my belief that the ranges calculated according to my method represent fairly accurately the actual ranges of the mice.

The number of refuges which an individual deer-mouse might have within its home range is indicated by the data on some of the mice. An adult female during July, 1938, went to 4 refuges within its home range. Two were spermophile holes, and two apparently had been dug by the mouse. The mouse went to one of the spermophile holes on two different occasions. In August the same individual went to 6 different refuges within its home range, which then partially overlapped its July range. However, all of the refuges were different from those used during July. Four of the 6 were spermophile holes, and two apparently had been made by the mouse. One of the spermophile holes was used twice. The refuges were fairly evenly distributed over the home range, so the mouse when it was in any part of its range had a possible refuge near at hand. An adult male in August went to 5 different refuges, of which 4 were within its calculated home range, and one was about 15 feet beyond. Four of the 5 were spermophile holes, and one apparently had been made by the mouse. One of the spermophile holes was used twice. The August home ranges of the male and female discussed above overlapped broadly, but neither went to a refuge to which the other went. During August the home ranges of two other adult females overlapped this area. One of these females went to three different refuges, one of which also was used by the male discussed above, but on a different date. The other female went to one refuge, which also was used by the first female, but on a different date. Thus in this particular area there were at least 13 separate refuges used by the resident deer-mice. In most cases only one individual went to a particular refuge, but two females went to the same refuge at different times, and to another refuge a male and female went at different times.

In July, 1938, an area of sand with an extremely sparse cover of panic grass *(Panicum)* and scattered oak seedlings and clumps of prostrate juniper was trapped. Several of the deer-mice went directly to clumps of juniper. When other of the deer-mice in this area were released they traversed barren situations as rapidly as possible, and often instead of finding a hole they stopped in the shelter of a seedling or clump of grass. Field notes on two representative individuals indicate the type of behavior in this area:

Adult ♀ 2044, July 22, 1938, went 12 feet west from trap 96 to base of oak seedling 8 feet in height; sat up and nonchalantly washed face.

Young adult ♀ 2042, July 18, 1938, ran 10 feet from trap 64 to shrubby oak; could hardly be routed from beneath it; ran around and around base of oak; finally ran out and under clump of panic grass; when routed out there ran back to base of oak.

As long as the mouse remained beneath the oak seedling it would be relatively safe from attack by a winged predator, which probably would, on the other hand, have little difficulty in catching it if it ventured out onto the barren sand. The reluctance of the mice to leave such cover as that of the oak seedlings seemed to indicate their recognition of its protective value.

In this sandy area, too, was taken the only deer-mouse that entered even the edge of the woods when released from the trap. This individual, a young, adult male, was unable to find a hole in the grassland and went 96 feet to the edge of the woods. At the wood's edge the deer-mouse climbed three feet up the trunk of a large tree but fell off. Then it hid among the dead leaves on the ground. En route to the woods, the mouse stopped under juniper seedlings and other cover, but I chased it away from such cover each time it stopped. The deer-mouse did not go directly to the woods, but apparently encountered the woods by accident due to my chasing it there.

Twenty-nine wood-mice were followed to some refuge in the blue-grass association a total of 43 times. Many other times the wood-mice went into nearby woods, but they were not followed to their refuges there. The wood-mice went only 10 times to holes in the blue-grass association that they apparently had themselves made. Eleven times they went to stumps in the blue-grass association, and one wood-mouse went to a log there. The wood-mice went 16 times to mole tunnels, and 5 times they went to abandoned spermophile holes. It is obvious from these data that the wood-mouse, when it ranges into the blue-grass association, sometimes utilizes the same types of abandoned mammal burrow used by the deer-mouse. In areas where the two species occur together there is, therefore, the strong possibility of competition between them. The wood-mice tend to utilize stumps and logs when those are available in the blue-grass association, but they certainly are not limited to this type of refuge.

In general, these observations on the released mice show that each deer-mouse and each wood-mouse generally knows the terrain within its home range and has little difficulty in finding a hole or other refuge at a time of apparent danger. Both the deer-mouse and wood-mouse use some holes that they apparently have themselves made, and both also utilize the abandoned holes and tunnels of other mammals when these are available. As both species tend to use the same type of refuges there possibly is competition for refuges and home sites when the two animals occur in the same area. In areas of sparse cover the deer-mice sometimes take refuge beneath tree seedlings and other scattered bits of cover from which they can hardly be forced away. Most of the deer mice that were followed after being released from the traps went to refuges within their calculated home ranges. A deer-mouse may have 6 or possibly more refuges within its home range. Where several individual mice occupy the same area, the abundance of refuges may be still greater.

*   *   *   *   *   *   *

References

ALLEN. D. L. 1939—Michigan cottontails in winter. Journ. Wildlife Management **3**: 307–322, 6 pls.

BLAIR, W. F. 1940a—Home ranges and populations of the meadow vole in southern Michigan. Journ. Wildlife Management **4**: 149–161, 1 text fig.

————1940b—Home ranges and populations of the jumping mouse. Amer. Midl. Natur. **23**: 244–250.

————1940c—Notes on home ranges and populations of the short-tailed shrew. Ecology **21**: 284–288.

BURT, W. H. 1940—Territorial behavior and populations of some small mammals in southern Michigan, Misc. Publ. Univ. Mich. Mus. Zool. **45**: 1–58, 2 pls., 8 text figs., 2 maps.

CHITTY, DENNIS. 1937—A ringing technique for small mammals. Journ. Anim. Ecol. **6**: 36–53, 5 figs.

DALKE, P. B. and P. R. SIME. 1938—Home and seasonal ranges of the eastern cottontail in Connecticut. Trans. Third North Amer. Wildlife Conference: 659–669, 4 text figs.

DICE, L. R. 1938—Longevity in Peromyscus maniculatus gracilis. Journ. Mammalogy **14**: 147–148.

DICE, L. R. and H. J. LERAAS. 1936—A graphic method for comparing several sets of measurements. Contr. Lab. Vert. Genetics **3**: 1–3, 1 text fig.

HAMILTON, W. J. 1937a—Activity and home range of the field mouse, *Microtus pennsylvanicus pennsylvanicus* (Ord.). Ecology **18**: 255–263, 2 text figs.

————1937b—The biology of microtine cycles. Journ. Agric. Research **54**: 779–790, 2 text figs.

16

Reprinted from *Contrib. Lab. Vert. Biol.*, (67), 8–9, 11–12, 16–19, 21–23 (Apr. 1955)

Social Behavior, Social Organization, and Population Dynamics in a Black-Tailed Prairiedog Town in the Black Hills of South Dakota

JOHN A. KING

LIFE HISTORY

Published accounts of prairiedogs have dealt with many phases of their life history. Their food habits have been studied by Kelso (1939) and by Merriam (1901), their burrows by Merriam (1901) and by Scheffer (1937), their growth by Johnson (1927), their reproduction by Anthony and Foreman (1951), and their ecologic responses by Scheffer (1947) and by Osborn and Allan (1949). More general accounts have been given by Bailey (1926, 1931) and by Seton (1929). Other general notes concerning their life history have been reported from time to time, particularly in the Journal of Mammalogy. Some information has been collected on other species of Cynomys, primarily by Longhurst (1944) on C. gunnisoni and by Stockard (1929, 1930) on C. leucurus. The life history summary presented in this study is intended to aid in the interpretation of the population dynamics and social structure of these animals.

Food Relations

The vegetation in this prairiedog town was much modified by the feeding of the animals, by their burrows, and by their clipping of the tall vegetation. These activities completely denuded some areas of all vegetation, particularly around the burrows. In the center of their areas of activity forbs had replaced the original grass cover. Around the periphery of the town much of the original vegetation remained. The natural vegetation varied somewhat from place to place in the study area because of the heterogeneity of edaphic features.

Kinds of vegetation available. — The principal herbaceous species found on Wards A and B in Shirttail Canyon are listed below;[2] those species eaten by prairiedogs (as ascertained by observation only) are marked with an asterisk:

Poaceae: wire grass (Aristida longiseta)*, spear grass (Stipa viridula), Texas crab grass (Schedonnardus paniculatus), blue grama grass (Bouteloua gracilis)*, tall grama grass (Bouteloua curtipendula), buffalo grass (Buchloë dactyloides), brome grass (Bromus japonicus)*, wheat grass (Agropyron smithii), witch grass (Panicum sp.), big bluestem (Andropogon scoparius). Cyperaceae: sedge (carex filifolia). Iridaceae: iris (Sisyrinchium angustifolium). Polygonaceae: knotweed (Polygonum ramosissimum)*, knotweed (Polygonum buxiforme). Cheno-

[2]Identifications were kindly made by William Watson, chief ranger, Wind Cave National Park. The names used are those of Rydberg (1932) or in a few instances McIntosh (1931).

Editor's Note: Figures 1 and 3 and Tables 1, 3, and 4 have been omitted, owing to limitations of space.

podiaceae: goosefoot (Chenopodium album). Amaranthaceae: pigweed (Amaranthus retroflexus)*. Nyctaginaceae: umbrella wort (Allionia linearis). Portulacaceae: whitlow wort (Paronychia sp.). Brassicaceae: wall flower (Cheirinia inconspicua). Grossulariaceae: cinquefoil (Potentilla sp.). Fabaceae: white sweet clover (Melilotus officinalis), few-flowered psoralea (Psoralidium tinuiflorum), silvery psoralea (Psoralidium argophylum), locoweed (Geoprumnon plattense), prairie clover (Petalostemon purpureus), locoweed (Oxytropis plattensis). Oxalidaceae: yellow sorrel (Xanthoxalis corniculata)*. Simarubaceae: milkwort (Polygala alba). Euphorbiaceae: spurge (Chamaesyce sp.)*. Malvaceae: scarlet mallow (Sphaeralcea coccinea)*. Cactaceae: cactus (Neomamillaria sp.). Onagraceae: evening primrose (Onagra albicaulis), evening primrose (Meriolix serrulata), butterfly weed (Gaura sp.). Asclepiadaceae: milkweed (Asclepias pumila). Boraginaceae: stickweed (Lappula sp.), stickweed (Lappula occidentalis)*, puccoon (Lithospermum gmelini), (Lithospermum linearifolium), (Onosmodium occidentale). Verbenaceae: vervain (Verbena bracteosa), vervain (Verbena hastata). Lamiaceae: savory (Clinoposium sp.). Solanaceae: nightshade (Solanum nigrum)*. Scrophulariaceae: mullein (Verbascum thapsus)*, beardtongue (Pentstemon sp.). Plantaginaceae: plantain (Plantago purshii). Compositae: blazing star (Liatris punctata), brownweed (Gutierrezia sarothrae), golden aster (Chrysopsis villosa), goldenrod (Solidago pallida), goldenrod (Solidago sp.), fleabane (Erigeron flagellaris), mare's tail (Leptilon canadense), black-eyed susan (Rudbeckia hirta), cone flower (Ratibida columnifera), sage (Artemisia gnaphalodes), sage (Artemisia frigida), thistle (Cirsium sp.)*, fetid marigold (Boebera papposa)*. Cichoriaceae: skeleton weed (Lygodesmia juncea)*.

Species of plants other than those indicated above are eaten by prairiedogs. Kelso's study (1939) of 247 stomachs of prairiedogs, 228 of which were taken from four localities in Montana, shows that the preferred item of food is grass, which totaled 51.5 per cent of the volume in the stomachs analyzed. The amount of grass consumed by prairiedogs in the Shirttail Canyon town probably is not so high as the percentage quoted by Kelso indicates, since the prairiedogs seem to rely mostly on forbs for food, owing to the paucity of grass in some parts of the town.

A variety of forbs thrive on the area, most of them providing food for the animals. Certain plants on the study area, Artemisia for example, although probably not eaten by the prairiedogs, fail to dominate the other vegetation because the animals keep them clipped to within a few inches of the ground.

The parts of the plants selected for food are largely those of high energy or high water content. Usually, the succulent, meristematic tissue at the base of the grass blades or the seeds and fruits of forbs are eaten first. Considerable quantities of crude fiber and cellulose are also ingested and appear to be selected at times. More often the fibrous parts are taken along incidentally with the parts containing more nutritious food. Thistles (Cirsium), eaten in abundance during the drier part of the season, were selected perhaps for their high water content.

Quantities of vegetation available. — Types of vegetative cover, in

1948, were measured along a 450-foot line transect oriented north and south approximately through the middle of Wards A and B. The amount of ground covered along the transect by each genus of plant was measured to the nearest inch, and its percentage of cover was calculated for every section of 50 feet.

The data from the line transect show a difference in percentage of bare earth between Wards A and B and also a downward gradation in percentage of bare earth toward the outer edge of Ward A (Table I). Ward B has a higher percentage of bare earth in most of the 50-foot sections than does Ward A. The two 50-foot sections of Ward A (350-400 ft., 400-450 ft.) at the outer edge of the prairiedog town contain the highest percentages of grass cover. Forbs predominate in Ward B and in the 50-foot sections of Ward A (250-300 ft., 300-350 ft.) adjacent to Ward B.

The air-dry weights of vegetation taken from 20 randomized clipped quadrats (Table II) further illustrate the difference in the relative abundance of grasses and forbs between Wards A and B. Over ten times as much grass by weight is present in Ward A, at the edge of the town, as in Ward B, near the middle. In contrast, the quadrats in Ward B have a greater weight of forbs than those in Ward A. The total weight of the vegetation per quadrat is over twice as great in Ward A as in Ward B. Ward B, situated closer to the middle of the town, was occupied by prairiedogs before 1938, whereas Ward A, at the edge of the town adjacent to a large grass field, became inhabited by prairiedogs after 1938.

Table II

Air-dry weight in grams of clipped vegetation on random square-yard quadrats. Weight of vegetation differs in the two wards. Ward A is at the edge of the Shirttail Canyon town.

Quadrat	Weight of Grasses, Grams	Weight of Forbs Grams	Total Weight, Grams
Ward A			
Ji	89.8	6.5	96.3
Ik	20.0	0	20.0
Ii	111.0	0	111.0
Oh	24.0	0	24.0
Jl	18.0	0.5	18.5
Pg	22.3	7.0	29.3
Qj	4.5	4.7	9.2
Ml	17.7	22.5	40.2
Jh	140.7	9.0	149.7
Hg	61.0	0	61.0
Totals	509.0	50.2	559.2
Ward B			
Qp	5.5	10.5	16.0
Op	16.5	13.0	29.5
Np	11.5	18.0	29.5
Pp	1.0	72.5	73.5
Pn	2.5	18.5	21.0
Mn	13.5	5.0	18.5
Jp	0	5.0	5.0
Ho	0	21.0	21.0
Hn	0	7.0	7.0
In	0	7.5	7.5
Totals	50.5	178.0	228.5

Frequency indices from a series of uniformly spaced square-foot quadrats along two belt transects placed at right angles to each other across the middle of Ward A show a rough zonation of the vegetation. The presence or absence of a species of plant was recorded for each quadrat at ten-foot intervals along the transects. In all, 86 quadrats were examined. The data, presented in Tables III and IV and in Figure 1, demonstrate a rough zonation of vegetation in which forbs predominate in those zones of Ward A which are proximal to Ward B and diminish in abundance in the outer zones where grasses predominate. This arrangement of the vegetation in zones is somewhat similar to that in the zones found by Osborn and Allan (1949) in an abandoned prairiedog town in Oklahoma. Since Ward A represents only a section of a prairiedog town, it does not contain the complete concentric zones of an entire town, such as that described by Osborn and Allan.

Effects of feeding upon vegetation. — The distribution of kinds and quantities of plants in a prairiedog town is due to the feeding of the prairiedogs. Intensive utilization of the original grass cover by the prairiedogs results in a depletion of the grass in the middle of the town. Forbs replace most of the grass and provide food for the inhabitants. Some grass persists, however, in the middle of a town after prairiedogs have inhabited it for a long period.

Grasses cannot withstand the feeding of the prairiedogs as well as can the forbs. Consequently, forbs flourish in the new habitat created for them through the destruction of the grass. Probably no species can resist constant destruction by feeding. As a result, the species of forbs preferred by the prairiedogs decrease in abundance, and less preferred species increase. Domination of the vegetation by a less preferred species is probably restricted by the feeding habits of the prairiedogs. First, their food habits are modified with changes in the vegetation; the more abundant a less preferred species becomes, the more it is used for food. Thus, the area is opened for reinvasion by other species. Second, the feeding range of each individual usually contains several different vegetation types (Fig. 1). A prairiedog may for a time feed primarily on one type of vegetation and neglect a less preferred type, which may be replaced in the following year by the preferred species. For example, the prairiedogs fed about the edges of Ward B at a time when part of the ward had become almost entirely covered by edible trailing fleabane (Erigeron flagellaris). In the following year competing species of plants partly replaced the fleabane. The new plant species provided food for the prairiedogs, and they again fed in the middle of the area. These shifts in the feeding range and the resulting changes in vegetation are analogous to the human practice of fallow farming.

Another analogy to human farming methods is offered by the cultivation of the soil by the burrowing of the prairiedogs. Subsoil is thrown out on the surface of the ground and is exposed to invasion by forbs, which, on the topsoil, cannot compete with the grasses. The earth exposed about the burrows frequently is covered by plants not found abundantly elsewhere in the region. Fetid marigold (Boebera papposa), scarlet mallow

229

(Sphaeralcea coccinea), black nightshade (Solanum nigrum), and pigweed (Amaranthus retroflexus), for example, are species which occur
almost exclusively on the subsoils exposed about a prairiedog burrow.
These plants, which have been more or less "cultivated" by the prairiedogs, are often eaten and at times appear to be important food items.

Feeding behavior. — Prairiedogs spend most of the day eating and
foraging for food. The mean proportional time spent in feeding by 11
animals watched individually for the greater part of 11 days, or a total
of 82 hours, was 57 per cent of their active day (Table V). Other activities of the day included the making of social contacts (6 per cent),
moving about (6 per cent), staying in burrows (15 per cent), acting
alert (13 per cent), and miscellaneous activities (3 per cent). Since
all of the prairiedogs could not be watched on the same day, the percentages are not directly comparable. In general, young animals spent
a larger proportion of their active day feeding than did mature animals.
Frequently, the behavior of the prairiedogs was affected by a particular
activity or circumstance which reduced the amount of time spent in feeding. On days when the temperatures were over 100° F. or when the animals were severely frightened, they spent less time feeding and more
of the day in burrows. An adult male was building a new nest on the day
he was observed and therefore spent much of his time digging and carrying nesting material into the burrow. Strong winds often kept the prairiedogs more alert and cautious than they were on calm days. On the day
♂ 37 was observed, he was alert in guarding his range from a strange
individual. Despite these other factors which interrupt the amount of
time spent in feeding, it still represents the major part of activity.

Table V

Percentage of time spent in various activities by 11 prairiedogs observed individually for
an entire day. A total of 82 hours of observation during July and August of 1948 and 1950
on Ward A of the Shirttail Canyon town.

Number of Animal	Sex	Age	Total Minutes	Feeding	Inside Burrow	Alert or Standing	Social Contacts	Moving About	Miscel- laneous
6	♂	Ad.	436	50	18	8	5	10	9
10	♂	Ad.	761	74	6	11	5	5	0
16	♀	Ad.	338	43	25	14	9	5	3
17	♀	Ad.	421	62	17	8	4	3	5
24	♂	Ad.	341	57	25	6	5	4	3
37	♂	Ad.	576	33	8	42	5	12	0
51	♀	Ad.	559	48	11	30	5	4	3
88	♀	Im.	313	64	24	4	5	4	0
89	♂	Im.	401	71	17	4	4	3	1
90	♀	Im.	388	73	3	5	11	8	0
130	♂	Im.	369	57	14	15	8	5	2
Mean percentage				57	15	13	6	6	3

Prairiedogs find their food by smelling along the ground until a mor-
sel is discovered. Each kind of food is collected and consumed in a
characteristic fashion. Large leaves and fruits are bitten off at the base
and eaten while the animal sits up and holds them in its forepaws. A

prairiedog procures grass by pulling near the base of a blade with its incisors. The blade is held in one forepaw and run through the mouth until the lower part containing the soft meristematic tissue is in the mouth. This is bitten off and chewed while another blade is being pulled out. The operation is performed with great rapidity. Occasionally, the entire blade is consumed. Small pups, not strong enough to pull the grass out readily, often may be seen pulling and tugging at a blade, trying to loosen it from its base. Thistles are bitten off at the base of the leaf and are also eaten from proximal to distal end; thereby the prairiedog is able to masticate the spines on the leaves without being pricked. Because of the practice of eating the basal part of the vegetation first, much of the distal part is cast aside and wasted. Certainly, however, the most nourishing pieces are consumed. Seeds are stripped off the vegetation by running the stem along the mouth until the seeds are brought into it and bitten off. Many seeds are also picked up from the ground and eaten. Roots and tubers are obtained by digging, a process which makes little holes two to three inches deep and about as wide throughout the town. Such holes are most frequently seen during the early spring. During most of the summer scratching for food is too shallow to make a noticeable impression in the earth.

Prairiedogs seldom feed exclusively on one type of plant for any length of time before passing on to another. On August 24, 1950, for example, a young female within 19 minutes ate plants of nine different genera: Schedonnardus, Lappula, Verbena, Chamaesyce, Boebera, Aristida, Carex, Polygonum, and Lygodesmia. Although she spent much time finding and eating spurge (Chamaesyce) seeds, the largest quantity of food consumed was thistle, as entire leaves were devoured at a time.

Some grasses and forbs are clipped off near the base and permitted to fall and dry up without being eaten. Tall vegetation in the middle of the town is entirely removed, but at the edge it is only thinned. The removal of this vegetation is accomplished with great rapidity as the animals move haphazardly from one plant to another. Usually, the vegetation which grows tall enough to be cut down in this fashion is not selected for food.

Vegetation in the middle of the town rarely grows over a foot high because of the clipping and eating activities. In one place a small patch of mullein (Verbascum) had grown over three feet high. It was attacked by the prairiedogs off and on throughout the summer of 1948. Part was eaten; the rest was completely destroyed, and it did not grow up again in 1949 or 1950. The tallest plants in the middle of the town in 1950 were a few stems of needle and thread grass (Stipa), which grew out of a protecting clump of cactus. In the new areas of settlement the grasses were completely cut off only near the burrows, where there was the greatest amount of prairiedog activity. Beyond the immediate vicinity of the burrows, in the newly settled area, the extent of clipping was noticeable only at close range where the cut, dried stems were visible.

By clipping the tall vegetation prairiedogs keep much of the town free of vegetation which might obstruct the view. Since a clear view is important both in their survival from predators and in their social life, this behavior is quite adaptive. Why the presence of tall vegetation should induce this clipping response at some times and places and not at others is not understood at the present time.

Prairiedogs make runways through the tall grass between their burrows (Pl. II, Fig. 1) by trampling the vegetation under foot as they run from burrow to burrow and by the concentrated clipping of vegetation along these pathways. In the older part of the town runways were not visible, since most of the vegetation had been removed, and the animals wandered indiscriminately over the entire area. Even in the new section, where the grass was tall, the prairiedogs did not limit their movements to the runways, but went readily through the thickest grass. The runways were only a by-product of the frequent trips between burrows.

Burrows

The burrow is an important feature in the life of a prairiedog. Not only does it form a refuge from the external environment and a place to rear the young, but it also affects the social structure of the prairiedog society. The intimate relationship between prairiedogs and their burrow systems contributes to the distinct spatial patterns of their social organization. Since the burrows remain essentially the same from generation to generation, they have a stabilizing effect on the organization of the colony.

The exposure of the excavated soil and the denudation of vegetation from the mound and from the surrounding area have many important ecologic effects. The exposed subsoil is affected by weathering and by the encroachment of vegetation. The denuded soil also is especially subject to erosion by water and winds. The size of the denuded area surrounding a burrow is often increased by bison, which use the craters as wallows and consequently expose even more bare earth to the climatic elements.

Kinds of burrows. —The external appearance of the burrows varies considerably. Three general categories based on the excavated soil can be established: (1) a crater-like mound of soil, (2) a mere pile of soil, and (3) practically no soil. The crater-like mound is the largest and most characteristic. Sometimes the mound is as much as three feet high and seven or eight feet wide. The shape of the mound varies somewhat with the mode of its origin and with its age. A distinction between a rim crater and a dome crater is based on the shape and kind of soil in the crater. The rim crater is generally smaller and is carefully formed by the prairiedogs into a crested mound about the entrance of the burrow (Fig 2, top). It is made of the top soil scraped into a pile from the surface of the ground about the entrance and molded by the feet and noses of the prairiedogs to form a compact structure. Sometimes the dirt for this mound is scraped up from an area with a radius of as much as five or six feet. In contrast to the rim crater, the dome crater is often large. It is shaped like a dome, except for the burrow entrance in the middle (Fig. 2, bottom). The mound is composed of pale subsoils or clay which cannot be shaped as effectively by the animals as can the topsoil.

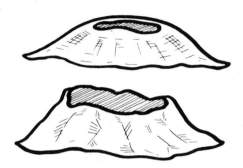

Fig. 2. Diagram of craters around the entrances of prairiedog burrows; top, dome crater; bottom, rim crater.

The differences between these two types of craters apparently depend upon the manner in which the prairiedogs construct their subterranean burrows. The subsoils exposed in the dome craters suggest that the burrow began at the dome crater and that the crater was formed from the excavated soil. When the opening to the burrow comes to the surface some distance away from the original opening, there is at first only a hole with no soil about its entrance. The prairiedogs then scrape the surrounding topsoil and form a rim-shaped crater. A subterranean connection then exists between the two entrances. This underground connection may be plugged with soil and reopened at intervals; hence a connection between the two entrances cannot be ascertained from the nature of the craters alone, except in new areas of only a few burrows. In an old town many other subterranean connections are made which obscure the direct relationship between any two burrow openings.

Those burrows without craters at the entrance are sometimes the result of degeneration of old unattended craters. At other times, a small pile of dirt is left at the entrance of a new, seldom-used burrow. Some of these burrows characterized by an unshaped pile of dirt at the entrance are frequently used and are only in a temporary state of disrepair, whereas others are seldom used. A burrow with little or no excavated soil about the entrance may be the opening of a new entrance from beneath, or, more often, a short, exploratory burrow from which the original dirt has been eroded.

The subterranean passageways of the burrows were not thoroughly explored in this study, although one burrow system was partly excavated. The frequently published diagram (Merriam, 1901) of Osgood's exca-

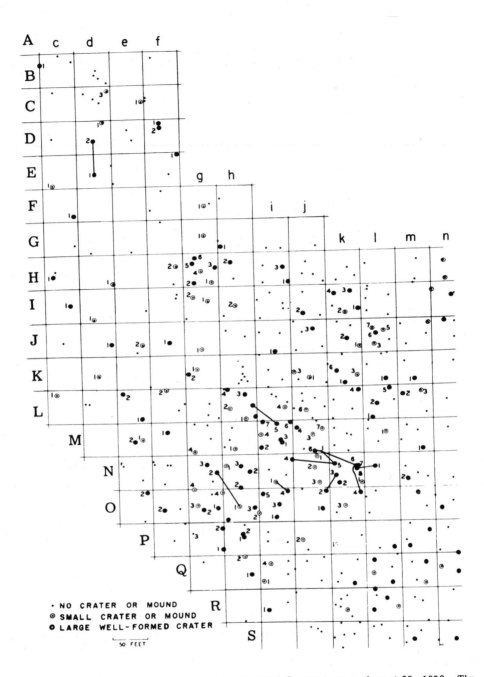

Fig. 4. Burrows on Wards A and B of the Shirttail Canyon town on August 25, 1950. The separation of Wards A and B may be seen by the scarcity of burrows in the longitude of n, with Ward B on the right. The grid pattern illustrates the 50-foot square quadrats used in observing the prairiedogs. Lines connecting the burrows indicate a subterranean connection. Some of the burrows are numbered.

vation of a black-tailed prairiedog's burrow in Nebraska represents a tunnel system very different from the one excavated in Shirttail Canyon. Burrows of related species (Foster, 1924; Whitehead, 1927) have more in common with the burrows studied here. Foster, for example, gave the average depth of the lateral passages as three and one-half feet with seven entrances to the burrow system. This contrasts sharply with the 15-foot depth and single burrow entrance described by Merriam. A diagram of the burrow I partly excavated is presented in Figure 3. Of the 36 feet of the tunnel uncovered, the maximum depth below the ground surface was three feet, except for the nest chamber which was one foot deeper. Only one entrance to the burrow system was found. The uncovered lateral passageways, however, might have led to other openings. The passageways followed a horizontal stratum of sandy loam directly below a stratum of coarse gravel and large rocks. Since most of the soils in Shirttail Canyon are distinctly stratified, tunnels dug in one stratum might easily intersect each other.

The existence of subterranean connections between the burrows was ascertained by watching a marked prairiedog enter one burrow and reappear at another entrance. Figure 4 shows the position of the burrows known to be connected on the study area. These subterranean connections probably are not permanent, but vary from season to season, since the prairiedogs plug them and later reopen them. If I blocked the entrance of a burrow after a prairiedog went down, the same animal usually appeared at another entrance, often with dirt on its claws, indicating that it had probably dug out an old earthen plug. Shortly after the pups came out of the burrows in May the adults frequently were observed digging at the old burrows. I assumed that they were reopening the connections which had been closed for the winter or while the young were in the nest.

The nest chamber in the excavated burrow was 18 inches in diameter and 14 inches in height and was about 12 feet from the burrow entrance at a depth of three feet below the soil surface. The nest was composed of fine, dried grass, which almost filled the chamber. Nesting material is carried into the nests throughout the year by animals of both sexes and of all ages. The animals pick up dried grass, stuff it into their mouths with their forefeet, and carry it into the burrows (Pl. II, Fig. 2). Frequently, nest grass is gathered at a distance from the burrow, often as far away as 150 feet. The carrying of nesting material into the burrows is stimulated to some extent by the discovery of a particularly favorable material or by a period of cold weather.

Numbers of burrows. — In all, 614 burrows were mapped in 1950 on Wards A and B, which together comprise 11.7 acres (Fig. 4). Of the three categories of burrows mentioned previously, 155 were crater burrows, 100 were mound burrows, and the remaining 359 were merely holes. The total number of burrows increased from 540 in 1948 to 614 in 1950. This increase was due partly to the construction of additional crater burrows (89 in 1948 and 155 in 1950) and partly to the digging of more hole burrows (302 in 1948 and 357 in 1950). The number of mound burrows, however, decreased (203 in 1948 and 100 in 1950). The subjectivity in the classification of the burrows may account for some of the difference in the numbers of each type of burrow between 1948 and 1950. Some increase in the number of crater burrows between 1948 and 1950 actually occurred through the development of mound burrows into crater burrows by the activity of the prairiedogs. The increase in number of hole burrows is largely a result of an expansion in the size of Wards A and B from 9.4 acres in 1948 and 11.7 acres in 1950.

The density of the burrows on Wards A and B was 57 per acre in 1948 and 52.5 per acre in 1950. This reduction in density resulted from an increase in the area of Ward A. The density of burrows in the area of expansion, which is included in Ward A, was much lower than in older sections of the town. Burrow density varies considerably throughout a prairiedog town, as well as from one town to the next (Merriam, 1901; Scheffer, 1937; Soper, 1938).

No reliable index can be used to estimate from the number of burrows the number of prairiedogs in a colony. The burrows are more or less permanent, but the number of prairiedogs is not; consequently, any such index would be unreliable, as can be seen by comparing the number of burrows per prairiedog throughout the three years of this study. There were 5.4 burrows per prairiedog in the study area in 1948, 9.6 burrows in 1949, and 4.2 burrows in 1950. These calculations do not mean that one prairiedog is limited even approximately to this number of burrows. The actual number of burrows used by a given individual depends primarily upon social factors and therefore varies according to the social organization of the group. A prairiedog may use any burrow within its home range.

The burrows of the prairiedogs were dug in soils of many different types and textures. Burrows were under large boulders, under roots of trees, and between large buried rocks. Most active burrows were on the floor of the canyon and on the terrace, where the fine-textured soil enabled the prairiedogs to construct deep, continuous passages. Generally, shallow burrows occurred in the marginal regions about the town, where the presence of rocks or impenetrable strata limited subterranean excavations. Prairiedogs apparently do not select certain soil types for their burrows, but begin to dig in almost any kind of soil. Since burrows cannot successfully be made in rocks or loose sand, only short exploratory tunnels are found in these soils. The lack of selection of soil types by prairiedogs agrees with Osborn's (1942) observations of prairiedog burrows in Oklahoma.

* * * * * * *

Literature Cited

ANTHONY, ADAM, and DARHL FOREMAN
   1951 Observations on the reproductive cycle of the blacktailed prairiedog (Cynomys ludovicianus). Physiol. Zool., 24 (3): 242–48.

BAILEY, VERNON
  1926 A biological survey of North Dakota. U.S. Dept. Agric., North Amer. Fauna, No. 49.
  1931 Mammals of New Mexico. Ibid., No. 53.
FOSTER, B. E.
  1924 Provision of prairie-dog to escape drowning when town is submerged. Journ. Mammal., 5: 266–68.
JOHNSON, GEORGE E.
  1927 Observations on young prairie-dogs Cynomys ludovicianus born in the laboratory. Journ. Mammal., 8: 110–15.
KELSO, L. H.
  1939 Food habits of prairiedogs. U.S. Dept. Agric., Circ. 529.
LONGHURST, WILLIAM
  1944 Observations on the ecology of the Gunnison prairie dog in Colorado. Journ. Mammal., 25: 24–36.
MERRIAM, C. HART
  1901 The prairie dog of the Great Plains. Yearbook of U.S. Dept. of Agric., 257–70.
OSBORN, BEN
  1942 Prairie dogs in shinnery (oak scrub) savannah. Ecology, 23: 110–15.
OSBORN, BEN, and PHILIP ALLAN
  1949 Vegetation of an abandoned prairie-dog town in tall grass prairie. Ecology, 30: 322–32.
SCHEFFER, THEO H.
  1937 Study of a small prairie-dog town. Trans. Kans. Acad. Sci., 40: 391–93.
  1947 Ecological comparisons of the plains prairie-dog and the Zuni species. Ibid., 49: 401–6.
SETON, ERNEST T.
  1929 Lives of game animals. New York: Doubleday, Doran and Co.
SOPER, J. DEWEY
  1938 Discovery, habitat and distribution of the black-tailed prairie-dog in western Canada. Journ. Mammal., 19: 290–300.
STOCKARD, A. H.
  1929 Observations of reproduction in the white-tailed prairie-dog (Cynomys leucurus). Journ. Mammal., 10: 209–12.
  1930 Observations of the seasonal activities of the white-tailed prairie dog, Cynomys leucurus. Papers Mich. Acad. Sci., 11: 471–79.
WHITEHEAD, L. C.
  1927 Notes on prairie-dogs. Journ. Mammal., 8: 58.

THE VICUÑA AND THE PUNA

Carl B. Koford

Museum of Vertebrae Zoology, University of California, Berkeley

INTRODUCTION

High in the central Andes of western America, above the limit of cultivated crops, lies a treeless pastoral zone, the puna. While scanning the bleak rolling grasslands of the puna a traveler may be startled by a prolonged screech. The cry attracts his gaze to a racing troop of fifty gazelle-like mammals, bright cinnamon in color—vicuñas! As they gallop up a barren slope he sees that a single large vicuña pursues them closely. The pursuer charges at one straggler, then another, as if to nip its heels. But suddenly the aggressor halts, stands tall with slender neck and stout tail erect, stares at a line of llamas in the distance, and whistles a high trill. Then it gallops away to join a band of several vicuñas, some obviously young, which graze close by. And these it follows as they file uphill, away from the approaching llamas and the somber Indian who trudges behind them.

The barefoot herdsman tells the traveler that the small retreating band is a family of vicuñas, females and young, protected from the rear by their leader, an adult male, and that the large fleeing troop consists entirely of males. The traveler sees no difference in the appearance of the supposed sexes. On reflec-

tion, however, he recalls that the vicuña is reputed to spend its life at heights reached only by the most lofty peaks in his own country, and that it bears a costly fleece which, centuries ago, clothed Inca royalty. To him the vicuña resembles, in size, actions, and habitat, the pronghorn antelope of the Great Plains. But inasmuch as the vicuña has a long neck and rather large feet, and lacks horns, it is more like a small humpless camel.

The vicuña and other llama-like ungulates, or lamoids, are of the camel family (Camelidae). The fossil record indicates that this family originated and developed in western North America, spread via land bridges into Asia and South America, and finally became extinct in its original homeland. In the Old World two species of camelids survive: the Bactrian camel and the Dromedary. These two (genus *Camelus*) are much larger in size than any of the four species that live in South America: the llama (*Lama glama*), alpaca (*L. pacos*), guanaco (*L. guanicoe*), and vicuña (*Vicugna vicugna*). Paleontological studies made in Argentina show that these species were distinct in the Pleistocene period. Yet, some authors consider that the llama was derived from the

Fig. 1. Trains of llamas carrying sacks of produce are a common sight in the puna. The tola shrubs and fescue grass are poor forage. Near Mazocruz, Puno; August 22, 1952.

guanaco, or that the alpaca is a hybrid between the llama and the vicuña, and others have expressed still different views (López Aranguren 1930; Gilmore 1950).

The four lamoids resemble one another in structure, but they differ in size, range, pelage, temperament, and usefulness to man. The llama is in common use by highland natives, from northern Peru to northern Chile and Argentina, for carrying burdens (Fig. 1). Trains of llamas march long distances over the inhospitable Andean highlands, descend westward to the arid coast, and eastward to the low humid jungles. Each llama may carry across its back a sack of dried potatoes or similar modest load. As do other domestic animals, llamas vary much in size, but the usual weight is about 200 pounds, considerably more than an alpaca and twice as much as a vicuña. Variable too is the color of llamas—black, white, and browns, in pure form, patchwork, or blends. Alpacas are usually of a solid color, as befits their primary use as producers of abundant long wool. Because they are capricious as compared to the stolid llamas, alpacas rarely carry burdens or walk great distances from their home pastures. They thrive only in Peru and Bolivia, and only at elevations above the level of Lake Titicaca, 12500 ft.

In contrast to the llama and alpaca, which have

been domesticated for hundreds of years, the vicuña and guanaco are wild animals. Guanacos look much like llamas, but the color of their pelage is a fairly uniform reddish-brown. Although guanacos range from the highest grasslands to sea level, they are rather scarce in the range of the other lamoids, and they are most abundant near the southern tip of South America. Adults are of little commercial value, but the hides of the very young are prized by the natives for the manufacture of robes.

The phylogenetic classification of the lamoids seems to be indeterminate, at least on the basis of structural criteria. Of the group, the llama and guanaco are the most similar, and the vicuña, although resembling the alpaca in some characteristics, is the most distinct. It has been variously classified as specifically, subgenerically, or generically separate from the other lamoids. The characteristic, unique among living artiodactyls, that best supports generic separation of the vicuña is its peculiar lower incisors. In *Vicugna* these are very long with parallel sides, enamel on but one face, and an open root (Miller 1924:2). In contrast to these rodent-like incisors, those of *Lama* are short and wedge-shaped, with enamel on both faces and a closed root.

Another characteristic that distinguishes the vicuña from the other lamoids is the bib of long hair which

FIG. 2. At the approach of an intruder, an adult male vicuña postures with head and tail high, making himself appear as large as possible. Huaylarco, Department of Arequipa; March 11, 1952.

hangs from its brisket. This hair is white, like the rest of the under parts, and contrasts strongly with the dark buffy-brown hue of the upper parts (Fig. 2). Often differences in the tone of the dorsal color enable an observer to recognize individual vicuñas in a group. Some color differences are due to age. But apparently the sexes are alike in color.

As their wool is the finest known, and as their hides and meat are prized by the natives of the Andean highlands, vicuñas have been hunted for centuries. In many parts of its range the species is threatened with extinction. Man, therefore, has a moral obligation to take action to conserve the species, and an economic interest in perpetuating the supply of vicuña wool. Further, the vicuña is of great scenic value as the most conspicuous wild animal of the dreary puna, and it is unexcelled in lithe beauty.

Until now, no methodical study of the life of the vicuña has been made. Nearly all published summaries concerning vicuña habits are based on repetitions of the statements, many true but some false, of J. J. von Tschudi, a naturalist who traveled in Peru more than a century ago and who published a large volume on the Peruvian fauna (1844-1846, in German). The essential facts heretofore known about vicuñas and the other lamoids are well summarized by Cabrera & Yepes (1940:256-269) and by Gilmore (1950:429-454). The agricultural knowledge of lamoids have been reviewed by Cardozo (1954), Romero (1927), and Maccagno (1932). The only significant account of recent observations of wild vicuñas is that of Pearson (1951:161-168).

The present study is primarily a survey of the behavior of the vicuña in its native habitat, with emphasis on social relationships. Ecology, distribution, history, utilization, and management have been investigated in moderate degree in order to fill out a broad picture of the animal, its environment, and its relations with man.

The basic reason for this kind of study is that a sound program for the conservation and management of any wild animal must be based on scientific knowledge of the habits and environment of the species. This paper is a contribution to that knowledge. But further, detailed accounts of the actions of wild animals are fundamental to studies of comparative behavior, especially parallel and convergent evolution in behavior. And, as studies of morphology have not clearly determined the phyletic relationships among the lamoids, studies of behavior may clarify our understanding of these relationships.

The vicuña is an ideal subject for study by observation in its native surroundings. Groups of vicuñas may be watched for hours with relative ease, for they dwell on open grasslands, and they are large, gregarious, active by day, and not overly wary. Important to the investigator was the fact that the Andean highlands are healthful in climate and accessible by automobile, yet rather sparsely populated by man.

I am deeply grateful to an anonymous donor to the University of California, who provided funds for these studies. In the field I was accompanied and aided by Mary Koford. Assistance in making arrangements for the field work and hospitality were generously provided by Jávier Ortíz de la Puente, Sr. and Sra. Francisco Paredes, Dr. and Mrs. David P. Duffie, Dr. Guillermo Mann, William Rudolph, Paul Crum, Erik Neilsen, and Miguel Urdangarín. Identification of grasses was undertaken by Dr. John Reeder, Yale University; that of other plants by Dr. Helen Sharsmith, University of California. Drs. A. H. Miller and O. P. Pearson, Museum of Vertebrate Zoology, read the manuscript and made numerous helpful suggestions in the course of the work.

METHODS

Early in April, 1951, I first saw vicuñas in the wild. On this occasion, several groups were observed in southern Peru near the highest point of the road that runs eastward from the city of Arequipa to the town of Puno, which lies at the edge of Lake Titicaca in the department of Puno. After several weeks of search for a suitable study area in Puno, we established camp just east of the highest pass on the road from Tirapata to Limbani, at the extreme head of the Amazon (Inambari River) drainage, about 65 mi north of Lake Titicaca. This place, Aricoma, is situated at an elevation of about 15500 ft in an area of damp grasslands and clear lakes, close to glacier-flanked peaks. There we remained for a period of 5 months (May 24 to October 16). In May, and again in August, we spent several days at a vicuña ranch, Calacala.

About the middle of the year Indians and livestock began to interfere with studies at Aricoma, and these annoyances increased. Therefore, we sought a less disturbed area in which to study birth, mating, and intergroup relationships. The site chosen was Huaylarco, which is located near the Arequipa-Puno road at a point about 55 mi west of Puno, at an elevation of 15000 ft, in an arid region of rolling hills and pampas. There we camped for a period of 6 months (November 7, 1951, to May 3, 1952).

On occasions we explored various other roads in southern Peru and briefly observed vicuñas in the vicinity. The principal sites of these observations were: the road over the cordillera between Chalhuanca and Púquio, about 200 mi northwest of Arequipa; between Macusani and La Raya, 120 mi north northwest of Puno; along the Crucero to Poto road, about 50 mi north of Lake Titicaca; near the road from Ocoruro to Imata, the latter 65 mi west of Puno; in the vicinity of Laguna Salinas, 25 mi east of Arequipa; near Lake Loriscota, 60 mi south of Puno; and, finally, adjacent to the road between Mazocruz and Tarata. This last road we traveled while driving from southern Peru to northern Chile.

In August, 1952, we entered the northernmost province of Chile, Tarapacá. The following month we camped for a week near Caritaya, 80 mi north northeast of Iquique. But we found vicuñas scarce there, so we drove on southward to the province of Antofagasta. There we stayed 5 weeks (October 2 to November 6, 1952) in the volcanic region about 55 mi northeast of Calama, and we saw vicuñas nearly every day. The span of time spent in the vicuña zone of the Andes was 18 months (April 1951 to November 1952), but as much time was spent in other tasks, only about 10 months were spent in field studies of the vicuña.

The principal method of study was to watch undisturbed groups of vicuñas from distant points with binoculars and a 20x telescope. Watching from long range, rather than from a close blind, is advantageous in that the observer can reach or leave his post without disturbing the animals, minor movements of hands or notebook pass unnoticed, and events occurring within a large area can easily be seen. Stone walls and native huts were often used as blinds, but where roads or terrain permitted I drove my car, a jeep with closed metal top, to the observation post. The car made a spacious and comfortable blind with cameras, maps, and other bulky equipment at hand.

An intimate acquaintance with the normal composition and behavior of many groups of vicuñas was necessary to the interpretation of actions observed. For understanding behavior the most valuable data, but the most tedious to obtain, were observations of a few well-known groups for long periods on successive days. In adjacent bands on successive days many factors of the environment that vary with place or time of year, such as weather, vegetation, and water, are relatively constant. Thus, the factors that cause variations in behavior can be recognized.

In the puna evening the cold is penetrating and writing or other work with the hands painful, except within a heated shelter. We found that a small coach trailer of standard design, modified to use kerosene

for heat and a storage battery for light, made a comfortable home. The trailer remained at one site for many weeks at a time, while the jeep alone was used for short, rapid, or difficult trips.

DISTRIBUTION, POPULATION, AND MORTALITY

OCCURRENCE

To try to outline the entire range of the vicuña is presumptuous. Naturalists have reached few parts of the zone where vicuñas dwell. Specimens are few, specific records are scanty, and allusions by natives and tourists are indefinite. I have attempted only to sketch the approximate limits of the geographic range of the vicuña (Fig. 3). For the most part, the range outlined follows the elevation contour, judging the limit of the vicuña zone by the level of the nearest known occurrence. Many of these levels were determined by observations made from some of the dozen good dirt roads that cross segments of

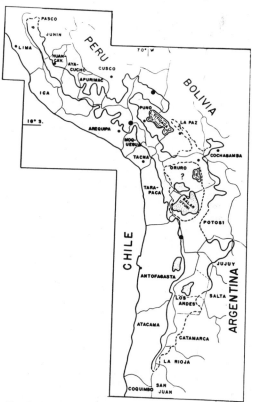

FIG. 3. Outline map of the present range of the vicuña. The limits of distribution are shown by the heavy solid line, broken portions indicating where the limits are most doubtful. The three large black dots indicate the principal areas of study, from north to south: Aricoma, Huaylarco, and northeastern Antofagasta. Smaller dots show locations of cities bearing the same names as the departments or provinces in which they are located.

the vicuña range. The accurate determination of the range of the vicuña and of the population densities in different regions would be a difficult but valuable project for future investigators.

The limits of the present and former range of the vicuña are discussed in the appendix. In general, the vicuña now occurs over an area 1300 mi in length, between the latitudes of 10° and 29° South. Although Allen (1942:410) considered the vicuña to be a vanishing species, perhaps already extinct in Chile, my review of the records suggests that the overall range may not have diminished within historical times. Doubtless there has been a decline in vicuñas in local areas heavily used by man. But in the last century or more, it appears that the activities of man and domestic animals have not greatly modified the general distribution of vicuñas. Insofar as stability of overall range of occurrence is a valid criterion of welfare, the vicuña does not seem to be approaching early extinction.

In elevation, vicuñas normally range as high as 16000 ft. Although they occur as low as 12000 ft, I estimate that at least two-thirds of the population lives above the 14000-ft level. In southern Peru I found vicuñas to be common only at sites above 14500 ft, and most abundant at about 15000 ft. The limits of range are apparently determined by availability of food and by freedom from disturbance, not by the rarity of the atmosphere.

HABITAT

Climate.—In the Southern Hemisphere summer commences in December, autumn in March, winter in June, and spring in September. In the puna summer is the wet season, as in the grasslands of central North America. At many puna localities more than ¾ of the annual rainfall of about 25 inches falls in summer. But because, even in summer, the air is cold at high elevations, summer in the puna is a time of snow and hail, and it is locally known as "invierno" (winter). At the end of summer plants are green and water is widespread. Then during autumn, vegetation and soil become dry, and drier yet in winter when skies are often cloudless.

In the United States spring is typically a season of awakening from dormancy for both plants and animals. Plants grow rapidly as the sun warms moist soils. Most animals produce their young. But in the puna spring brings warmth to ground already dry. Green food is scarce; few animals bear young. Then, with the coming of summer rains and snows, moisture and warmth favor abundant plant growth. Vicuñas and most other puna mammals are born in summer, although for humans the weather is then miserable. The climate of the Peruvian puna has been described by Bowman (1916), Weberbauer (1945), and Pearson (1951). Probably similar is the puna climate of Bolivia, Argentina, and Chile, although south of Lake Titicaca the highlands are generally more arid than in Peru.

The climate of the mountains northeast of Lake Titicaca, where studies were made at Aricoma on the eastern slope of the Cordillera de Carabaya, is less typical of the vicuña zone. This region is comparatively humid, for the trade winds which sweep westward over the damp Amazon jungles are lifted by the eastern slopes of the Andes and here drop their moisture. On the windward slopes of high peaks spectacular glaciers descend to a level of about 16000 ft. In 1951 at Aricoma, in the "dry" season the sky was cloudy nearly every day, evening fog was common, and snow fell in every month save July. But in spite of this dampness the grass, viewed from a distance, appeared brown and lifeless in winter and spring.

Temperature.—One might expect that even in the tropics the nights would be very cold at an elevation of nearly 3 mi. At Aricoma in winter and at Huaylarco in summer, I kept fairly complete records of daily maximum and minimum temperatures. With few exceptions, at neither locality did the nightly minimum fall below 18° F and on many occasions the minimum was above freezing. Winter minima were higher in humid areas, such as Aricoma, than in dry areas, such as Huaylarco. At Aricoma it did not freeze on some nights of the coldest months of the year, while at Huaylarco I recorded temperatures as low as 5° F in November, a few months after the coldest time of year. For a station west of Lake Titicaca at an elevation of 15500 ft, Bowman (1916:161) gives the absolute minimum for one year as 0.2° F (June). Evidently, then, below zero temperatures are rare in the puna.

For the same high station Bowman gives the maximum temperature for one year as 70.4° F (December). The highest that I recorded in the puna of Peru was 61° F (Aricoma, September). It seems, then, that in the vicuña zone the highest and lowest temperatures are much less extreme than in many cities of central North America. The yearly temperature range of about 70° F in the puna is about a third less than the yearly range at Chicago.

Neither is the daily temperature range in the puna extreme. The greatest daily range that I noted was 50°F (Huaylarco, November). For one month (January) the maximum daily range at Huaylarco was only half as great. Daily ranges are generally higher in spring than in summer, for in the latter season the sky is frequently overcast. Daily maxima in summer may be no higher than in winter.

As compared with many grazing mammals of North America, vicuñas need withstand neither extreme cold, extreme heat, nor extreme variations in air temperature. But of great effect at high altitude is solar radiation. In thin air bodies absorb heat rapidly from the sun, and they lose it quickly when the rays of the sun are blocked. Moreover, wind, dry air, and low atmospheric pressure are all factors that increase the rate of evaporation. Puna organisms, therefore, must be able to cope with rapid changes in solar heat and with rapid rates of drying. Doubtless the dense fleece of the vicuña is of great importance in protecting the animal from weather extremes.

Water distribution.—At Huaylarco streams were dry by late October, the middle of spring. Rains commenced about Christmas. Several days later streams flowed fast and lake levels were rising. One lake spread and covered many acres of a pampa where vicuñas had grazed a few days earlier. In December numbers of Indians mined the thick white layer of salt in the bed of Laguna Salinas, east of Arequipa. But in January this bed of salt was replaced by a wide shallow lake, colored pink by several thousand Chilean flamingos. After the middle of February storms were few at Huaylarco, and by the end of April many streams were dry. Even at Aricoma, a far wetter area, many streams, seeps, and ponds dried up during the winter and spring.

Vegetation and food.—In most of the region inhabited by vicuñas the dominant plants are perennial bunch grasses. The larger tufts are about 1 ft in diameter but, because of heavy grazing, they are usually less than a foot in height. On the better grazing sites at Huaylarco the crowns of plants covered no more than 50% of the ground surface. Groups of low evergreen shrubs, or "tola," alternate with grasslands in various proportions, from isolated bushes to continuous stands. The principal species of tola are *Lepidophyllum rigidum, L. quadrangulare,* and *Baccharis microphylla* (all family Compositae). On occasions I saw dozens of vicuñas and hundreds of

alpacas grazing on pampas where tola was dominant, though seldom browsed. In the areas of study, the lower altitudinal level of occurrence of vicuñas was often noted to be near the lower limit of *L. rigidum.* In the higher vicuña country mat-forming plants, cushion-plants, and rosette-plants, well-adapted to resist desiccation, are common. Many aspects of the phytogeography of the puna are discussed by Weberbauer (1945).

The principal food of vicuñas is grass. At Huaylarco, away from streams, the grasses most sought by vicuñas are *Calamagrostis brevifolia* and *C. vicunarum.* These are normally kept cropped to a height of 1-2 inches by grazing animals. They thrive well in the protection of tall coarse grasses, thorn bushes, and rocks, and near the edges of plant mats and dung piles. Commonly vicuñas feed on the grasses that border streams or grow on islets. Here the main forage grass is *Festuca rigescens,* which grows in compact blue-green tufts (Fig. 4). Pearson (1951:167) reported that this fescue was an important food of vicuñas at Caccachara, 60 miles southeast of Huaylarco. He also noted that *C. nitidula* was favored. Several other species of the genera *Calamagrostis, Festuca, Poa* and *Stipa* were grazed, in varying degree, by vicuñas at Huaylarco.

In many areas where vicuñas forage the most conspicuous plants are tufts of coarse bunch grasses.

FIG. 4. While a vicuña family stands in a cold running stream, one female feeds at a clump of bunch grass *(Festuca rigescens),* another urinates in the water, and a third spits at a yearling. The three females at the right appear to be pregnant. Huaylarco, Arequipa; February 25, 1952.

These are principally the sharp-tipped *Festuca orthophylla* and *F. dolicophylla*. Although these grasses are rarely nibbled by vicuñas, their roots are important in retarding wind and water erosion. And when snow blankets the ground, the stiff tufts break up the snow surface and expose adjacent herbage to grazing animals (Fig. 5). Further, the tall tough grasses protect short succulent plants from sun, wind, and the hoofs of livestock. So heavy is the grazing of livestock in some portions of the Peruvian highlands, however, that even the coarse grasses are often mowed to a stubble a few inches in height (Figs. 6 and 7).

Succulent forbs, mostly annuals that rise not more than half an inch above the ground surface, are relished by vicuñas. At Huaylarco a small legume, probably *Astragalus peruvianus*, was common and apparently sought by vicuñas. They pawed this plant from the soft ground and ate the upper roots, about a quarter inch in diameter, as well as the tiny leaflets and blue flowers. Near the end of the rainy season, short forbs grow in abundance between tufts of grass.

Viewed from a distance, many sandy pampas where vicuñas feed appear to have no vegetation (Fig. 8). Upon closer inspection, however, one finds that the ground is sprinkled with rosette plants, few larger than a pocket watch. In the early months of the year

the even distribution of these plants is obvious, for each rosette bears a white flower. Apparently because of this white speckling, one plain where vicuñas dwell is named the Pampa de Confital (Candy Plain). Most of the rosettes that vicuñas eat are species of *Nototriche* (family Malvaceae). On average grazing sites these plants grow in a density of about five per square yard. On poor grazing sites plants cover as little as one thousandth of the ground surface. The most conspicuous plant that dots many barren hills is a large rosette with hairy aromatic leaves, *Werneria* (family Compositae), but this does not appear to be eaten by vicuñas.

At many moist sites in puna grasslands there are spongy bright green carpets, some more than 100 feet in width, of *Distichia muscoides* (family Juncaceae). Vicuñas may crop the shoots of these soaking wet mats. At drier sites pale-green carpets of *Pycnophyllum* (family Caryophyllaceae), rarely more than 5 ft across, cover patches of rocky soil and lend a green aspect to otherwise naked ground. Although vicuñas seem not to crop these dry mats, they commonly eat blades of succulent grasses that grow up through them.

Vicuñas and other lamoids deposit their dung in discrete piles. Most of the plants that grow on or close to these piles are conspicuously different from the plants of the surrounding pasture (Fig. 9). Early in the wet

FIG. 5. When snow covers the ground vicuñas feed on uncovered vegetation or thrust their muzzles down through the soft snow. These three are troop males. Huaylarco, Arequipa; March 4, 1952.

FIG. 6. Most of the members of male troops are immatures of about the same size. These six are probably yearlings. Livestock has grazed the coarse fescue in the foreground to a low stubble. Huaylarco, Arequipa; January 11, 1952.

FIG. 7. In feeding on short herbage, a vicuña ambles forward slowly, muzzle close to the ground. This pregnant female, damp from recent bathing, feeds between thorn bushes hedged by passing livestock. Huaylarco, Arequipa; February 27, 1952.

season grass shoots grow up through the edges of dung piles, and vicuñas crop these shoots. But the majority of plants that grow on dung are low annual forbs. Late in the wet season brilliant green circular spots mark the location of dung heaps, and on many barren hills these spots are the only greenery that can be seen from a distance. Near the crests of dung piles at Huaylarco, a nettle about an inch high, probably *Urtica flabellata* (family Urticaceae), grew thickly. Apparently vicuñas did not eat this, although they seemed to nibble the green carpet of another forb, *Arenaria* (family Caryophyllaceae), that grew about two inches high around the edges of the mounds. In general, the forbs on dung piles are not relished by vicuñas. What, then, is the advantage of concentrating the dung and urine? One possibly advantageous effect is that the excretory chemicals influence the vegetation on but a small fraction of the pasture.

Food and distribution.—The upper altitudinal limit of the vicuña range is apparently set by the lack of herbage at higher levels. Few plants grow in the cold desert above 16000 ft elevation. Surrounding Lake Titicaca, below the normal range of the vicuña, and bordering the puna on the eastern slopes of the Andes, is a belt of rainy-green shrubs and grass steppes. The grass in this zone is much taller than the bunch grass of the puna, and there is little bare ground between tufts. Even though human culture

FIG. 8. On sandy pampas which appear barren from a distance, vicuñas find sustenance in small rosette plants. Cerro Chucca, near Pati, Arequipa, in the background. April 16, 1952.

FIG. 9. Although grass covers most of the surrounding ground, old dung piles are carpeted with short green forbs in summer. Here an adult male voids. The rocks are a temporary fireplace made by an Indian atop his fuel supply. Huaylarco, Arequipa; February 27, 1952.

did not prevent vicuñas from frequenting these lower grasslands, they might be unsuitable for grazing, for vicuñas feed mainly on short perennial herbs. Moreover, tall, thick grass would interfere with running. Nevertheless, a herd of domestic vicuñas seems to thrive near Lake Titicaca in the grass steppe zone at an elevation of 13000 ft.

In arid parts of the vicuña range it is probably also the lack of suitable food, principally green herbage, that determines the lower limit of occurrence of vicuñas. In Peru most of the western slopes of the

Andes are barren. In northern Chile there is little grass cover, away from streams, below the 12000 ft level. Compared to the dry western slopes at somewhat lower elevations the puna is moist, and some green food is available there at all seasons. It is this greenery of the higher pastures that attracts shepherds and their flocks from lower elevations during the dry winter.

That green herbs are the preferred food of vicuñas was apparent from observations of feeding, and confirmed by the examination of the stomach contents of three animals. Furthermore, in dry regions vicuñas are most abundant where the soil is moist and the vegetation succulent. Where vicuñas feed on small hairy plants, such as *Nototriche*, usually only the newest leaves are eaten. Roots that vicuñas eat are carrot-like in texture, not woody. But on the other hand, very dry food is sometimes taken. Many times I saw a vicuña chew down a single dry stalk of coarse fescue. One, while crossing the dusty bed of a dried up pond, picked up a wad of desiccated algae in its mouth and chewed this several times before dropping it.

In order of decreasing requirement for succulence in food, I would rank the lamoids: alpaca, vicuña, llama, and guanaco. The fact that this is also the order of increasing geographic range suggests that the seasonal availability of green food and the differing ranges of tolerance of the lamoids for dry food are important factors determining the limits of their distribution.

Soil.—The geology of the puna is complex. Evidently puna soils are fertile, but coolness and aridity limit plant growth. Damp sites are verdant. In the study areas the soil is pale in color, sandy in texture. Even in grasslands the soil is generally so loose that the footprints of birds and mice can be followed. Gravel is commonly mixed with the sand, and on barren pampas wind keeps the scattered gravel exposed as a moderate erosion pavement. Rocky slopes, cliffs, and rockslides are common. In walking and running on slopes and rocks, the lamoids are probably more adept than the Old World camels. The fact that the feet of lamoids are more deeply cloven than the feet of the other camels may be an adaptation for locomotion on hard or rocky ground.

POPULATION COMPOSITION

Specific meaning of terms.—Vicuñas are gregarious. They gather into two basic kinds of groups, here distinguished as *bands* and *troops*. A typical band is comprised of one adult male with several females and their young of the year. These *band females* and young together constitute the *family* of the male (the *family male* or *band male*). Year around the family of vicuñas occupies a certain piece of ground, the *territory*. The territorial boundaries are defended by the *resident male* (a family male on his own territory) against encroachment by alien males and *displaced bands* (family bands that are not on their own territories). Areas that are not defended are called *unclaimed* ground. It is well to note that the vicuña family is not a "harem," in the sense that the term has been used in studies of fur seals and red deer. In those species the harem exists only in the breeding season, and the females have no attachment to the male or his territory, except through having been captured by him and prevented by him from leaving (Dice 1952: 242-244). Males that are not members of bands gather into leaderless *troops* consisting of many individual *troop males*. Most of these males are yearlings and two-year olds. In the sense that they have not yet mated, most troop males are "bachelors," but as this term has special connotation for a certain age group of fur seals, I have avoided it.

In the study area, most vicuñas were born in March. The *young juveniles* become *old juveniles* at about six months of age (October 1, for convenience) and *yearlings* on January 1. Yearlings and two-year olds are recognized as *immatures* as contrasted to juveniles and adults.

Sex ratio.—In a polygynous society one might expect to find that females outnumber males in the total population. Because of the difficulty of quickly distinguishing the sex of a vicuña, the simplest method of determining the relative numbers of males and females is to compare the total number in bands (excluding juveniles, which may be of either sex) with the number in troops. For a 1:1 sex ratio and an average band size of 5 members, 42% of the non-

juvenile vicuñas in a region would be troop males. This percentage would be lower for smaller bands, higher for larger. Does approximately such a proportion exist?

While traveling by auto through the open vicuña country, one can count the numbers of vicuñas seen in groups and attempt to classify the groups. With fair assurance, groups containing juveniles are family bands, and groups that have 15 or more members, but no juveniles, are male troops. Unfortunately, it is difficult to determine whether smaller groups that have no juveniles are bands or troops. If all these groups are assumed to be bands, the results of the census will give a very low percentage for males. Making this assumption, in one count of 315 vicuñas the indicated proportion of males was only 19%. Doubtless many of the small groups that were counted as bands were in fact troops.

Another factor that lowers the validity of single counts from the road for determining sex ratio is that large troops are encountered by chance. Thus, too many or too few troop males might be assigned to the census area. The larger the area, the less the error from this source. At best, a single count from the road provides a gross index of comparison between areas, not a measure of sex ratio.

The study area at Huaylarco was approximately 1000 acres. On this area at a time when the total number of resident vicuñas in bands was 125, I saw a single troop of 75 males. These males, plus 17 known band males, made up 46% of the total number of vicuñas present (200). At that time (early January), 30 new yearlings of both sexes were still members of bands. If 8 of these yearlings were males, as is not improbable, the ratio of all males to females on the area was 100:100. In general, then, the sex ratio in vicuñas is approximately balanced, but because of the difficulty of recognizing small troops and of assigning troops to a certain census area, the exact sex ratio of a population cannot easily be determined.

Doe:fawn ratio.—It is also difficult to distinguish between old juveniles and yearlings, and between yearlings and two-year olds. Therefore, no reliable ratio of breeding females to juveniles (the "doe:fawn ratio" of wildlife management workers) was obtained for vicuñas, except in the study area at Huaylarco. For that area there were 46 juveniles per 100 reproductive females (two years of age and older) in April, 1952, when nearly all young had been born.

For the 20 bands studied at Huaylarco, a summary of the age composition is given in Table 1. Each band is designated by a letter and number indicating the location of its territory. These numbers are distances in kilometers from Arequipa on the road to Puno. In Table 1, for each band the number of individuals, of yearlings, and of juveniles is tabulated by half-month periods. The numbers given are the maximum counts made on at least two days of each period, so that temporary changes in composition of bands are not included. In the few cases where individuals were known to have left one band and

TABLE 1. Composition by half-month periods of 20 bands at Huaylarco.

Band	Month 1951				Month 1952							
	November		December		January		February		March		April	
W. 139.4			9-0-0	9-0-0	9-0-0	10-1-0	9-0-0	10-0-1	11-0-2	1-0-0	2-0-0	2-0-0
W. 139.0			2-0-0	2-0-0	2-0-0	2-0-0	3-0-0	3-0-0	2-0-0	11-0-2	10-0-2	
S. 139.0			8-1-0	8-1-0	8-1-0	10-1-0	7-0-0	9-0-2	11-0-2	13-0-4	15-0-4	16-0-5
E. 139.0	11-5-0	11-5-0	10-4-0	10-4-0	10-4-0	8-0-0	6-0-0	9-0-3	10 0-4	10-0-4	10-0-4	10-0-4
E. 138.8					3-1-0	2-0-0	2-0-0	1-0-0	3-1-0	2-0-0	3-1-0	3-1-0
W. 138.6							1-0-0	1-0-0	1-0-0	1-0-0	1-0-0	
E. 138.4	6-1-0	6-1-0	6-1-0	6-1-0	6-1-0	4-0-0	2-1-0	2-1-0	2-1-0	2-1-0	2-1-0	
E. 138.2	5-0-0	5-0-0	5-0-0	4-0-0	3-0-0	4-1-0	4-2-0	5-3-0	5-2-0	5-2-0	5-1-0	5-1-0
W. 138.2	8-3-0	7-2-0	6-2-0	6-2-0	6-2-0	4-0-0	4-3-0	4-3-0	4-3-0	4-3-0	4-3-0	4-3-0
W. 137.7	8-3-0	10-4-0	11-5-0	11-4-0	13-5-0	11-5-0	9-2-0	8-2-0	7-1-0	7-0-0	7-0-0	7-0-0
E. 137.5	16-5-0	16-5-0	10-4-0	15-5-0	15-5-0	14-4-0	3-2-0	3-2-0	3-2-0	3-2-0	3-2-0	
W. 137.3	5-2-0	5-2-0	4-1-0	4-1-0	3-1-0	3-1-0	2-0-0	2-0-0	3-1-0	3-1-0	5-0-2	5-0-2
E. 137.0	4-1-0	4-1-0	4-2-0	4-1-0	4-1-0	4-1-0	4-1-0	7-1-0	12-0-4	11-0-4	11-0-4	
S. 135.6							5-0-0	4-0-0	7-0-1	10-0-2	9-0-2	9-0-3
N. 134.5	10-2-0	10-2-0	8-2-0	9-1-0	9-1-0	9-1-0	11-0-0	9-0-0	20-0-4	22-0-6	23-0-8	23-0-8
N. 133.5					1-0-0	1-0-0	1-0-0	3-2-0	1-0-0	1-0-0	1-0-0	2-0-0
S. 133.2			12-4-0	12-4-0	12-4-0	10-3-0	12-4-0	11-1-0	14-1-1	23-0-6	26-0-9	25-0-9
N. 133.0								7-0-0	6-0-2	5-0-0	7-0-1	
S. 133.0	13-4-0	15-5-0	14-5-0	14-4-0	18-5-0	15-2-0	13-0-1	13-0-1				
S. 132.8	5-0-0	7-0-0	6-0-0	5-0-0	5-0-0	6-0-0	4-0-0	3-0-0	6-0-1	6-0-1	6-0-1	
Totals:												
All	91	108	116	119	125	124	107	119	139	144	142	86
Adult ♂♂	11	12	15	15	17	18	20	20	20	20	19	11
Non-juv. ♀♀	54	65	70	76								
Adult ♀♀					78	85	74	75	83	82	79	48
1951 juveniles	26	31	31	28								
1952 yearlings					30	21	12	15	11	9	8	5
1952 juveniles							1	9	25	33	36	22
Average band	8-2½-0	9-2½-0	8-2-0	8-2-0	7½-2-0	7-1-0	5½-1-0	6-1-½	7-½-1	7-½-1½	7½-½-2	8-½-2
Ratios:	Per 100 non-juvenile females				Per 100 reproductive females (older than 1 year)							
Adult ♂♂	20	18	21	20	22	21	27	27	24	24	24	23
1951 juveniles	48	48	44	37								
1952 yearlings					38	25	16	20	13	11	10	10
1952 juveniles							1	12	30	40	46	46

*Numbers indicate: total in band—old juveniles (in 1951), or yearlings (in 1952)—young juveniles (born 1952).

joined another, the loss and gain are ascribed to separate half-month periods. The totals for each sex and age class, and their proportion to 100 non-juvenile or reproductive females, are given in the lower part of the table. At Huaylarco at the end of 1951, the ratio of juveniles to older females (including yearlings) was 37 to 100 for 15 bands of known composition. Excluding four bands that did not contain juveniles, the ratio was 47 to 100. This ratio was much higher than one based on miscellaneous roadside counts of bands at various localities in southern Peru. On trips made during the period from April to September, 1951, I counted the members of 30 bands that contained juveniles. These counts gave a ratio of only 17 juveniles to 100 older females.

On some trips I counted all the vicuñas, both in bands and in troops, that I could see from the car, distinguishing juveniles from older vicuñas in the counts. These counts seemed to indicate a very low reproductive success. For example, one day in May, 1952, in a 55-mi segment of road northeast of Púquio I counted 534 vicuñas of which only 7% were juvenile. And one day in October, 1952, between Ocoruro and Pati I counted 333 vicuñas of which but 5.5% were juvenile. Assuming a balanced sex ratio, these counts give ratios of juveniles to older female vicuñas of only 14:100 and 11:100.

The first time that I drove through the Huaylarco area, in April, 1951, I counted 186 vicuñas of which approximately 6% were juvenile. Another count in October gave 176, with nearly the same proportion of young. But early in November, a more careful census indicated that 14% of 270 vicuñas in the Huaylarco area were juvenile. A few weeks later, after intensive study, I found that of the vicuñas in resident bands, 27% were juvenile. Now, including males in troops and assuming a sex ratio of 1:1, the proportion of juveniles to all vicuñas at Huaylarco at that time (December) was 18%. This percentage is much higher than any obtained by a single count. Evidently single counts give misleading results; they usually indicate a much lower percentage of juveniles than are actually present. Such counts may be of value, nevertheless, as indexes for comparing reproductive success in the same population in different years.

POPULATION DENSITY

Maximal densities.—It is not difficult to count the vicuñas in resident bands of a given area. At Aricoma on the best sites, where good pasturage was nearly continuous, the average density of band vicuñas in winter was about 1 animal per 10 acres. At Huaylarco, in early February, 1952, just before the first young were born, the average density of band vicuñas on the study area of about 1000 acres was 9.5 acres (horizontal measurement) per animal. Some of the smallest territories supported 1 vicuña per 5 acres for a period of many weeks but such high densities were exceptional.

The number of males in a troop can also be counted easily. But because a troop wanders far, its presence on a census area is fortuitous and the proper acreage to assign it is doubtful. I made many counts, however, that included both bands and troops. For example, from the road at Huaylarco, I once (November 8, 1951) counted all the vicuñas seen in a strip 30 chains wide and 11 km long. For the total of 270 vicuñas counted, the average density was approximately 6.5 acres per animal.

These average densities are probably near the carrying capacity of the vicuña range. In the Andean highlands the grazing capacity of good sheep pasture is generally taken as 1 hectare (about 2.5 acres) per animal. Allowing for the facts that vicuña range is generally less productive than good sheep range, that a vicuña is larger than a sheep, and that livestock grazes the vicuña range during part of the year, a density of 1 vicuña per 10 acres may be all that the range can sustain permanently.

Variation in population density.—Of course if unsuitable or unclaimed ground is included in the census area estimated densities are much lower than when only territories are included. Pearson (1951:121) estimated a density of 120 acres per vicuña on a continuous area of approximately 15 sq mi at Caccachara, in southern Puno. As the topography was rough, doubtless much of the included area was not pastured by vicuñas. In arid regions where green food is scant population densities may be much lower than those illustrated. In the Antofagasta study area the population density was estimated at 50 per 100 sq mi— more than 1000 acres per vicuña. Probably similar low densities predominate in most of the vicuña range in Chile, Argentina, and southern Bolivia.

From one area to another the concentration of vicuñas differs widely. I counted 428 vicuñas while driving 48 miles of the road between Chalhuanca and Púquio, Peru, in May, 1952. Considering the area searched, the density was about 35 acres per animal, including juveniles. In driving from Ocoruro southward to Pati in October, 1951, I counted 309 vicuñas in a strip 50 mi long and 0.5 mi wide. The average density was 50 acres per animal, but much of this strip was unsuitable for vicuñas. In one 7-mi segment of the strip the density was 20 acres per vicuña. Driving from Crucero to Poto in the winter of 1951, I saw an average of less than 2 vicuñas per mile along the two most favorable stretches, 14 and 25 mi in length. The density there was in the neighborhood of 160 acres per animal. For large areas where there are many vicuñas, the average density in southern Peru is estimated to be less than 50 acres per animal.

Counts from an automobile without regard to the width of strip surveyed give a gross index of vicuña population densities. Using this method I saw an average of 15 per mi while driving 5 mi across a tola flat at Chilliculco (Puno, Peru) on May 1, 1951; an average of 4.5 per mi while driving 18 mi across semi-desert pampas near Lake Loriscota on May 3, 1951; and 5.9 per mi while driving 19.5 mi from Imata to Pati on January 27, 1952. Such counts may give misleading results unless standardized as to length of run, time of day, and other factors. Unknown previous disturbances, such as the passing of a dog, can greatly influence the number of vicuñas in an area.

For Peru alone the total number of vicuñas has been stated to be approximately 1,000,000 (Cardozo 1954:194, from other source). In my opinion this figure is much too high. The perimeter of the vicuña range in Peru (Fig. 3) includes about 38000 sq mi, but mountain peaks, barren soils, and human activity within this region make an estimated half of it unsuitable for vicuñas. Thus, in Peru there may be only 19000 sq mi, about 12,000,000 acres, of land inhabited by vicuñas. If there were 1,000,000 vicuñas in Peru the average density would be about 12 acres per animal. But such high densities exist only at favored and restricted sites. On the basis of a more probable average density of 50 acres per vicuña, there are slightly less than 250,000 vicuñas in Peru.

The ground included in the outline of the entire range of the vicuña (Fig. 3) has an area of approximately 100,000 sq mi, two-thirds of the area of California. If half of this ground is occupied by vicuñas, they utilize 32,000,000 acres. About 60% of the vicuña range is in the dry and salty zone south of the latitude of Salar Uyuni, Bolivia (latitude 20° S), where average population densities are thought to be low, perhaps 1000 acres per vicuña. On the basis of these estimates, there are approximately 400,000 vicuñas in all, more than half of them being in Peru.

MORTALITY

Death rate.—At the end of the season of birth many adult females at Huaylarco were without young. Yet a few months earlier nearly all females except yearlings appeared to be pregnant. The ratio of only 46 juveniles per 100 reproductive females in April (Table 1) suggests that about half of the young perished during the late fetal or infant stage. The cause of this apparently high mortality was not discovered. Probably productivity was even less at other localities, where bands were smaller and the proportion of young to adults was lower than at Huaylarco.

In the wild, dogs, Indians, and condors probably kill many newborn young. But on the vicuña ranch at Calacala, where these predators are of minor importance and stillbirths are rare, reproductive success is equally low. There, 200 breeding females produce about 70 juveniles, according to Sr. Paredes. He blamed foxes for the high loss of young. I suspect that, in some years or under poor range conditions, fertility may be low among the younger breeding females (two-year-olds). Future wildlife workers might well concentrate on this problem of reproductive success.

In the bands at Huaylarco in April, the proportion of juveniles to yearlings was 46 to 10. If these figures are typical of all years, there is a loss from bands of about ¾ of the juveniles during their first year of life. This loss is caused not only by death, however, but by the departure of all male and many female yearlings from the bands. Thus, the mortality rate of juveniles is difficult to judge. Comparison of the ratio of juveniles to band adults in early April (29%) with the ratio in late November (25%) suggests that mortality in juveniles between 2 and 7 months of age is low.

Theoretically, in a stable population annual mortality in vicuñas older than yearlings would be equal to the number of new yearlings. On the basis of figures for Huaylarco, this number would be about 38 per 100 reproductive females, in January, or 17% of all vicuñas older than yearlings (assuming a 1:1 sex ratio). One should not trust these figures too far, because the basic assumptions that the population is stable and that the data for one year are typical of all might be wrong. For significant analysis of age composition and mortality it is necessary to have accurate counts of one population, by age classes and including troop males, for a period of several years.

Disease and injury.—To determine the causes of mortality in any wild population is difficult. Fresh carcasses are hard to find, for dogs, foxes, condors, caracaras, and other carnivorous animals may consume the body of an adult vicuña within a week, and of a juvenile within a day. Hunters who kill vicuñas leave little evidence behind.

In the higher parts of their range, vicuñas may have few parasites. None of the vicuñas that I studied appeared to be suffering from disease. I made a gross examination of three vicuñas that had been shot, but I found no abnormalities, parasitic worms, or ectoparasites. Pearson (1951:167) tells of a pregnant female that died in the wild "apparently from an intestinal inflammation." Allen (1942:412) reports that lungworm, scab, and flukes have been found in vicuñas, but he does not state whether these were wild animals. Romero (1927:151) mentions mange and itch ("la sarna") as lamoid diseases that probably spread to vicuñas. Garcilaso de la Vega (1871:378) tells of an outbreak of the itch in llamas in the year 1544. He relates that the disease spread to vicuñas,

but implies that it did not have a serious effect on the population.

James Mitchell, for many years a veterinarian on Peruvian haciendas grazed by llamas, alpacas, sheep, and vicuñas, told me that all of these animals were infected with scab, and many with tapeworm cysts in the liver and other organs. Another veterinarian of central Peru, D. F. Watson, wrote me (1954) that vicuñas found dead on sheep ranches usually have scabies and many internal parasites. Apparently, then, where vicuñas graze sheep pastures they may become heavily infected with parasites. Possibly infection follows the weakening of the animal from other cause. Tschudi (1844-46:232-233) relates that sometimes a single vicuña would allow itself to be approached, and it could be caught by a horseman with lasso or bolas. The Indians told him that these animals were tame because they were infected with worms. Some carcasses examined by Tschudi were found to have cysts of intestinal worms in the pancreas and liver, and roundworms in the colon. Evidently, under certain conditions parasitism may reduce the vitality of vicuñas.

In a four-month period at Huaylarco, I noted 10 injured vicuñas. About half of these had a lame or broken leg. Some wounds were surely made by gunshots. Others were probably caused by collisions with automobiles, or by falls on rough ground. Nearly all the injured vicuñas were males and, because an injured male cannot well defend a territory, most of them were members of troops. One adult male in a troop appeared to be well fed and grazed normally although one of its legs, broken just above the ankle, dangled limply.

The family male of one band which I was studying became seriously injured. I found him lying close beside the road. Half an hour later he rose, with great effort. For several minutes he remained at that spot, sometimes raising or lowering his head very slowly. Then he walked a short distance, very unsteadily, for one leg was unable to bear his weight, and again he lay down. Half an hour later he got up again, faltered up a gentle slope, crossed the road, drank from a puddle, nibbled some grass, and then lay down, almost tumbling sideways in the act. When I walked toward him he allowed approach to a distance of 50 yards before he rose and trudged away. Three days later he could walk fairly well.

In the puna, there is much snow and hail during the season of birth. Severe storms might cause a heavy loss of juveniles, on occasions. At Calacala young vicuñas, as well as lambs, are sometimes found dead after snowstorms.

Life span.—A female vicuña born at the New York Zoological Park in 1904 lived for 24 years (R. M. McClung, letter, 1953). This is the greatest longevity known for a vicuña. One of 3 still living in that zoo is at least 15 years old (a female, purchased in 1940). Many other zoo specimens have survived more than 10 years, and as experience is apparently of great survival value in the wild, some wild vicuñas may live at least 10 years, though the average life span is doubtless much less.

GENERAL ACTIVITIES AND REACTIONS

To anyone who watches vicuñas for a few hours it soon becomes obvious that they remain in discrete groups whose principal activities are grazing and resting. From time to time individuals void or scratch. When one walks too close to another, the latter may spit at it. And if the observer approaches the vicuñas, one of them gives a call and the group runs off.

RESTING IN DAYLIGHT

Lying down.—A vicuña spends more than half of its life lying belly down with its legs tucked under its body. In this prone position the wrists and ankles are flexed double and the heels nearly touch each other. This is the characteristic posture for resting and sleeping, for females during copulation, for newborn young before they can stand (Fig. 10), and for sick and dying vicuñas. In play-fights, lying prone is often the final defensive action. On occasions while lying prone a vicuña extends one wrist, then the other, so that the forefeet lie close together in front of its chest. A few minutes later it doubles the feet under its chest again. Sometimes a resting vicuña lays its head on the ground, perhaps on a rock or mound. Commonly, in this pose only the chin touches the ground, the neck being extended far forward and arched a few inches above the ground surface. Even newborn young may rest thus.

FIG. 10. Vicuñas are born in the wet season, when snow or hail falls nearly every day. This infant, about 20 minutes old, lies prone in typical camel fashion. Huaylarco, Arequipa; March 4, 1952.

The prone position is warm, for the slender legs are exposed as little as possible and the long brisket hairs protect the forelegs from cold. That the pose is steady is attested by the observation that, for several days after death the body of one vicuña remained in the characteristic prone position. One native belief is that a vicuña sleeps with its throat behind its forefeet so as to guard against the attacks of foxes.

As a vicuña prepares to lie prone, it noses the ground and makes two or three false starts by flexing

one foreleg. Then it doubles one wrist, then the other, placing the two knuckles down side by side. By flexing the hindlegs, it lowers its knees to the ground. Next, as the elbows drop and the wrists slide forward, the chest descends. Lastly the hindfeet shuffle forward so that the heels bear on the ground. The whole process may be performed in slow jerky fashion, or too rapidly for the observation of details. Apparently all the members of the camel family lie prone in approximately this manner. Rising is normally rapid. The trunk heaves forward as the straightening hindlegs raise the posterior body. Then one wrist is lifted, the forefoot planted, and the other foreleg likewise extended.

Vicuñas prefer to rest on dry bare soil, but sometimes they lie prone on mud or snow. When a group is resting the spacing and orientation of individuals is irregular. Frequently the male of a band lies at a short distance from his family, and the juveniles may rest together a short distance from the adults, but the group does not assume a formation of any seemingly defensive advantage. Juveniles often graze while lying prone, but adults rarely do. All, however, chew cud while resting. Upon becoming alert a prone vicuña stops chewing, raises its ears, and elevates its head a few inches. Then, as alarm subsides, it resumes chewing and lowers its ears and head. Prone

TABLE 2. Schedule of daylight resting and voiding for one band[a].

Band member	Male	Female 1	Female 2	Female 3
Duration of rest periods[b]	6:00- 6:12 6:38- 6:41 7:28- 7:32 7:43- 8:06 8:50- 9:17 9:39-10:02 11:55-12:43 1:03- 1:28 3:26- 3:55	6:00- 6:23 7:22- 7:32 9:59-10:18 11:32-11:52 12:28-12:43 1:12- 1:32	6:00- 6:15 6:22- 6:31 9:07- 9:17 9:52-10:01 10:02-11:01 12:33-12:43 3:33- 3:58	6:00- 6:05 6:15- 6:25 7:23- 7:35 8:28- 8:47 10:18-10:48 12:00-12:43 2:22- 2:35
No. periods	9	6	7	7
Total minutes	194	107	137	132
Ave. duration	22	18	20	19
Range (min.)	3 to 48	10 to 23	9 to 59	5 to 43
Time of voiding	6:16 6:52 8:18 10:28 1:35 2:19 3:15 4:15 4:54	6:23 10:52 2:11 3:20	6:21 11:04 12:12 3:05 3:20	6:25 11:04 3:20
Voidings	9	4	5	3
Interval between voidings: minimum maximum	36 min. 3 h. 07 m.	1 h. 09 m. 4 h. 29 m.	15 min. 4 h. 43 m.	4 h. 16 m. 4 h. 40 m.

[a]Band *B*, Aricoma, August 2, 1951.
[b]Only periods of more than 3 minutes are included.

vicuñas occasionally yawn with ears back, muzzle raised high, and mouth held wide open for a second or two.

When the sky is heavily overcast during the day vicuñas may lie prone for long periods, but in this weather they seldom lie down on their sides. It was often observed that soon after sunshine broke through the clouds several prone vicuñas lay down on their sides, as if sunlight stimulated the action. Lying on the side is an unwatchful position, seldom maintained more than 15 minutes at a time. While the vicuña lies on its side, its neck and head may be held up or they may lie sideways on the ground. The flexed legs usually sprawl, though the forelegs may be tucked under the chest. On hillsides vicuñas lie with body horizontal across the slope, legs downhill.

Resting schedule.—At Aricoma, most of my observations were of two bands called A and B. The resting schedule of the members of band B for 12 hours of daylight is given in Table 2. These data show that each animal took 6-9 rests for an average period of about 20 minutes, the longest periods being about one hour. On the average, each vicuña rested nearly 2.5 hours, about 20% of the 12-hr period. After 4:00 p.m. these vicuñas did not rest during daylight. In fact, only one third of the 29 rest periods started after 12 noon. In this example, the male rested half again more than the average for females. If there had been many transgressions of his territory by other vicuñas, the resident male would have had less time for rest than the females. Of the intervals between rests 40% were over one hour in length, the longest being 2 hours and 50 minutes.

Another day, during the 5 hours preceding noon I watched a band of 6 members. There occurred 15 individual rest periods of 10-70 minutes each, not counting periods of less than 3 minutes. The average per animal was 2.5 rest periods, each 33 minutes in length.

The frequency and length of daytime rest periods are irregular. The amount of time spent in rest is greater at some localities than others, presumably because less time is needed to ingest sufficient food at the former sites. But in general, vicuñas in bands rest for a period of one-quarter to one-half hour at intervals of 1-2.5 hours. In Fig. 11, a bar graph of the data of Table 2 without regard to individual

identity, a tendency for the group to rest at intervals of about 2 hours is apparent.

In male troops too, there is a tendency for a large proportion of the members to lie down at intervals of about 2 hours during the day, the first rest period usually occurring about 8:00 a.m. For example, in one troop of 30 males, 20 were lying down at 7:55 a.m., 27 at 9:30 a.m., and 14 at noon, fewer being down in the intervals between these times. Resting periods of a troop are not clear-cut, however, because the members lie down or arise a few at a time, not all at once. In one troop of 19 undisturbed males 9 were lying down at 8:30 a.m., then 15 by 8:44, 16 by 8:55, and 17 by 9:05. But by 9:18, 11 males had risen, and all were up 4 minutes later. The rest of troop males is often interrupted by the attacks of family males. But soon after the attack is withdrawn many members of the troop lie down, as if a rest were overdue. Among the members of troops the action of lying down is moderately contagious; if one lies down others soon follow suit.

Reaction to storms.—Vicuñas take little notice of falling rain. They allow snow to accumulate on their backs and necks, without shaking it off. And they do not hesitate to walk in snow when it is several inches deep. If the wind is moderate, grazing continues in snow, hard rain, or light hail. But when these are driven nearly horizontally by the wind, many vicuñas lie prone facing downwind.

Before lying down, the members of a group may run a short distance, often from a flat to the base of a hill as if to reach a sheltered spot, and they gather close together. While lying prone during the storm the vicuñas in a group may be close together or scattered. One band of 11 lay down within a space of 15 x 5 yds. But the members of another large band lay down in approximate file, strung out for a distance of 50 yds, no more than 2 vicuñas lying together.

All members of a band may lie prone within a minute, and sometimes within half a minute, after the start of hard hail. In one observation, all 18 males of a troop quickly lay down as hail commenced. Then when the hail increased in intensity a few vicuñas rose, walked forward, and lay down again; others did likewise. The group thus grew more compact, so that the sides of some vicuñas nearly touched

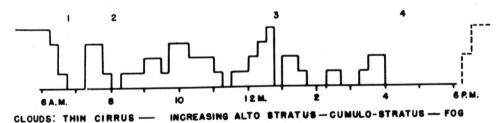

CLOUDS: THIN CIRRUS — INCREASING ALTO STRATUS—CUMULO-STRATUS — FOG

Fig. 11. Daylight resting periods of a band of 4 members. Height of bars indicates maximum number of vicuñas resting at one time during each quarter-hour period. Numbers above bars indicate events as follows: (1) another group neared; (2) band walked a quarter-mile; (3) mild disturbance by man; (4) light snowfall.

others. Doubtless, within a compact group vicuñas are somewhat protected from wind and hail by the bodies of their fellows. While the hail falls the prone vicuñas hold their ears back horizontally, heads lowered to the level of their backs, and they remain nearly motionless even though their heads and necks become caked with ice.

How long will vicuñas remain prone during a hailstorm? In one observation, all 10 members of a band remained down for a period of an hour. Then the hail turned to wind-driven snow and 3 vicuñas rose and grazed, but the other 7 stayed prone a half hour longer. Vicuñas differ in their readiness to lie down when a storm commences, or to rise after it is over. The last member of a group to lie prone is often the first to rise and graze when the hail abates. It may be the hungry individuals which are least inclined to remain inactive for long. Frequently one member of a band, perhaps usually the male, continues to graze while the others lie down. Sometimes all members of one group remain down while a band close by continues to graze. When finally the hail stops or turns to rain, the prone vicuñas rise a few at a time. Upon rising, some shake off part of the accumulated ice. Several minutes after the storm all are up and grazing.

RESTING AT NIGHT

Characteristically, bands of vicuñas spend the night resting on their daylight feeding grounds. On many of the territories that I studied, a lower flat area was favored for late afternoon grazing. From these flats the vicuñas walked uphill shortly after sunset in order to bed down. On many occasions I censused bands of vicuñas in the late afternoon as they grazed on the lower parts of their territories, and then again early in the next morning when they were still resting on the higher slopes.

Time of bedding down.—The time that vicuñas bed down was found to be fairly regular; about an hour after sunset at Aricoma in winter, and a half hour after sunset at Huaylarco in summer. For example, here is an observation made at Aricoma one evening in August. At 5:05 p.m. the sun set at the site where the 10 members of bands A and B were grazing. At 6:03 p.m. 3 members of band A were lying down. Seven minutes later, with the Southern Cross visible in the sky, only 2 of the 10 vicuñas remained standing. Vicuñas normally bed down by the time that bright planets become visible to the naked eye. Evidently the time when foraging stops depends upon light intensity. In an observation made one evening in July, the 7 members of a band were diligently grazing on a flat at sunset, 5:03 p.m. The band started uphill to bed down at 5:30 p.m. At that time the light from the zenith was 1.0 ft-candle and from the brightest point of the sky 3.2 ft-candles, as measured with an ordinary photoelectric exposure meter. Variations in the schedule of bedding down are evidently caused by hunger, the shadow of mountains, and clouds. In a few observations in which heavy clouds obscured the setting sun, bands of vicu-

ñas bedded down about one-quarter hour earlier than on clear days.

As twilight is short in the tropics, all the vicuñas in an area usually bed down within a period of less than half an hour. Bedding down is delayed little, or not at all, by bright moonlight. One evening in July, I watched a band of 5 members which was still grazing one hour after sunset, casting shadows by the light of the full moon. But all were lying prone by 6:25 p.m.

On occasions when a band is excited, its members may be active when darkness precludes observation. In early January, I was observing a band of 18 grazing vicuñas when the sun set behind clouds on the horizon at 5:55 p.m. A quarter hour later the male of the band attacked a nearby troop. Then, at 6:37 p.m., when Orion was visible and the light from the brightest point in the sky was only .4 ft-candles, I noted that 3 members of the band were lying down. Soon 8 were down. But about 10 minutes later I saw by a flash of my spotlight that half of the band was still up, and several vicunas were running about.

Site of bedding down.—Vicuñas usually rest for the night on dry sites of moderate slope, well downhill from ridgetops. Many bands normally bed on upper slopes of their territories, but downhill from still higher barren slopes. These bedding sites are relatively free from disturbance because there are seldom vicuñas on the unclaimed ground upslope, and because livestock, men, and carnivores usually travel the lower slopes and canyon bottoms. With moderate consistency, the same general sites are used repeatedly for bedding down. One band of 10 members bedded down within 50 yds of the same spot on 3 consecutive nights.

Bands that inhabit flat territories may remain all night far from slopes or cover. For example, one evening at 6:00 p.m. I saw two bands, totaling 21 adult and 10 juvenile vicuñas, lying prone on a wide pampa, 200 yds from the nearest slope. At 5:30 a.m. the following morning they were still lying at the same site. Cover is not avoided when present, however, for one band often bedded down near a certain group of boulders a few feet high. No definite bedding depressions, such as those formed by deer, are made by vicuñas.

Although bands normally bed down on their own territories, they sometimes move to unclaimed ground to spend the night. Perhaps this move is made only when the band is disturbed on its territory in the evening. On one occasion, at dawn I discovered a band resting on a narrow trail across a shale slide, 200 yds uphill from the nearest grassland. It is probable that the group had been disturbed by horses, which grazed in the vicinity. On another occasion, I watched a large band graze on its flat territory near the foot of a steep barren ridge for a long period, until 5:30 p.m. Then the vicuñas started to walk up the slope, out of their territory. Twenty minutes later all had reached the crest and were walking away, farther from their territory. On this occasion I did not detect any cause for alarm. In a similar observation, a band grazed on a pampa until half an hour after sunset, and then walked to the foot of a slope 300 yds distant and out of their territory. They ascended to a barren hilltop and walked on over it. In this instance, the vicuñas might have been slightly alarmed by my car, for a few days later the same band remained all night on the pampa in their territory, some adults lying down only 20 minutes after sunset. I suspect that vicuñas may tolerate during daylight some mild disturbance, such as the presence of a car, which they will not tolerate in the evening, when they commence to bed down. In studies of the California condor, I noted such a lowering of the threshold of tolerance for disturbance at the time that the birds commenced to roost for the night.

Activity at night.—Vicuñas rest nearly all night, it seems, usually lying prone. As when resting in daylight, they face various directions in no particular formation. Members of the same band may lie 3-15 yds apart, and a large band may bed down within an area 30 yds in diameter. The band male tends to lie a short distance from his family. Perhaps he usually lies lower on the slope, for on two occasions when I alarmed a band mildly at night only the downhill member rose and gave the alarm trill. By day it is typically the male which acts thus. The members of a troop also bed down together. At dawn one morning I noted a troop of 24 males lying on a hillside, spread a distance of 100 yds up and down a slope. A cluster of 7 lay 25 yds uphill from a group of 6, and the others were scattered farther uphill.

Bands often move a short distance during the night, for in the early morning they are seldom found at the exact sites where the first few members lay down the previous evening. For example, one evening 3 members of a large band lay prone at a certain spot at 6:05 p.m. At 4:20 a.m. the next morning the group was lying 200 yds distant from that spot. Movements of short distance during the night may be incident to voiding or to nursing young. Long moves are probably caused by disturbance. On a few occasions, I observed that a familiar band was not on its territory at dawn although it had bedded down there the previous evening. Presumably there had been a disturbance at night, for on each occasion the band returned to its territory soon after sunrise.

Rising in the morning.—Sunlight, evidently, is the principal factor that determines the time of rising in the morning. Yet, vicuñas do not seek sunshine from shadow. At Aricoma I often started watching the activities of family bands in the early morning, before the first vicuñas had risen. Typically, the earliest risers stood up soon after sunshine struck the highest peaks, half an hour before it struck the bedding site, and the latest risers got up a few minutes after the rays of the sun struck them. An example of the rising of a band at Aricoma is as follows: 6:00 a.m., all 4 members of band B still prone. 6:05, immature female stood, grazed. 6:12, male rose, grazed. 6:15, immature female lay prone; one adult female rose. 6:21, adult female lay prone again; second adult female rose. 6:25, immature female rose

again. 6:31, one minute after sunshine struck group, first female rose again; all four were up. Here, as in most observations, the interval of time from the rising of the first to the last member of a large group was about half an hour.

Within this interval there is much variation in the time of rising of two bands in the same area, or of the same band on successive days. In one observation at Aricoma, 4 members of band B were standing at 6:05 a.m., but all 6 members of A were still lying down. Two members of A stood briefly, then again lay prone. At 6:30 a.m. sunshine first struck this band. All members were up five minutes later. The following morning all 11 members of these two bands were on their feet nearly half an hour earlier, at 6:10 a.m.

The observed times of rising at Aricoma in winter were about 1.5 hours later than at Huaylarco in summer. Part of this difference was due to the difference in season, and part to the fact that surrounding peaks blocked early sunshine at Aricoma. But in addition, to satisfy their hunger vicuñas apparently had to graze for more hours at Huaylarco than at Aricoma, because forage was less abundant at the former site. Vicuñas also spent less time in resting during the day at Huaylarco than at Aricoma.

Observations at Huaylarco indicated that during a period of about 4 months, from early summer to mid-autumn, the average time of morning rising became later by about 1 hour. For approximately 30 band vicuñas on a certain pampa, the usual time that all were first on their feet was about 5:00 a.m. in December and 6:00 a.m. in March. This difference was probably due to changing pasture conditions and changing demands of young, rather than to any change in the position of the sun.

The time of rising may be delayed by cloudy weather. In an observation made one morning in January, when the sky was overcast and snow was falling lightly, only one vicuña of a group of 12 had risen by 6:30 a.m., an hour and a half after the usual rising time. Rising in the morning takes place at much greater light intensities than bedding down for the night. Before they rise in the morning vicuñas are clearly visible at long distance, yet they often continue to graze in the evening when human sight fails.

Although I found no significant difference due to sex or age in the time of rising in the morning, some observations suggested that troops tend to start foraging earlier than bands. Troops would profit from early grazing, for they are subject to many disturbances that interrupt feeding during the day.

A mannerism that often occurs shortly after first rising in the morning, but rarely at other times, is a cat-like stretch. In stretching, a vicuña leans forward with hindlegs slanting far back, extends its neck forward and down so that the chin nearly touches the ground, and holds this pose for a few seconds.

VOIDING

Dung piles.—Lamoids defecate and urinate on communal dung piles. Vicuñas of both sexes and all ages use the same piles, normally visiting the closest at hand. Displaced bands freely use dung piles that are situated on the territories of other bands. Alpacas and llamas use the same piles that are used by vicuñas. One typical large pile was .1 ft thick at the center and 5 yds in diameter. The center third of the surface was solid feces, the peripheral two-thirds carpeted with short nettles and other plants. Average pellets are 11-15 mm long, a third less in diameter, and irregularly ellipsoid in shape. Dry pellets are firm but can be mashed by a hard pinch of the fingers.

On flat ground where there are many vicuñas the regularity of the spacing of dung piles, about 50 yds apart, is striking. On poor range the piles are widely spaced, perhaps only one, a mere sprinkling of pellets, occurring every quarter mile near frequented pathways. The spacing, size, and freshness of dung piles could be useful in estimating the abundance of vicuñas on different sites, providing allowance were made for the contributions of alpacas and llamas.

Manner of voiding.—The male vicuña typically defecates in the following manner: He walks to a dung pile, sometimes as far as 50 yds distant, raising his tail high when he reaches a point a few yards from it. Upon reaching the pile he noses it for several seconds. Next he puts his forefeet atop the pile and treads them alternately a number of times, at a rate of about two steps per second. Then he turns about so that his anal region is downslope and spreads his hindfeet wide. In the defecation posture the head is lightly lowered, the ears are horizontal, and the tail points out or slants upward (Fig. 9). During a period of 5-10 seconds, as many as 7 fist-sized clumps of pellets fall.

Urination starts as defecation stops. The male urinates for a period of about half a minute, the fine urine stream passing posteriorly between the backs of his knees and falling behind his ankles. The stream surges and his extended tail pulses up and down simultaneously, apparently in time with the beating of his heart (130-150 pulsations per min). A male may remain in the voiding stance for 1.5 minutes. Then he walks away from the dung pile and lowers his tail slowly. Immediately after voiding a vicuña sometimes turns about, noses and treads the pile, turns once more, and again voids. Walking in water stimulates urination (Fig. 4). As one troop of 20 males crossed a stream, 7 stopped and urinated in the water.

Females differ from males in the manner of urination. After defecation, at the start of urination, the female lowers her hind quarters about 4 in. Then the angle at her heel joint is about 110°, as contrasted to about 160° in the male. Also the urine stream of the female is much thicker than that of the male, and it flows for a period about half as long. The stream of the female spurts strongly for approximately one second at irregular intervals of about a second. Opportunity for the distant observer to determine the sex of individual vicuñas is greatest when they void. Not only can differences in the manner of urination be observed, but when the tail is raised the bulging testes and black-tipped penis of the male may sometimes be seen.

Juveniles void in the manner of adults of the same sex. While standing alongside its mother, a juvenile but 40 hours old spread its hindlegs as if to defecate, but no feces fell. The youngest one actually seen to defecate was about 4 days old. In the adult manner, this infant nosed a dung pile, turned around, and voided a single pellet. Another, 7 days old, extruded but 2 pellets. In the few days before young commence to graze they probably have little digestive waste to eliminate.

Frequency.—The interval between successive defecations may be as long as 5 hours. Adult males void more frequently than females—perhaps twice as often. This difference is probably due to the fact that males react more strongly than females to the scent or sight of dung piles. One male defecated 3 times during a period of 10 minutes, then again 10 minutes later. For a band which I watched for 12 hours during daylight, the average interval between defecations was about 3 hours for the females, and half as long for the male (Table 2). Voiding is infrequent late in the day, when vicuñas are absorbed in feeding.

Although there is great irregularity in the intervals between defecations, all members of a band are fairly sure to defecate within a period of half an hour after first rising in the morning. Yet, vicuñas are in no hurry to void upon rising; 20 minutes may elapse between the time that the first member of a band rises and the time that the first defecates. An example of the schedule of the first voidings of the day by 4 members of a band follows: 6:00 a.m., all prone. 6:05, immature female rose. 6:12, male rose. 6:15, adult female rose. 6:16, male defecated. 6:21, adult female defecated. 6:22, other adult female rose, defecated. 6:25, immature female defecated; male defecated again.

As the amount of feces dropped in the first evacuation of the day is not greater than subsequently, and as vicuñas do not defecate at once upon rising, it is probable that they void from time to time during the night. Shortly after dark on a moonlight night, one member of a band which had been bedded down for half an hour was seen to rise, walk to a pile, defecate, return to its bedding place, and lie down.

All visits to dung piles are not for the relief of rectal pressure. Commonly a male, and sometimes a female, will defecate on two separate piles within a few minutes. When walking near a dung pile, and especially when nosing it, a vicuña of either sex may raise its tail and even void a few pellets. Probably the scent of dung stimulates defecation, but this action also appears to be contagious. On occasions, many or all members of a band defecate within a period of a few minutes. Several vicuñas in turn may void on a pile, one waiting nearby until another leaves it. Sometimes two or more vicuñas void on a pile at the same time. In all, the attraction of a dung pile to passing vicuñas is great. As one band of 5 members filed past a pile the first two stopped and voided, the third passed by, the fourth nosed the pile, and the trailing male voided on it. Even when a troop of males is retreating under attack, several members may pause at a dung pile and add to it. When an adult male is running to challenge another, or as he returns from an encounter, he usually stops at one or more dung piles and defecates.

SPITTING

In popular accounts the spitting of lamoids has been publicized as a hose-like stream of vile-smelling saliva, accurately directed at the white of the eye of an enemy. Some books state that the whole contents of the stomach may be thrown at the attacker. But the spitting of vicuñas is no more than a quick forceful expulsion of air, which incidentally sprays out saliva and small fragments of whatever masticated food is in the mouth. When I scratched the neck of a domestic vicuña, it spat at my chest. So sudden was the blast that I was startled, but only a few bits of grass and dewlike droplets of clear liquid clung to my sweater. Often I heard the sudden *puff* of spitting at a distance of 100 yds. At some distance the blast is normally invisible, but on a cold day the warm moist expelled air may condense as a cone of fog.

Spitting or blowing is an act that shows "displeasure" insufficient to merit a kick. The spitter holds its muzzle tilted up and its ears far back. At least to a man, the throwing back of the ears serves as a warning that a blast might follow. The vicuña receiving the blast stands stiffly with its head turned away from the other and its ears back. Sometimes the two remain poised for more than half a minute. A kick or a chase of a few steps often follows a blowing attack. Occasionally an annoyed vicuña spits forward into the air while the disturber stands behind it. Spitting encounters are common between females of a band, between band females and juveniles, and between troop males, but rare between an adult male and female, or two family males.

GROOMING

Rolling.—A vicuña dusts its fleece and scratches its neck, side, and legs, by rolling vigorously on the ground (Fig. 12). Usually before rolling the animal noses the ground and paws the soil a few times. Then it drops to the knuckles of its wrists, flops down onto its side, and thrashes a few times. In each thrash the forelegs are flung forward, the hindlegs backward, and the head, neck, and tail tossed dorsally, all at once. Then the vicuña flops over onto its other side and thrashes again. In one observation, a vicuña repeated, 5 times in quick succession, the whole act of lying down, rolling, and standing up. Often a vicuña remains prone for several minutes after rolling, or it rolls just before standing up. Even juveniles a few days old may roll energetically.

Vicuñas prefer to roll in dust, but commonly they roll on dry turf, and occasionally on bunch grass or tola. Rarely do they roll on mud, or when their wool

FIG. 12. In rolling on bare ground a vicuña thrashes its neck and legs forward and backward. This female gave birth a few days later. Huaylarco, Arequipa; February 27, 1952.

is wet from rain or bathing. At Huaylarco, they often rolled in shallow pits of dust about 6 ft in diameter. These pits were apparently formed by the rolling of llamas and alpacas, however, more than by the actions of vicuñas. The flat-bottomed dusty beds of dry pools are also choice sites for rolling. As a group of vicuñas walks across such a dry pool, several members may lie down and roll in the dust. But rolling is also a means of scratching, and vicuñas roll their backs hard against the uphill banks of dust pits.

The action of rolling is contagious, even between bands. Within a period of a few minutes all members of a band may roll. In one troop of 19 males, seven rolled at the same time. In bands of 12 and 15 members, 40 yds apart, 5 of each band rolled simultaneously. Soon after one vicuña walks to a dust bowl and rolls, others may follow suit.

Members of male troops may roll even though a challenging family male stands but 50 yds distant. In fact, troop males and displaced bands tend to roll soon after being chased by a territorial male, as if rolling were a sign of relief from tension. The same explanation might account for the fact that a family male sometimes rolls immediately after chasing other vicuñas, and that vicuñas often roll after watching a fox.

Scratching and shaking.—Vicuñas scratch themselves many times a day. The tip of one hind toe is delicately used to scratch from 10-20 strokes at various parts of the forelegs, shoulders, ribs, or lowered head. In attempting to scratch, juveniles a few days old sometimes lose balance. A common action of both adults and young is the rubbing up and down of the rear surface of one metatarsus with the anterior surface of the ankle of the other hindleg. Possibly this action rubs loose scales from the external metatarsal gland. The rapidly nibbling incisors are also versatile in scratching the skin, or in cleaning the wool of the back, legs, belly, buttocks, and other parts. Grooming with the mouth often continues for 3 minutes, 10-20 nibbles being taken at each spot.

While lying down a vicuña may rub its neck and head on bare soil, rock, grass, or shrubs. Occasionally one pauses in walking and brushes its head or neck back and forth through a bush or a tuft of grass. At other times, especially just before rolling, a vicuña lowers its neck and one shoulder to the ground, then scrapes them along the surface or through a bush by pushing a few steps forward with the hindlegs.

Commonly after scratching, rolling, or bathing, a vicuña shakes its head. In addition, after rolling or bathing it may shake its body with a vigorous lateral twisting motion. Sometimes the ears are waggled back and forth independently to shake off a snowflake or perhaps a fly, and the tail is occasionally switched.

COMMUNICATION

Voice.—When a band is approached by a man, dog, alien vicuña, or other disturber, an alarm trill is given by the resident male. In typical form this call consists of approximately 5 separate notes, delivered in a period of about 4 seconds. The first note is often near middle C, and each succeeding note starts a tone lower than the one before it. Each chirp slurs down two or three full tones and diminishes in volume. Commonly the final note slurs down very slowly and trails off into inaudibility. A diagram of the call is given in Fig. 13.

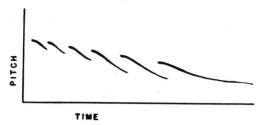

TIME

FIG. 13. Diagram of the alarm trill of a male vicuña. Several whistling notes are given in rapid succession, each starting a tone lower than the one before and slurring down, diminishing in volume.

There are many variations in the trill. On occasions the sound is a pure whistle. Again it is thin, stringy, squeaky, screeching, hoarse, or rasping, with no musical quality. Perhaps only 2 notes can be heard, or as many as 10. In some variations the first few notes, or all of them, are monotone, or the descending steps are half tones. No notes, or only the terminal one, may slur down and fade in volume. The delivery may be fast (more than 2 chirps per sec) or slow and mournful (one per sec). In some observations, I noted the call as *chee chee choo choo choo,* or as a series of *chucks, chirts,* or *cheets.* The trill often reminded me of the sound made by rubbing a cork back and forth on glass. The whole call may be repeated again and again. Probably there are differences correlated with sex and age.

When giving the alarm trill a male does not vary

his facial expression or open his mouth, but with the sounding of each note his nostrils dilate and his abdomen contracts. It appeared to me that the male trilled while forcing air out through his nostrils.

The calling male faces the source of alarm and vicuñas close by then turn and face him, unless they see the disturber. Immature troop males call, but less readily than family males. Female and young vicuñas rarely trill, but they can do so. Once as I walked near 2 juveniles about 8 months old they squealed at me in different pitches.

The effect of the alarm call on other vicuñas reaches far beyond the family of the male, but if the trill is heard in the distance grazing vicuñas may respond only by raising their heads for a moment. Yet, in one observation, persistent calling at me by an adult male evidently alarmed a distant band and caused it to move a quarter of a mile away. Alpaca-vicuña hybrids at Calacala were heard to trill in about the same manner as vicuñas, but alpacas and llamas were not heard to call. Guanacos have a call similar to that of the vicuña, it is said.

The fact that vicuñas have large ears suggests that their sense of hearing is well developed, but my observations in daylight indicated that the hearing of vicuñas is no more acute than that of man. In daylight good hearing would seem to be of much less value to a vicuña than good sight, but at night the most important sense for detecting predators might be hearing.

Smell.—For a vicuña to obtain olfactory information of others or their excrement, it is apparently necessary for it to nearly touch the object with its muzzle. It would seem that, for the detection of predators or alien vicuñas, it would be advantageous to have a sense of smell that would function at long range. This sense, however, seems to be poor in the vicuña. On occasions, when I was hidden behind a rock wall several vicuñas approached me to a short distance, evidently unaware of my presence although the breeze bore my scent toward them. In some instances, intruding males remained on the territory of a hostile male, hidden from him by high ground, for a considerable period before he saw them and chased them off. And vicuñas were never seen to react to a fox or dog until it was actually in sight. In fact, they sometimes walked toward a poorly seen dog or man until they saw it clearly, and then they fled. For fleet animals that live in open country and which are rarely attacked by terrestrial native carnivores, a highly developed sense of smell might be of much less value than keen sight.

LOCOMOTION

Gaits.—The basic gaits of the vicuña are the diagonal walk and the transverse gallop (Howell 1944: 228, 233). In walking the order of movement of the feet is: left front, right hind, right front, left hind. The feet of one side move forward at almost the same moment, the hindfoot being placed on the ground an instant before the forefoot of the same side is planted. When a grazing vicuña ambles forward slowly

FIG. 14. As vicuñas gallop across a pampa at high speed, their limber necks keep their heads on a steady course. Near Lago Suche, Moquega; 1952. Photograph by O. P. Pearson.

its feet move in this same order, at intervals of several seconds. In a slow walk the feet move through one cycle during a period of 1–2 seconds.

Vicuñas gallop (Fig. 14) in approximately the manner of the swifter antelopes and deer, which also have a light build, slender legs, curved rump, and tucked-in hindquarters (Howell 1944:241). The supple neck enables the head to travel in a smooth path while the shoulders surge up and down. In galloping at moderate speed, about 20 miles per hr, the legs move through two cycles per sec, each cycle moving the animal forward a distance of about 5 yds. A speed of "at least 30 miles per hour was maintained for some time" by vicuñas chased with an automobile at an elevation of 15000 ft (Hall 1937:472).

Howell (1944:65) suggests that the cursorial ability of camels has been developed for moving long distances between food and water, rather than for fleeing from predators. Perhaps of greater evolutionary importance in vicuñas is the speed and agility of adult males in defending their territories efficiently, and the ability of all vicuñas to avoid the attacks of hostile males. Whatever the reason for the evolution of speed in vicuñas, it is now of great value in fleeing from dogs and horsemen.

Jumping and climbing.—Vicuñas can jump fairly well. At Calacala I saw a domesticated adult female, accompanied by a young juvenile, jump to the top of a 4.5-ft high stone wall, then down on the other side. Sr. Paredes told me that a wire fence 1.8 m (5.9 ft) high was needed to hold vicuñas. They easi-ly jump about 7 feet from one stream bank to the other, and they often hurdle 2-ft high rock walls during normal movements.

Vicuñas climb rocky slopes with agility. At Aricoma I saw them ascend steep shale slides, sending plates of rock clattering down slope. Running adult males thread their way among large rocks, and they mount boulders about 3 ft high in order to view pursuers. In one observation, a troop of 19 males walked 20 yds across a rough solid rock ridge of moderate slope, although they could easily have detoured on grass. The padded soles of the feet apparently do not slip on rock surfaces. In fleeing from men, dogs, or other predators, agility in climbing rough slopes may be of greater value to a vicuña than high speed on level ground.

Trails.—On the better grazing sites vicuña trails are generally inconspicuous, but where vegetation is sparse trails are well-defined. On sandy barren ground pathways are about 8 inches wide and worn an inch into the soil. At close range these paths are obscure, but in strong cross lighting they are easily seen from a distance. At Aricoma, vicuña trails slant across the steepest shale slides above the highest grasslands, and some trails lead over high barren passes. Vicuñas also use llama trails, which in grasslands are conspicuous. Commonly these trails are hard-packed, about 12 inches wide, and worn several inches into the soil. A dozen llama trails may run approximately parallel, a few feet apart, with normal grass cover between them.

* * * * * * *

LITERATURE CITED

Allen, G. M. 1942. Extinct and vanishing mammals of the western hemisphere. Amer. Committee Internat. Wild Life Protection, Special Publ. **11**. xvi + 620 pp.

Acosta, J. (de) 1880. The natural and moral history of the Indies. Transl. E. Grimston. Ed. C. R. Markham. London: Hakluyt Society. vol. **1**. xlviii + 295 pp.

Bowman, I. 1916. The Andes of southern Peru. Amer. Geographical Soc., N. Y. xi + 336 pp.

Burt, W. H. 1943. Territory and home range concepts as applied to mammals. Jour. Mamm. **24**: 346-352.

———. 1949. Territoriality. Jour Mamm. **30**: 25-27.

Cabrera, A. & J. Yepes. 1940. Mamíferos Sud-americanos. Buenos Aires: Compañia Argentina de Editores. 370 pp.

Cardozo, G. A. 1954. Los Auquénidos. La Paz, Bolivia: Editorial Centenario. 284 pp.

Carpenter, C. R. 1934. A field study of the behavior and social relations of howling monkeys. Compar. Psychol. Monog. **10**: 1-168.

Cieza de León, Pedrò 1864: The travels of Pedro de Cieza de León, A. D. 1532-50, . . . Transl. and ed. by C. R. Markham. London: Hakluyt Society. xvi + lx + 438 pp.

Darling, F. F. 1937. A herd of red deer. London: Oxford Univ. Press. x + 215 pp.

———. 1952. Social behavior and survival. Auk **69**: 183-191.

Dice, L. R. 1952. Natural communities. Ann Arbor: Univ. Michigan Press. x + 547 pp.

Garcilaso de la Vega, Inca 1871. First part of the royal commentaries of the Yncas. Transl. and ed. by C. R. Markham. Vol. 2, 553 pp.

Gilmore, R. M. 1950. Fauna and ethnozoology of South America. In Handbook of South American Indians, Bull. 143, Smiths. Inst., Bureau Amer. Ethnology **6**: 345-264.

Hall, F. G. 1937. Adaptations of mammals to high altitudes. Jour. Mamm. **81**: 468-472.

Howell, A. B. 1944. Speed in animals. Chicago: Univ. chicago Press xii + 270 pp.

King, J. A. 1955. Social behavior, social organization, and poplation dynamics in a black-tailed prairiedog town in the Black Hills of South Dakota. Contrib. Lab. Vert. Biol., Univ. Mich., **67**. 128 pp.

López Aranguren, D. 1930. Camélidos fósiles Argentinos. Anales de la Sociedad Científica, Argentina **109**: 15-35, 97-126.

Maccagno, L. 1932. Los auquénidos peruanos. Lima: Dirección de Agricultura y Ganadería. 64 pp.

Madueño, C. G. 1912. Monografía de la vicuña. Santiago, Chile: Pan-Amer. Sci. Cong. **16**(2): 5-30.

Miller, G. S., Jr. 1924. A second instance of the development of rodent-like incisors in an artiodactyl. Proc. U. S. Natl. Mus. **66**, article 8. 4 pp.

Nice, M. M. 1941. The role of territory in bird life. Amer. Midland Nat. **26**: 441-487.

Osgood, W. H. 1943. The mammals of Chile. Field Mus. Nat. Hist., Chicago, Zool. Ser. **30**. 268 pp.

Pearson, O. P. 1951. Mammals in the highlands of southern Peru. Bull. Mus. Comp. Zool. **106**: 117-174.

Romero, E. C. 1927. Llamas y alpacas, vicuñas y guanacos. Buenos Aires. 208 pp.

Stroock, S. I. 1937. Vicuña. The world's finest fabric. New York: S. Stroock and Co., Inc. 23 pp.

Tinbergen, N. 1952. "Derived" activities; their causation, biological significance, origin, and emancipation during evolution. Quart. Rev. Biol. **27**(1): 1-32.

Tschudi, J. J. 1844-46. Untersuchungen über die fauna Peruana. St. Gallen. xxx + 262 pp.

———. 1847. Travels in Peru, during the years 1838-1842. Transl. T. Ross. London: David Bogue. xx + 506 pp.

Weberbauer, A. 1945. El mundo vegetal de los Andes Peruanos. Lima, Ministerio de Agricultura. xix + 776 pp.

18

Reprinted from *Auk*, **68**, 471–482 (Oct. 1951)

REMOVAL AND REPOPULATION OF BREEDING BIRDS IN A SPRUCE-FIR FOREST COMMUNITY

BY ROBERT E. STEWART AND JOHN W. ALDRICH.

IN 1949, while engaged in population studies of birds in northern Maine, the authors accumulated considerable information concerning population dynamics of birds inhabiting the Spruce-Fir forest community. This information was obtained in connection with investigations of the effective control by breeding birds of an infestation of the spruce budworm, *Choristoneura fumiferana* Clem.

Field studies were conducted during June and July on two separate areas; a square 40-acre tract which was used as an experimental area, and a rectangular 30-acre tract that served as a control area. The two areas were 1.25 miles apart and were both within one-half mile of the shore of Cross Lake, located about 14 miles southeast of Fort Kent, Aroostook County, Maine. In the experimental area an attempt was made to eliminate or to reduce drastically the bird population by use of firearms, while in the control area the natural bird population was allowed to remain unmolested. Both skins and stomachs of the birds killed were preserved for the Fish and Wildlife Service collections. The spruce budworm populations were studied in both areas before, during, and after the removal of the birds in order to see if any differential developed between the two areas in their total populations. The entomological portion of the field study was conducted by Philip B. Dowden and V. M. Carolin of the Bureau of Entomology and Plant Quarantine of the U. S. Department of Agriculture. Stomach contents of the birds collected were analyzed by Robert T. Mitchell of the Fish and Wildlife Service. The results of the various phases of this work concerning the effective control by birds of spruce budworms will appear in a separate report. The present paper includes an analysis only of the data on removal and repopulation of birds in the experimental area.

The Spruce-Fir forest in the experimental area was somewhat varied. The canopy in certain sections of the forest had been partially opened by selective lumbering, resulting in a fairly dense understory growth of conifers. The other portions of the forest with a closed canopy contained little or no understory. Balsam fir, *Abies balsamea*, and black spruce, *Picea mariana*, were the primary dominant trees throughout the forest. The majority of the trees ranged between 40 and 60 feet in height. Small sections of mixed coniferous-deciduous brushy undergrowth were present and the area was crossed by a state highway.

The population of territorial males in the area was determined prior to the removal of birds, by using the spot-mapping method (see Breeding-Bird Census, Audubon Field Notes, 4 (2): 185–187). Twenty-five census trips totaling 55 man-hours were conducted on the area from June 6 to June 14, inclusive. Most of the census trips were taken during early morning hours between 6:00 and 10:30 a.m. A few trips were also made in the evening for the determination of thrush populations.

Beginning on June 15 and extending through July 8, an attempt was made to remove all birds on the experimental area by shooting with 16-gauge shotguns. Shells used were loaded with very fine (No. 12) shot so that birds would not be too mutilated for specimens. By using this method the total population on the area was greatly reduced, although the actual degree of reduction could not be readily appraised. The chief difficulties were that the breeding territories were completely disrupted during the period when the original occupants were being removed and at the same time new adult males were constantly invading the area. Most of the adult females were so inconspicuous and secretive that it was not possible to census them accurately. In all, a total of about 130 man-hours were spent on the area collecting birds. The number of birds removed from this 40-acre tract during the entire collecting period was 455 (420 adults and 35 birds in juvenal or immature plumage).

Nearly all of the birds in the Spruce-Fir forests appeared to be at or near the peak of their breeding cycle during this study. With the exception of the American Crow, Golden-crowned Kinglet, and Red-breasted Nuthatch, it is doubtful if any species, at the onset of collecting, had produced nestlings. Nestlings, as well as three or four groups of fledglings, of the Golden-crowned Kinglet and Red-breasted Nuthatch were observed, but they were practically all removed from the area soon after they were first encountered during the first few days of collecting. The constant pressure of collecting and the resultant turn-over in adult birds completely prevented the development of other nestling populations in the study area during the collecting period.

Small, roving groups of adult non-breeding birds were found in the area from time to time. The most common of these were the American Robin and Cedar Waxwing, both species being occasionally represented by from one to nine individuals. These birds were noted with increasing frequency during the latter part of the collecting period, from about July 2 to July 8. Other non-breeding birds occurring more rarely included the American Goldfinch, Pine Grosbeak, and Chipping Sparrow. A few wandering immature birds were also seen during the last few days of collecting. These had undoubtedly moved in from outside areas after having become independent of their parents. An attempt was made to collect all of these birds whenever they were encountered.

After the first nine days of collecting (June 15–June 23) the resident males of all species except the Myrtle Warbler and Olive-backed Thrush had been reduced to 15 per cent or less of the original numbers. Continued collecting coupled with a steady influx of new males maintained these species at about this level for the remainder of the collecting period. The number of male Magnolia Warblers was reduced to the 15 per cent level by June 23 even though only 16 males (67 per cent of original population) had been taken by that time. Possibly some of the males had deserted owing to the disturbance of walking and shooting in the area; the Magnolia Warbler, being an understory species, would probably be affected by such disturbance more than tree-top birds would be. The number of male Myrtle Warblers was reduced to about 25 per cent and maintained this approximate level thereafter. Apparently a proportionately larger number of non-territorial males of this species was present to fill evacuated territories. The Olive-backed Thrush was found to be a very difficult species to collect. It was much more wary and secretive than the other species and was seldom seen, although its songs and call-notes were evidence of its continual presence. Although a total of 18 adult Olive-backed Thrushes was taken during the collecting period, the rate of collection was so low and repopulation so fast that the resultant voids in the population were filled by new males almost as soon as the voids were produced. As a result, the population level for males of this species was reduced only slightly below (probably about 80 per cent of) the pre-collecting population.

The approximate reduction in total number of adult males of all species was from an average of 148 to 28 or, in other words, from a density of 370 males per 100 acres to 70 males per 100 acres (reduction of 81 per cent). This density of 70 males per 100 acres is the approximate level that was maintained from June 24 to July 8. With the exception of wandering non-breeding birds, including the Arctic Three-toed Woodpecker, Cedar Waxwing, Black-throated Green Warbler, Chestnut-sided Warbler, Pine Grosbeak, American Goldfinch, Chipping Sparrow, and most of the Robins, practically all of the adult males taken appeared to be on their breeding territories at the time they were collected.

Most of the new males which took possession of the evacuated territories probably arrived during the night or early morning, as practically all of them when first observed seemed to be actively establishing territory, and no evidence of any late arrivals during the day was noted. The males of many species appeared to be completely eradicated time after time, only to be replaced by other males on following days. The characteristic behavior pattern of most of the new arrivals was quite distinct from the behavior displayed by males with territories that had been maintained over a longer period. The songs of these newcomers were louder and uttered more frequently, while at the same time the birds were much more active in covering and inspecting the areas within their territories. It is probable that most of these new arrivals were unmated. Lack (1946) in his study of the English Robin found that "In the late spring nearly all the best robin song comes from cocks which are unmated." ". . . almost immediately after the cock has obtained a mate, its song declines to a rather moderate intensity and remains so, except during fights, unless the mate is lost, in which case the cock again comes into loud song."

Due to the variable nature and uneven growth of the Spruce-Fir forests in the area, the required habitat niches for the various resident species were not uniformly distributed. Because of this, the pre-collecting distributional pattern of breeding territories was quite different for each species. It was therefore of interest to note that subsequent to the collection of the original territorial males, the new males of each species almost invariably established territories in the same places that were occupied by their predecessors. Since an abundant food supply was available for birds throughout the area, these observations substantiate the importance of "the niche" in explaining local distribution of breeding birds.

Prior to the collecting period, the number of territorial males on the shooting area was 148, as determined by the spot-mapping method from June 6 to June 14. The number of adult males collected on the area during the period June 15 to July 8 was 302 + (Table 1). Thus, over twice as many males were removed from the area as were present before the collecting was started. The rapid influx and establishment of new territorial males, following the removal of the former occupants, account for the large number of males collected and are indicative of the amount of population pressure that was present in this community. It would appear that this pressure was due to severe competition, between individual males of the same species, for suitable areas to establish their territories. Apparently, the appropriate habitat

TABLE 1

PRE-COLLECTING POPULATION AND COLLECTION TOTALS OF ADULT MALES

Species	Pre-collecting population, territorial males June 6-14	Total males collected June 15–July 8	Collecting periods		
			June 15–23	June 24–July 1	July 2–8
Tetraonidae					
Ruffed Grouse, *Bonasa umbellus*	+	1	1		
Picidae					
Yellow-shafted Flicker, *Colaptes auratus*	1	2	1		1
Arctic Three-toed Woodpecker, *Picoides arcticus*	V	1		1	
Tyrannidae					
Yellow-bellied Flycatcher, *Empidonax flaviventris*	4	3 (4)	2		1
Corvidae					
Blue Jay, *Cyanocitta cristata*	1.5	2	2		
American Crow, *Corvus brachyrhynchos*	+	1 (1)		1	
Paridae					
Black-capped Chickadee, *Parus atricapillus*	2	3 (4)	3		
Brown-capped Chickadee, *Parus hudsonicus*	2.5	3 (5)	3		
Sittidae					
Red-breasted Nuthatch, *Sitta canadensis*	3.5	4 (1)	4		
Troglodytidae					
Winter Wren, *Troglodytes troglodytes*	2	1	1		
Turdidae					
American Robin, *Turdus migratorius*	2	14 (4)	4	4	6
Hermit Thrush, *Hylocichla guttata*	2	4	2	2	
Olive-backed Thrush, *Hylocichla ustulata*	7.5	11 (2)	4	2	5
Regulidae					
Golden-crowned Kinglet, *Regulus satrapa*	4	4	3	1	
Ruby-crowned Kinglet, *Regulus calendula*	2	2	2		
Bombycillidae					
Cedar Waxwing, *Bombycilla cedrorum*	V	6 (2)			6
Vireonidae					
Solitary Vireo, *Vireo solitarius*	3	4 (1)	3	1	
Red-eyed Vireo, *Vireo olivaceus*	1.5	7	2	3	2
Parulidae					
Tennessee Warbler, *Vermivora peregrina*	3	1	1		
Nashville Warbler, *Vermivora ruficapilla*	4	6	5		1
Parula Warbler, *Parula americana*	1	4	2		2
Magnolia Warbler, *Dendroica magnolia*	24	36	16	10	10
Cape May Warbler, *Dendroica tigrina*	12	23	13	10	
Myrtle Warbler, *Dendroica coronata*	8.5	29	11	9	9
Black-throated Green Warbler, *Dendroica virens*	V	1		1	
Blackburnian Warbler, *Dendroica fusca*	9	15	7	2	6
Chestnut-sided Warbler, *Dendroica pensylvanica*	V	1		1	
Bay-breasted Warbler, *Dendroica castanea*	35	81	34	21	26
Oven-bird, *Seiurus aurocapillus*	1.5	1	1		
Canada Warbler, *Wilsonia canadensis*	.5	1	1		
Fringillidae					
Purple Finch, *Carpodacus purpureus*	2.5	6	3	1	2
Pine Grosbeak, *Pinicola enucleator*	V	3			3
American Goldfinch, *Spinus tristis*	V	2			2
Slate-colored Junco, *Junco hyemalis*	5	13 (6)	6	2	5
Chipping Sparrow, *Spizella passerina*	V	1 (1)			1
White-throated Sparrow, *Zonotrichia albicollis*	3.5	5 (3)	5		
TOTAL	148	302 (34)	142	72	88

( ) undetermined sex   V visitor in area
+ indicates that less than .25 of one territory was present on area.

niches for most species were not numerous enough to satisfy the territorial requirements of all the males present.

The indications are that a large, surplus, "floating" population of unmated males must have been present in the Spruce-Fir forests of this region.   Since there was no evidence that spring migration was still continuing at the time the collecting was initiated, it was assumed that the birds which invaded the area were largely unmated wandering males that had been in the general region for some time.   Presumably these males were searching for suitable sites to establish territories, and whenever unoccupied habitat of the right type was found they lost little time in taking over and proclaiming their ownership by song. This is in line with the belief of Nice (1941) that "birds which fail to obtain territory, form a reserve supply from which replacements come in case of death of owners of territory."   In a study of the territories of Marsh Tits, Southern (1950) stated that "the speed with which deaths among the breeding population were replaced suggests that there was definitely a non-breeding population throughout the spring."

It is probable that at least some of the new males that were collected were those that had established territories in areas adjacent to the collecting area.   Upon the removal of the original territorial males in the collecting area, some of the males in outside areas could have taken advantage of the reduction in population pressure and expanded the boundaries of their territories into the collecting area.   In a study of banded English Robins, Lack (1946) found that "when the owner of a territory disappears, the owners of the neighboring territories expand into the vacant site almost at once, and often the ground is fully occupied within twenty-four hours."   In a comprehensive study of Ruffed Grouse, Bump, *et al.* (1947) discovered that the birds "tend quite rapidly to move into a covert whose population has been depleted out of proportion to those surrounding it."   In the present study, however, the fact that more new males first appeared in the center of the collecting area than on the periphery would indicate that most of them were wandering, unmated birds rather than neighbors expanding their territories.

The ratio of males collected to the number of territorial males on the area prior to the collecting period was found to vary considerably with different species.   In the case of birds belonging to Old World families such as the kinglets, chickadees, and nuthatches, there was very little difference between the two figures, indicating that their populations were comparatively stable.   However, for most of the thrushes, vireos, warblers, and fringillids, the number of males collected was from one and one-half to three times as great as the number present in the area before collecting. The Red-eyed Vireo showed the greatest difference of all with nearly five times as many males collected as were formerly present in the area (the relatively high numbers of American Robins and Cedar Waxwings that were collected were due to the presence of wandering non-territorial birds). These striking differences in ratios of individuals present before collection to those collected for the different species would seem to indicate that the surplus of males was much greater for some than for others. In the case of Brown-capped Chickadees, American Robins, Golden-crowned Kinglets, and Red-breasted Nuthatches, it is possible that because of their earlier period of nesting they might have passed that stage in their breeding cycle at which they were interested in establishing territories and, therefore, the number of individuals present before collecting was relatively similar to the number collected.

Differences in the ratio of males collected to number of males present in the area before collecting were quite pronounced even among species that were closely related and that had similar habits. The five most abundant species in the area were warblers, all belonging to the genus *Dendroica*. The number of males collected for every male that was present in the area before collecting is indicated for each of these species, as follows: Magnolia Warbler, 1.5; Blackburnian Warbler, 1.7; Cape May Warbler, 1.9; Bay-breasted Warbler, 2.3; and Myrtle Warbler, 3.4. Since these species are about equally conspicuous in the field, it is unlikely that such noticeable differences in the ratios can be attributed to variations in the ease with which they may be collected.

Those species that showed a definite surplus of males could be considered as having *supersaturated* populations (see Kendeigh, 1947). In other words for these species, the carrying capacity of the area, insofar as breeding territories were concerned, was insufficient to take care of the needs of all males present. Possibly, the presence of the infestation of spruce budworms could have accounted for this. The great abundance of food resulting from the budworm infestation might have attracted such a large number of males to the area that the territorial requirements of all of them could not be satisfied. The importance of territorial behavior in limiting bird population in a given habitat is stressed by Nice (1937) who stated, however, that "climate and other factors may keep the numbers of a species in a region so low that territorial behavior has no chance to limit populations." Thus, while various factors may affect the numbers of breeding birds occurring in an area, the true balancing agent between bird populations and environment is the competition for territories among individual males of the same species.   This is in accord with the findings of Nicholson (1933) who wrote "for balance, it is essential that the action of a controlling factor should be governed by the density of the population controlled, and competition seems to be the only factor that can be governed in this way."

It would seem probable that when certain elements, such as food supply, in the habitat vary, the saturation point for each species will vary also.   For example, the saturation point for most insectivorous birds in a Spruce-Fir forest without a budworm infestation would probably be much lower than it was in the area studied.   This belief is based on the assumption that the minimum territorial requirements are affected by the relative ease with which food can be obtained. Kendeigh (1947) has pointed out that in the case of the Tennessee, Cape May, and Bay-breasted warblers there appears to be a direct correlation between the abundance or occurrence of these species and the severity of spruce budworm infestations.   The principle of balance between changing animal populations and environment was expounded by Nicholson (1933) who stated that ' if a population is in a state of balance with the environment, its density must necessarily change in relation to any changes in the environment."   This principle is further clarified by Elton (1936) who maintained that "the chief cause of fluctuations in animal populations is the instability of the environment."

The number of adult breeding females was also greatly reduced during the collecting period, although the actual degree of reduction could not be determined from the data available.   The small number of females recorded at any time was somewhat puzzling.   Although female birds are generally much more difficult to observe in the field than are males, the exceedingly small number actually seen and collected over a three-week period would, assuming that most males had been mated, indicate that a preponderance of the females deserted the area shortly after their mates had disappeared.   The close attachment to the nests and general inconspicuousness of most females undoubtedly explains in part why such a small number was seen and collected during the first few days.   The presence of large numbers of conspicuous territorial males at that time would naturally tend to overshadow the count of the females.   However, the continued recording of low numbers of females following the initial period of collecting is thought to be due, in large part, to an extensive withdrawal of females from the area.   After the majority of the males had been collected, any type of bird activity, slight though it might be, became much more noticeable; thus the presence of any appreciable number of females would surely have been noted. At least 130 man-hours were spent in the area during the collecting period and practically all of the females seen were collected (Table 2). The females collected had well developed brood-patches on their breasts, indicating that they either were or had been sitting on eggs. If females had remained undetected in the area and continued to sit on eggs throughout the period, many of them would have been feeding young before the period was over and, therefore, would have become much more conspicuous at that time. The females of the later-breeding species did not become conspicuous at any time, and there was no evidence of young produced on the shooting area.

Certain species showed a much greater disparity in the proportions of the sexes seen and collected than others (Table 2). The greatest differences in the ratio of the sexes were found in the warblers (Parulidae) and vireos (Vireonidae). In the case of the warblers, the males of which comprised about 67 per cent of the total adult male population of all species, only 29 (13 per cent) of 228 adults collected proved to be females. For most birds, other than the warblers and vireos, the females were observed nearly as often as the males and were almost as readily collected from the start. It would seem to be significant that several species, of which almost equal numbers of males and females were collected, were earlier breeders which were caring for young at the time collecting began. This fact could be responsible for the females, being more conspicuous and more readily collected. It is well known also that females desert their nests and territories more readily before incubation becomes well advanced.

Variations in the sex ratios of the birds collected may be indicative of actual differences in sex ratios of certain of the species represented. The sex ratios, of birds collected, for the seven most abundant species are indicated as follows (ratio of males to females): Blackburnian Warbler, 15–1; Bay-breasted Warbler, 9–1; Magnolia Warbler, 6–1; Cape May Warbler, 6–1; Myrtle Warbler, 6–1; Slate-colored Junco, 2.5–1; Olive-backed Thrush, 2–1. It is unlikely that such marked differences as these could be due entirely to differences in comparative conspicuousness between males and females of the species listed.

SUMMARY

The number of territorial male birds in a 40-acre tract of Spruce-Fir forest in northern Maine was 148 during the period June 6 to June 14, 1949. Birds were removed from this area by shooting from June 15 to July 8, inclusive, with the intention of reducing the population to and keeping it at as low a level as possible. By this means the number

TABLE 2

NUMBERS OF ADULT MALES AND FEMALES COLLECTED

Species	Males	Females	Undeter-mined sex[1]
Tetraonidae			
Ruffed Grouse	1	1	
Picidae			
Yellow-shafted Flicker	2		
Arctic Three-toed Woodpecker	1		
Tyrannidae			
Yellow-bellied Flycatcher	3		4
Corvidae			
Blue Jay	2	1	
American Crow	1		1
Paridae			
Black-capped Chickadee	3	1	4
Brown-capped Chickadee	3	4	5
Sittidae			
Red-breasted Nuthatch	4	6	1
Troglodytidae			
Winter Wren	1	1	
Turdidae			
American Robin	14	11	4
Hermit Thrush	4		
Olive-backed Thrush	1	5	2
Regulidae			
Golden-crowned Kinglet	4	4	
Ruby-crowned Kinglet	2	1	
Bombycillidae			
Cedar Waxwing	6	2	2
Vireonidae			
Solitary Vireo	4	1	1
Red-eyed Vireo	7		
Parulidae			
Tennessee Warbler	1	1	
Nashville Warbler	6		
Parula Warbler	4		
Magnolia Warbler	36	6	
Cape May Warbler	23	4	
Myrtle Warbler	29	5	
Black-throated Green Warbler	1		
Blackburnian Warbler	15	1	
Chestnut-sided Warbler	1		
Bay-breasted Warbler	81	9	
Oven-bird	1	3	
Canada Warbler	1		
Fringillidae			
Purple Finch	6	3	
Pine Grosbeak	3	1	
American Goldfinch	2	3	
Slate-colored Junco	13	5	6
Chipping Sparrow	1		1
White-throated Sparrow	5	5	3
TOTAL	302	84	34

[1] Many birds lodged in tree-tops at time of collecting and were not examined closely.

of territorial males was reduced to approximately 19 per cent of the original by June 24. Continued collecting coupled with a steady influx of new birds maintained this low level until July 8. A total of 455 birds (420 adults and 35 young) were removed from the area during the entire collecting period.

Following the collection of the resident males, new males entered the area either during the night or early morning. The behavior of the new arrivals differed from the older residents in that they were much more active and vocal, singing more vigorously and more frequently. The importance of habitat niches in controlling the distribution of most species was indicated by the fact that most of the new arrivals established themselves in the same places that had been occupied by former residents of the same species.

For most species, over twice as many adult males were collected on the area as were present before the collecting started. This was due to the rapid influx and establishment of new territorial males following the removal of the original occupants. The carrying capacity of the forest from the standpoint of suitable sites to establish territories was not sufficient to accommodate all the males present, resulting in a surplus population. These surplus birds served as a reserve supply, replenishing areas that had been depleted. Some of the additional males collected were probably those that had territories adjacent to the shooting area, since the reduction in population pressure in the area due to collecting would permit an expansion of their territories into the shooting area. However, it is believed that the majority of evacuated territories were filled by entirely new birds. A few species were represented in the area only by wandering non-breeding birds.

The ratio of males collected to the number of territorial males on the area prior to collecting varied considerably with different species. The competition for territories among individual males of the same species appears to be the balancing agent between the size of bird populations and environment.

The number of adult females of later-breeding species collected was much less than the number of males. Although the general inconspicuousness of most breeding females undoubtedly explains in part why such small numbers were collected, there is evidence that a majority of the females in these species deserted the area after their mates had disappeared. The greatest differences in the ratios of the sexes collected occurred in the warblers (Parulidae) and vireos (Vireonidae).

In general, the later-breeding species, such as warblers, vireos, and thrushes, showed a greater disparity in the sexes of collected specimens than did the earlier breeders, such as the chickadees, nuthatches and kinglets.   This condition seemed to be correlated with the fact that these earlier breeders did not refill evacuated territories to nearly the extent that the later breeders did.   This could indicate a smaller reservoir of non-territorial birds and thus a lower population pressure in these species; or it could be the result of a more advanced condition of the breeding cycle during which the instinct to establish territory is absent or diminished in intensity.

LITERATURE CITED

BUMP, GARDINER, ROBERT W. DARROW, FRANK C. EDMINSTER, WALTER F. CRISSEY. 1947.   The ruffed grouse.   (New York State Cons. Dept.), pp. xxxvi + 915, 4 col. pls., figs. 1–171, tables 1–186, 127 sketches.

ELTON, CHARLES.   1936.   Animal ecology.   (Macmillan Co., New York), pp. xxx + 209, 8 pls., 13 figs.

KENDEIGH, S. CHARLES.   1947.   Bird population studies in the coniferous forest biome during a spruce budworm outbreak.   Dept. Lands and Forests, Ontario, Canada, Biol. Bull. 1: 1–100, figs. 1–32.

LACK, DAVID.   1933.   Habitat selection in birds with special reference to the effects of afforestation on the Breckland avifauna.   Journ. Animal Ecol., 2 (2): 239–262.

LACK, DAVID.   1946.   The life of the robin.   (H. F. & G. Witherby Ltd., London), pp. xvi + 224, pls. 1–8 (1 col.), figs. 1–6, rev. ed.

NICE, MARGARET MORSE.   1937.   Studies in the life history of the song sparrow. I.   Trans. Linn. Soc. New York, 4: vi + 247, 3 pls.   (1 col.).

NICE, MARGARET M.   1941.   The role of territory in bird life.   Amer. Midl. Nat., 26 (3): 441–487.

NICHOLSON, A. J.   1933.   The balance of animal populations.   Journ. Animal Ecol., 2 (1): 132–178.

SOUTHERN, H. N.   1950.   Marsh-tit territories over six years.   Brit. Birds, 43 (2): 33–47.

U. S. Fish and Wildlife Service, Patuxent Research Refuge, Laurel, Md., and Washington, D. C., August 24, 1950.

IV
Behavior of Populations

Editor's Comments on Papers 19 Through 23

Originally behavior was considered as an individual activity that allowed an animal to obtain from the habitat the maintenance requisites. However, the increase in knowledge showed first that the animals might have important effects on factors in the habitat. Then around 1950 it became apparent that the animals achieved a higher level of behavior when considered as a population and that the population became an ecological factor. Thus a new dimension was added to the complexity of analysis.

Lack's paper (1940) presents the beginning of this shift by describing the behavioral aspects of population change. However, it still treats the birds as individuals and hardly mentions their impact on habitat. But the paper was influential in helping people to think at the population level.

New ideas developed in Errington's (1945) paper on bobwhite quail. After a very old-fashioned description of factors, he describes the changes in numbers of bobwhites and the relation to habitat, which includes the bobwhites themselves as an ecologic factor. From the ideas on weather Errington gradually came to the idea of "basic security" from the hazards of winter weather. He noted that while the individual behavior of a bird was very important for survival, the covey also was an important unit. Thus the ideas of density and competition arose and the concept of a capacity developed. Note, however, that much of the competition was considered to be with other species (pheasants) rather than with the same species (quail). The inverse relation of rates of gain is noted, and the ideas of "inverse law" and "threshold" imply that behavior is an ecologic factor. This paper had wide influence in ecologic circles as well as in the profession of game management.

Perhaps an anecdote about Errington will illustrate how some advances occur in science. First note that Errington did not use mathematical notations in his discussion of the "inverse law." It happened that Thomas Park, then editor of *Ecology*, sent to Errington for review a paper submitted by Emlen, Stokes, and Winsor (1948) on increases of rat populations. These authors noted the application of the logistic to their data and presented a very simple mathematical discussion. Errington realized that his "inverse law" was simply a version of the logistic. The mathematics is simply to divide the first derivative of the logistic by N, resulting in an equation of a straight line with a negative slope.

Davis (1949) proposed that behavior through competition could be an ecological factor that determined changes in population. It would act at the population level to regulate the distribution of resources. Although at the time the intricacies of social organization of rats were not known, behavior clearly controlled the population forces of birth and mortality rates. The proposition that behavior could control populations seems to be original with this publication, although many other authors (such as Errington and Lack) gave hints. Also the concept of "density dependence," developed by entomologists, was very influential but did not include behavior as a factor (and hence is not included in this volume). Thus this paper seems to be the first to suggest that behavior can act in a density-dependent manner to regulate populations. The formulation of the principle was in very applied terms and perhaps therefore did not attract the attention of theoretical ecologists. A *mechanism* for this relation was suggested by Christian (see below), and much later (1959) Wynne-Edwards proposed the same idea but distracted the ecologists and behaviorists by his introduction of "epideictic displays" as a means of communication. It is important to note that the paper by Davis did not suggest a particular mechanism for the dependence on density.

A paper that provided much detail about the relation of competition and birth rates was Kluijver's monograph (1951) on the great tit. The part excerpted here was very influential because it has such marvelous data (30 years in length) and paid attention to ecologic aspects, often to the point of convincingly discarding them. The data were collected from the ecologic orientation of the early 1930s, which assumed that weather was a major factor in the control of birth rates. The extensive comparisons possible with data from so many years permitted the startling conclusion that weather was not important in determination of the second broods, which made the difference in population changes. Thus, although the habitat or type of woods affected population changes, the really significant factor is population density—which, however, can act at different levels (see Errington). The rejection of weather as a significant controlling factor freed the population ecologist to examine behavior as a density-dependent factor in regulating populations.

Christian's paper (1950) on the pituitary–adrenal system, while largely speculative, opened up the physiological study of the mechanisms that altered the birth and death rates in consequence of behavioral actions. In Christian's view (letter of August 1973): "Basically the article seems to be remarkably valid today . . . Much of Selye's G.A.S. is no longer correct . . . The role of behavior is clearly indicated in the article as is the inhibition of reproduction." After publication of this paper Christian began graduate study with D. E. Davis at the Johns Hopkins School of Hygiene and Public Health. Christian, because of his inherent ability and excellent preparation, produced promptly a set of studies, one of which (1956) received the Mercer Award of the Ecological Society of America. Subsequent papers showed that behavior acting through the pituitary–adrenal feedback mechanism restricted a population within the resources of the habitat. The influence of this paper and subsequent papers by Christian, and Davis and Christian ramified into many areas of physiology and behavior. For example, the role of behavior in resistance to disease now could be investigated in the context of specific mechanisms. Behavioral and endocrine effects on neonatal development (remember the stunting of infant mice) blossomed into one of the most active topics of research today. And, the understanding and control of population fluctuations (at least in rodents and deer) has greatly improved because a mechanism for density dependence is now known.

The framework was now complete. Behavior had been recognized as an ecologic factor in determining how resources were used and the physiologic repercussions were suggested.

References

Emlen, J. T., Jr., O. W. Stokes, and C. P. Winsor. 1948. The rate of recovery of decimated rat populations of brown rats in nature. *Ecology 29* (2): 133–145.

Wynne-Edwards, V. C. 1959. Population ecology. The control of population-density through social behaviour: A hypothesis. *Ibis 101* (3/4): 436–441.

Christian, J. J. 1956. Adrenal and reproductive responses to population size in mice from freely growing populations. *Ecology 37:* 258–273.

An Influential Book

Richdale, L. E. 1957. *A Population Study of Penguins.* Oxford University Press, New York, 195 pp.

Reprinted from *Ibis*, **88**, 299–307, 324 (Apr. 1940)

The Behaviour of the Robin:
Population Changes over Four Years

DAVID LACK

INTRODUCTION.

The main object of the present study, extending from January 1935 to October 1938, has been an investigation of the aggressive and sexual behaviour of the Robin, *Erithacus rubecula melophilus* Hartert, accounts of which have been published elsewhere (Lack (1939), (1940)). In 1935 over 11 acres, in 1936–38 over 20 acres, almost all the resident Robins were colour-ringed individually, and young birds hatched in the area were also ringed. Hence, incidentally, considerable data were collected on population changes. But it should be emphasized that, had population changes been a primary object of the present enquiry, a larger area could have been covered in greater detail. The present population study does not compare with that of Nice (1937) on the Song-Sparrow (*Melospiza melodia*), but it seems worth publishing because extremely few species have been investigated to the degree of detail of the present study. Further, the Robin has also been investigated by Burkitt (1924–26) at Iniskillen in Ireland, to which account repeated reference will be made. The present study was carried out in an area of gardens, quarries, and woodland on the Dartington Hall Estate near Totnes, Devon, some 330 miles south and somewhat east of Iniskillen. The results are closely similar to those of Burkitt.

Trapping.

The methods of trapping and ringing are described in Paper I. (Lack, D. (1939)). The Robin seems much the easiest of all British birds to trap in winter. Thus after two winter months of intensive trapping and ringing of all the small Passerine birds that would enter the traps, the Robin was the only species in which every resident had been ringed. The Robin, seems " curious ", that is to say, it soon comes to inspect any strange object in the territory, and, for catching a Robin, the more conspicuous the trap is the better. This " curiosity " often results in the Robin finding food, but seems primarily correlated more with a general awareness of what is happening in the territory. It may be noted that mated females are trapped much less frequently than mated or unmated males, and also less frequently than unmated females holding individual autumn territories.

Summary of Life-history.

To make what follows clearer, the life-history is briefly summarized : for details see Paper I. After the moult in July, the juveniles, and a little later the adult Robins, sing, fight, and take up territories, this applying to all males, and to at least some females. Song wanes in October, but territories are maintained. Pair-formation occurs from mid-December onwards, resident females leaving their territories to pair up with males, while there is an influx of other females from mid-December throughout the early spring. Nesting commences near the end of March, and there are a succession of broods until June, after which the adults become inconspicuous and later moult.

Sex Identification.

The sex of the Robin cannot be distinguished by plumage or measurement, but only by various differences in behaviour which become evident in the spring, described in Paper I. The sex of birds which leave before their first spring is therefore unknown, but individual colour-ringing followed by observations in spring means that the sex of all other individual Robins is certainly known.

Note to Maps and Tables.

The maps of the territories in different years refer to the bulk of the area investigated, but not to the whole, since, in an additional area where all the Robins were colour-ringed, the territories were not accurately mapped. For each aspect of this paper as many data were collected as possible, which means that the total number of Robins involved may be different in each case, *e. g.* the percentage of unmated males could be assessed over a wider area than that in which the areas of the territories were calculated, and similarly for all the other aspects.   This will account for apparent discrepancies in the totals in different tables.

I. Population Data.
Number of Adult Robins ringed.

Between January 1935 and June 1938 the number of adult Robins (*i. e.* excluding nestlings) ringed in the area was 64 males, 38 females, 17 resident birds unsexed because they left before their first breeding season, and 5 birds· unsexed because they were not seen after trapping (trespassers from a distance, wandering individuals) *.

Breeding and Autumn Populations.

Throughout this paper the breeding populations have been calculated for 1 April ; this is a little early, but most pairs have begun to nest, spring movements have normally ceased, and the leaves are not so thick that the birds are hard to see. There is a small decrease in adult birds during April and May.   The autumn–winter population should be calculated about 1 October, since by then nearly all territories have been staked out, but the leaves are still thick on the trees, and it is not possible to trap most of the new arrivals until November ; hence 1 December was selected, by which time all residents have usually been ringed, and this is before any spring movements occur.   The population on 1 December seems to be very little different from what it is on 1 October. In 1938, the autumn–winter population was calculated for 7 October, since I left the area after this.

TABLE I.—*Number of resident Robins in part of the Area.*

[Note.—Calculated to the nearest Robin, since territorial boundaries are not the same each year.]

	1935.	1936.	1937.	1938.
Breeding season (1 April)	12	10	15	11
Autumn–winter (1 December) † . .	13	13	12	10

* The sex ratio is discussed later, and should not be estimated from the above figures ; first, a few males and more females, paired with ringed mates, themselves escaped ringing ; secondly, the different degree to which the two sexes are resident, and thirdly, the unsexed autumn residents, need to be considered.

† 7 October in 1938.

y 2

Table I. shows that the number of resident Robins during the breeding season is roughly the same as in the late autumn ; Burkitt had similar data. This statement is not, however, true if only very small areas are considered, as clearly brought out by the maps (see pp. 304 & 305).

Sex Ratio.

The sex ratio is most fairly estimated by considering the total number of residents in the area at some particular date. As before, 1 April and 1 December were selected. A rather larger area was investigated than that for which the population densities are known, and a rather different area in each year, which accounts for the difference in total figures. (Only the ratio is required.)

TABLE II.—*Sex Ratio. Breeding Season* (1 *April*).

	Males.	Females.
1935	10	6
1936	14	9
1937	18	16
1938	24	19
Total.........	66	50
Percentage	57	43

TABLE III.—*Sex Ratio. Autumn Residents* (1 *December*).

	Males.	Females.	Unsexed.
1935	14	3	8
1936	20	7	5
1937	12	6	2
Total.......	46	16	15
Percentage ..	60	21	19

Burkitt's figures are very similar. Adding up three summers, he had 29 mated pairs and 7 unmated males, giving 55 per cent. males and 45 per cent. females. His autumn populations are harder to estimate, but if the results of his first autumn are included, the ratios are very similar to those in Table III. ; if the first autumn is omitted (it seemed rather unsatisfactory), the percentage of unsexed autumn residents is definitely lower than that in Table III.

The sex of the "unsexed" autumn residents, which materially affects the autumn figures, is discussed in the section on migration.

Size of Breeding Territories.

Burkitt states that breeding territories consisted of about 180 yards of hedgerow or 1½ acres. He did not observe great individual differences in size. However, he does not seem to have studied this aspect in detail.

In the present study the areas of the territories were calculated by careful mapping on graph paper and then " counting the squares "; this seems the best method for such irregular areas. Since the boundaries of territories are being continually modified, the boundaries on 1 April and 1 December are considered.

The average of 29 territories of breeding pairs works out at 6580 sq. yards, or 1·36 acres, a figure close to Burkitt's. The two smallest territories (male 85 and female 64 in 1937, male 66 and female 111 in 1938) were each about 2000 sq. yards or 0·4 acre. Only one other territory was less than 3000 sq. yards. Fourteen of the territories were between 3000 and 6000 sq. yards, six between 6000 and 9000 sq. yards, and six between 9000 and 16,800 sq. yards, the last being 3·26 acres. But in the case of territories temporarily exceeding 9000 sq. yards, the whole area was not maintained long ; the males gradually sang less in the distant parts and neighbours gradually encroached. Two acres seems about the maximum size of a stable breeding territory, which is four or five times as large as the smallest breeding territory. As can be seen from the maps, there were wide variations in the size of territories in the same area in different years, and in different but similar areas in the same year, and the same male or pair might occupy a territory of very different size in different years.

In 1937 the average size of territory for the eleven breeding territories measured was 4410 sq. yards ; the birds were appreciably more dense than in other years.

Text-fig. 9.

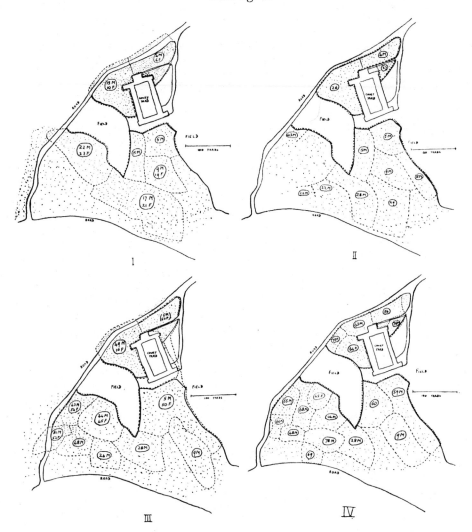

MAPS OF ROBIN TERRITORIES.

To show the variations in size and shape in different years, and also
how long each territory retained its occupant. M.=male. F.=female.
Where neither is put, the sex is in doubt.

(NOTE.—For large size and peculiar shape of territory, particularly
note 5 M and 50 F in Map III., and for small size 85 M and 64 F, also
48 M and 48 *a* F, in Map V.)

I.	II.	III.	IV.
Mid-April 1935.	Mid-October 1935.	Mid-April 1936.	Mid-October 1936.

Text-fig. 10.

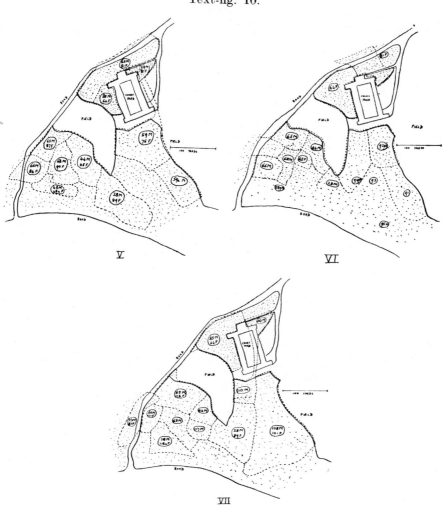

V.   VI.   VII.
Mid-April 1937.   Mid-October 1937.   Mid-April 1938.

(All the maps reproduced by kind permission of the Editor of the P. Z. S.)

Size of Territories of Unmated Males.

The average size of eleven territories of males unmated throughout the summer was 4460 sq. yards. Eight of the eleven territories were between 2000 and 4000 sq. yards, two being 6500 sq. yards and one 10,500 sq. yards. Hence the territories of unmated males are appreciably smaller than those of mated males. This is *not* because the males with larger territories more readily obtain mates, but because after pair-formation the pair usually enlarge their territory, and since an unmated male is much less aggressive than a mated one, this expansion is usually at the expense of unmated neighbours. In 1937 there were no unmated males, hence no territories which were readily compressible, which helps to account for the smaller average breeding territories in that year.

Size of Autumn Territories.

Burkitt considers his single autumn and winter territories were about half the size of the breeding territories, *i. e.* about 0·75 acre. At Dartington the average size of forty-one such territories was 3600 sq. yards, or 0·74 acre, which is very close to Burkitt's figure.

Kirkmann (1911) reported marked inequalities in the size of different winter territories ; Burkitt states that he did not find this. At Dartington the winter territories showed variations in size comparable with those of the breeding territories. The two smallest territories were each 800 sq. yards, the larger territories were not infrequently 6000 sq. yards. Hence while the *average density* of the Robin population seems almost identical at Dartington and Iniskillen in both summer and winter, the local variations in size of territory at Dartington itself are relatively very great. Howard's claim (1935) that in general bird territories are specific in size is therefore true of the Robin only between rather wide limits.

During the moult, Robins do not usually defend their territories, though they tend to remain in the same area. For four days during this period one moulting adult was not observed to move out of an area of 400 sq. yards, but this degree of restriction is exceptional.

Shape of Territories.

The territory is the area in which the Robin sings. Hence, where territories border open ground without trees, bushes or other song-posts, the open ground usually forms the boundary of the territory. Incidentally a road is not wide enough to achieve this. Where there is no such natural boundary, the boundaries between territories are determined by encounters with neighbours and do not follow natural landmarks.

Occasionally, Robins have curiously-shaped territories which necessitate crossing open ground to get from one part to another. Thus in 1936 male 5 and in 1937 male 59 regularly flew across 60 yards of open ground between two areas of wood, and in 1938 male 110 had even farther to go, but this bird used the intervening building for singing from (see maps). Burkitt notes a Robin which owned parts of three parallel hedges.

Food-trespassing.

Though a Robin sings and fights only in its own territory, it frequently trespasses for food. This accounts for the frequent records of several Robins caught in the same trap in the same season—Collennette (1931), Barnes (1934), and others ; the latter occurrences do not constitute a refutation of territory as I once thought possible (Lack (1933)).

In the twenty-four days between 14 March and 6 April, nine males and three females entered one trap. This is exceptional, for the trap was on the border of three territories, one of which changed owners four times during the period concerned. A more typical record is of eight Robins (five males and three females) caught in a trap in one territory in thirty-three days between 7 February and 12 March.

The greatest trespass recorded was a male trapped 350 yards from his territory. Another Robin was sometimes seen to fly across 80 yards of open ground to feed in another territory. Most trespasses were recorded in next-door territories.

REFERENCES.

BARNES, J. A. G. (1934).—Winter Territory of Robins. *Brit. Birds,* xxvii. pp. 267–268.

BARON, S. (1935). Robins changing Mates between Broods. *Brit. Birds,* xxix. p. 178.

BURKITT, J. P. (1924–26).—A Study of the Robin by means of Marked Birds. *Brit. Birds,* xvii. pp. 294–303 ; xviii. pp. 97–103, 250–257 ; xix. pp. 120–124 ; xx. pp. 91–101.

COLLENNETTE, C. L. (1931).—Some Notes on Robins. *London Naturalist* for 1930, pp. 135–136.

DROST, R., and SCHÜZ, E. (1932).—Vom Zug des Rotkehlchens, *Erithacus r. rubecula* (L.). *Vogelzug,* iii. pp. 164–169.

HOWARD, H. E. (1935).—' The Nature of a Bird's World.'

KIRKMAN, F. B. (1911).—' The British Bird Book ', vol. i. pp. 430–443.

LACK, D. (1939).—The Behaviour of the Robin.—I. The Life-history, with special reference to Aggressive Behaviour, Sexual Behaviour, and Territory. II. A Partial Analysis of Aggressive and Recognitional Behaviour. *Proc. Zool. Soc.* A, 109, pp. 169–200.

LACK, D. (1940).—Observations on captive Robins. *Brit. Birds* (in the press).

LACK, D. & L. (1933).—Territory reviewed. *Brit. Birds,* xxvii. pp. 189–190.

MICHENER, H. & J. R. (1935).—Mocking-Birds, their Territories and Individualities. *Condor,* xxxvii. pp. 97–140.

NICE, M. M. (1937).—Studies in the Life-History of the Song-Sparrow. —I. A Population Study of the Song-Sparrow. *Trans. Linn. Soc. New York,* iv.

TICEHURST, N. F. (1938).—' The Handbook of British Birds ', ed. H. F. Witherby, ii. p. 203.

Reprinted from *Ecol. Monogr.*, **15**, 3–21, 32–34 (Jan. 1945) with the permission of the publisher, Duke University Press, Durham, N.C.

SOME CONTRIBUTIONS OF A FIFTEEN-YEAR LOCAL STUDY OF THE NORTHERN BOBWHITE TO A KNOWLEDGE OF POPULATION PHENOMENA

PAUL L. ERRINGTON

INTRODUCTION

This paper attempts to bring up to date certain findings from a population study of the bobwhite quail *(Colinus virginianus)* in the vicinity of Prairie du Sac, Wis., 1929-44.

The investigation is one of several involving long-term, intensive observations of this species on definite areas in Wisconsin and Iowa. Because of the accuracy and continuity of winter censuses there and the relative clearness of the ecological picture displayed by some types of "key" data, Prairie du Sac is regarded in many ways as the most informative of the areas.

Although the early studies at Prairie du Sac have been well covered in the literature, data acquired over the greater part of a decade remain to be reported. These do more than merely pyramid evidence on phenomena that were adequately understood after the first years of work. It is true, for example, that virtually as much was known about starvation in wintering bobwhites by the mid-thirties as now but a number of concepts of population mechanics have been importantly modified by evidence more recently becoming apparent.

The present paper, despite the increasing reference to many species of animals that may be noted as treatment progresses, is essentially restricted in scope to the bobwhite. Even what may appear to be outright digressions are purposely included for the bearing they have upon the main subject matter. Nevertheless, the paper also is prepared in the hope that it may take us closer to principles of populations in higher vertebrates, generally, especially as relates to automatic controls and adjustments in rates of loss and gain.

The original Prairie du Sac data, here summarized, accrue from about 1,000 man-days of field work.

The 1929-32 studies were conducted throughout the year, chiefly by me but also with the help of Albert J. Gastrow, an experienced outdoorsman, resident in the area. This was done in connection with a program established at the University of Wisconsin by the Sporting Arms and Ammunition Manufacturers' Institute and the Bureau of Biological Survey, formerly of the U. S. Department of Agriculture. Thereafter, the area's investigations with which we are concerned in this paper consisted of winter (October or November to April) censuses and observations by Gastrow and were financed through Iowa cooperation (Iowa State College and the former Iowa Fish and Game Commission, with the aid of funds contributed by J. N. ("Ding") Darling), 1932-34; Wisconsin Alumni Research Foundation and the Iowa cooperators, 1934-35; Wisconsin Alumni Research

Foundation, 1935-36; National Association of Audubon Societies, 1936-37; Thomas E. Brittingham, 1937-38; Aldo Leopold, 1938-39; Wisconsin Conservation Commission (Pittman-Robertson funds), 1939-40 and first part of 1940-41; Aldo Leopold, second part of 1940-41; Wisconsin Conservation Commission (Pittman-Robertson funds), 1941-44.

In 1933, the year following my change of residence from Wisconsin to Iowa, the 1929-32 findings on wintering of bobwhites on the Prairie du Sac and other Wisconsin areas were published *(18)*. The Prairie du Sac studies continued under my supervision until the spring of 1935 and furnished much of the basis for a long report in 1936 *(33)*. During later years, while continuing to examine each season's data as they became available, I made no visits to the area, and the credit for keeping the program alive should be given very largely to the efforts—sometimes personal sacrifice—of Aldo Leopold, Professor of Wildlife Management at the University of Wisconsin. From 1941 to 1943, according to our opportunities, Leopold and I worked over the data with a view toward joint publication; by 1944, however, Leopold's schedule permitted no further expenditure of time on the project, and he invited me to go ahead on my own responsibility with the preparation of the manuscript.

Not here reported are supplementary studies at Prairie du Sac, begun in 1941 by members of the staff of the Department of Wildlife Management of the University of Wisconsin. Data from these will be written up for separate publication by the persons responsible for them.

Numerous individuals and organizations other than those already mentioned have given major help, some of which will later be acknowledged in appropriate places in text or bibliography. Among others deserving thanks for their cooperation, especially in the early years, are E. D. Ochsner of Prairie du Sac and W. B. Grange, now of Babcock, Wis.

THE AREA

The 4,500 acre study area (Fig. 1) lies on the east side of the Wisconsin River, at the corner of Columbia and Dane counties, opposite the village of Prairie du Sac. It includes the 3,200 acres referred to in other publications and in addition some bordering land. It is situated in the "dairy belt" of south-central Wisconsin, and has hills partly covered by hardwoods and valleys in pasture or cultivation.

Here, the bobwhite lives on the northern edge of its established range. To the northward are found only frontier populations of this "farm game" bird, usually sparse and discontinuous, though sporadically abundant *(50, p. 45)*.

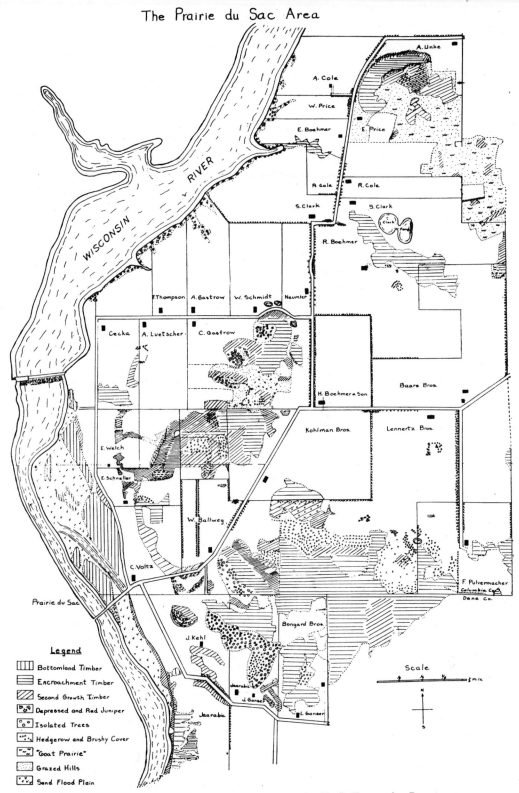

The Prairie du Sac Area

Legend
- Bottomland Timber
- Encroachment Timber
- Second Growth Timber
- Depressed and Red Juniper
- Isolated Trees
- Hedgerow and Brushy Cover
- "Goat Prairie"
- Grazed Hills
- Sand Flood Plain

Scale

FIG. 1. Study area east of Prairie du Sac, Wis. Map drawn by H. C. Hanson for Department of Wildlife Management, University of Wisconsin.

According to the evidence I have, local farming practices did not change perceptibly during the studies. Most woodlots were pastured and some were slashed. Trees and shrubs resistant to grazing by cattle encroached on sandy tracts and steep slopes. Frequent debrushing of roadsides produced its own plant and animal successions. Professor Leopold and Cyril Kabat, a graduate student who has done much work on the area, judge, on the basis of the 1934-35 cover map (*33*, pp. 394-395) and the recollections of Gastrow, that cover conditions for the bobwhite have long been deteriorating.

No artificial feeding, cover improvement, control of hunting or other form of bobwhite management was attempted. Hunting seasons on this species opened during the autumns of 1932 to 1935 and, on the Dane County part of the area, again in 1942 and 1943, but negligible numbers of the local birds were shot. Hunting of rabbits and squirrels and trapping of fur-bearers was similar to that usually carried on throughout the region. There were no "anti-vermin" campaigns; sometimes a hunter or a farmer shot a hawk or an owl. Foxes were hunted on occasion for sport or pelts. On the whole, resident wildlife was neither subject to excessive exploitation (or persecution) nor given more than passive encouragement.

For research on a species as relatively sedentary as the bobwhite, the Prairie du Sac area has many advantages. Nearly half of its periphery is bounded by a river sufficiently wide to discourage crossing by these birds at any season. On other sides, extensive tracts of habitat unfavorable for wintering bobwhites (deep woods and bare fields) help to reduce the chances of covey ranges being partly in and partly out of the area—one of the worst sources of confusion in many studies. The variables due to ingress and egress of wintering coveys are also lessened by the area's large size, as the majority of the coveys resident may engage in "normal" movements without crossing the borders. In most winters, tracking snows come often enough and last long enough to facilitate census-taking and "reading of sign."

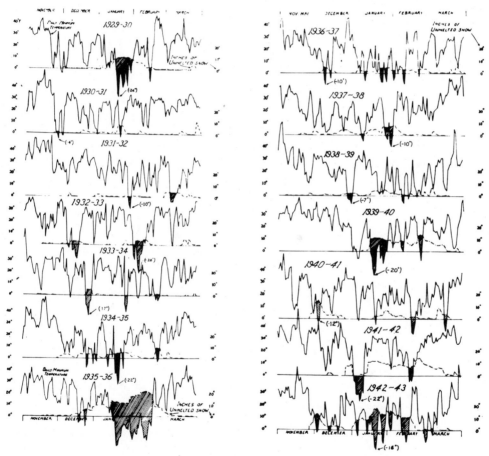

FIG. 2. Daily minimum temperatures and depths of snow on ground at Madison, Wis., winters, 1929-43. Prepared by Aldo Leopold from U. S. Weather Bureau data.

Furthermore, the area is one of those on which H. L. Stoddard had spent his early years as a naturalist. After completion of the Cooperative Quail Investigation of southeastern United States in 1929 *(66)*, he served as Biological Survey supervisor for a number of newly initiated cooperative studies of wild gallinaceous birds, including this one of Wisconsin bobwhites. The local program at Prairie du Sac was begun upon his advice and under his excellent guidance in the field.

WEATHER BACKGROUND

No precise weather data are available from Prairie du Sac for the period of study, so we must substitute records of the U. S. Weather Bureau for Madison, which is about 20 miles southeast of the area.

Figure 2 depicts temperature and snow depths for each winter except 1943-44, with times of simultaneous cold and snow being hatched for emphasis.

Figure 3 is an extension through 1942 of Wade's *(67)* use of an index of drought intensity proposed by Munger *(59)*, the base of each triangle covering a time in which no rain exceeded a twentieth of an inch per day.

Figures 2 and 3 are not claimed to register influence of weather upon the bobwhite.

As will later be shown, the winter of 1931-32, which looks mild on the chart, was terminated by a lethal emergency, and the winters of 1932-33 and 1934-35 look more similar when charted than one might expect from a far greater mortality associated with weather in 1934-35. Likewise, the mild-appearing winters, 1936-38, brought ice-storms and other ungraphed emergencies. The extremely severe winter of 1935-36 registers very well, but the "Armistice Day Blizzard" of November 11-13, 1940, looks unimpressive, although it killed many animals ordinarily weather-proof.

In Figure 3, the intensity of the 1934 drought appears far to surpass that of 1936. The latter summer does not look appreciably drier than most of the other summers, yet Wade *(67*, p. 560) states that "the damage done in 1936 was at least as great as in 1934."

POPULATION STATUS OF WILD HIGHER VERTEBRATES OTHER THAN BOBWHITES

It is axiomatic that the ecology of a wild animal cannot be thoroughly studied without reference to the associated biota. There are also practical limits to the scope of an investigation that can be carried on intensively by a few persons.

In this, an intensive study of the northern bobwhite, the associated species considered especially worthy of attention fell into three categories: (a) predatory species, (b) so-called buffer species, and (c) species possibly competing with the bobwhite as phyletically related forms.

PREDATORS

Presented in Table 1 are the available figures on winter densities of the principal large raptorial birds, foxes, and certain small and medium-sized fur-bearers.

The number of great horned owls *(Bubo virginianus)* regularly wintering on the area varied from four to eight and averaged six, or about one per square mile. In most springs, a nest was found for each two owls, but, in 1937, 1940, and 1944, no nests were found, and it is unlikely that any were overlooked. Food habits and prey relations of the species have recently been discussed at some length *(34)*.

The number of apparently wintering red-tailed hawks *(Buteo jamaicensis borealis)* varied from four to eight and averaged five. Field evidence indicates that these were mainly permanent residents, and, in springs when nesting data were procured, there were usually nests corresponding to every two wintering individuals listed in the notes. Failure to record nests in 1937 is laid to nonbreeding, but it is possible that late nests may have been missed.

The gray fox *(Urocyon cinereoargenteus)* has denning and ranging habits that simplify winter census work. In the early years of the study, density calculations were checked against numbers of animals known to have been taken in particular neighborhoods by hunters and trappers. Maximum winter densities of around five per square mile are suggested by the data both from Prairie du Sac and from another area near Blue Mounds, 18 miles to the south-

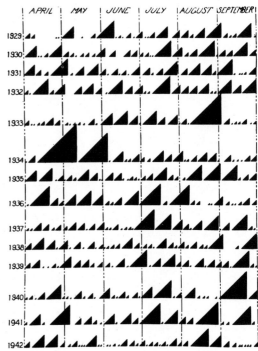

FIG. 3. Drought intensities, summers, 1929-42, plotted after Munger's method from rainfall recorded at Madison, Wis. Prepared by Aldo Leopold from U. S. Weather Bureau data.

TABLE 1. Censuses and estimates of certain predator populations, Prairie du Sac, Wis., 1930-1944.

Winter	Horned Owl Birds	Horned Owl Nests	Red-tail Birds	Red-tail Nests	Gray Fox Nov.	Gray Fox Apr.	Red Fox	Skunk	Mink	Weasel	Raccoon	Opossum
									(General winter estimates only)			
1930-31..	4	2	4	2	16	15	0					
1931-32..	6	3	4	2	21	17	2 transients		No estimates before 1933-34			
1932-33..	6	2	?	?	27	23	2 killed					
1933-34..	4	2	6	?	18	18±	0	45	0	4	10-12	?
1934-35..	7	3	8	2-3	20±	20±	0	40	1	2	10	12
1935-36..	7	3	6	2-3	30	15	0	50	3	6	0	12
1936-37..	5	0	4	0?	30±	28±	Species present at edge of area	45	2	12	10	0
1937-38..	6	2	4	2	30±	30±	2 or 3, of which 1 was killed	75	2	6	8	?
1938-39..	5	2	4	2	10	10±	8	15	3	6	8	1
1939-40..	4	0	4	2?	5	5	5	50±	2	?	3	1+
1940-41..	8	4	4	2	6	5	4 on area most of winter; 7 killed	50	2	4	3	?
1941-42..	7	3	5	3	?	2	5 on area by Feb.; 4 others killed	75	0	6	4	12
1942-43..	6	?	6	?	2	2	13 early winter; 7 by spring .	30+	2	4	7	20
1943-44..	7	0	5	?	11	8	17 on area by late Feb.; 28 others killed	70	5	5	15	10

west. The recent decline may have had a counterpart many years ago, for, after previous absence, the species was said to have reappeared locally in 1925 (50, p. 221).

Population figures for the red foxes (*Vulpes* sp.), which range beyond the area, should not be regarded as censuses, though Gastrow is an expert "still hunter" of red foxes and his opinion as to numbers should be close to the facts.

The figures on striped skunk (*Mephitis* sp.), mink (*Mustela vison*), weasels (*Mustela* spp.), raccoon (*Procyon lotor*), and opossum (*Didelphis virginiana*) are usually based upon Gastrow's winter estimates.

Not listed in Table 1, dogs (*Canis familiaris*) and house cats (*Felis domestica*) were about as abundant as they ordinarily are in midwest farming communities. Stomach contents of the few feral cats collected were similar to those of gray foxes. Gastrow estimated that there were five badgers (*Taxidea taxus*) resident in the winter of 1941-42; eight alive by the end of the winter of 1942-43; and 15 in 1943-44.

Of miscellaneous raptorial birds, barred owls (*Strix varia*) occurred regularly near the area but not often on it. Screech owls (*Otus asio*) were also to be found in some places, and there were variable populations of long-eared owls (*Asio otus wilsonianus*) and short-eared owls (*Asio flammeus*). Cooper's hawks (*Accipiter cooperii*) and marsh hawks (*Circus cyaneus hudsonius*) were often seen in summer and were irregular winter visitors. Bald eagles (*Haliaeetus leucocephalus*) wintered along the river; Gastrow has seen as many as seven in one day. Snowy owls (*Nyctea scandiaca*) and saw-whet owls (*Cryptoglaux acadica*) appeared infrequently in winter. Three goshawks (*Accipiter gentilis atricapillus*) stayed on the area for about a week in 1935-36 and another was recorded for early winter, 1943-44.

"BUFFERS"

Before 1941-42, when H. C. Hanson, a graduate student of Professor Leopold, made careful censuses of small mammals and birds (unpublished), no numerical data on "buffer" densities were obtained; great changes in populations were noted, however.

Cottontails (*Sylvilagus floridanus*) were considered "normal" for the area up to and including the winter of 1935-36. For 1936-37, Gastrow judged the cottontail population to be only about one-third "normal," and low densities were reported for other southern Wisconsin counties (53). Recovery at Prairie du Sac was visible by the winter of 1937-38; for the next two winters, numbers were again "normal"; and in 1940-41, Gastrow commented that he never before had known the rabbits to be so numerous. In 1942-43, they again were abundant; in 1943-44, much less so.

The white-tailed jackrabbit (*Lepus townsendii campanius*) has long been found on a sand prairie west of Prairie du Sac but occurrence on the area itself has been irregular.

The principal mice were meadow mice (*Microtus* spp.) and deer mice (*Peromyscus* spp.) of which the former showed much more pronounced fluctuations. It was plain that meadow mice were exceedingly abundant in 1929, 1938, and 1942, and there also was evidence (from their representations in the diets of local predators) that they may have reached another peak in 1934. The barn rat (*Rattus norvegicus*) spread into fields from farm yards in the warmer months but seemed unable to winter far away from buildings.

Populations of fox and gray squirrels (*Sciurus niger rufiventer* and *S. carolinensis*) appeared to be sparse but rather stable. Other resident sciurids included the woodchuck (*Marmota monax*), chipmunk (*Tamias striatus*), flying squirrel (*Glaucomys volans*), and the two common ground squirrels, *Citellus tridecemlineatus* and *C. franklini*.

WILD GALLIFORMES

Semiannual population figures for ruffed grouse (*Bonasa umbellus*) and ring-necked pheasants (*Phasianus colchicus torquatus*) are given farther on in this paper, in Figures 12 and 6, respectively. Neither grouse nor pheasant data are uniformly accurate, being for some seasons no more than estimates based upon miscellaneous "sign." The grouse may be regarded much more as true residents of the area than may the more mobile pheasants, which drifted irregularly across the land boundaries.

Judging from the annual survivors as noted in spring, the ruffed grouse populations wintered with fair security during the first 12 years at or under levels of 18 to 20 birds. In 1941-44, they varied from 30 to 38 (the spring density for 1944 was 32, not given in Fig. 12), or at levels approximately twice those for the earlier years. These grouse were legally protected throughout the study and, like the bobwhites, suffered little shooting. Unlike the bobwhites, they were not very susceptible to winter emergencies.

Large-scale pheasant stocking in Wisconsin began in 1928, but at Prairie du Sac, these exotic birds have not as yet shown any strong ascendancy—unless the 1941-42 density be so considered. A local farmer released 30 to 80 artificially propagated pheasants in August of each year since 1939.

Of other wild Galliformes, a flock of about 74 prairie chickens (*Tympanuchus cupido pinnatus*) was seen on the area in January, 1933.

SEMIANNUAL CENSUS DATA FOR THE PRAIRIE DU SAC BOBWHITE POPULATIONS

The data for the area's spring and fall populations of bobwhites, presented in Figure 4, were obtained by direct enumerations of local birds (for description of techniques, see *18*, pp. 1-11; *33*, pp. 310-333), which I have endeavored to sum up as for comparable times of each year.

Some trends may be detected in the gross fluctuations, but others are so obscured by variables as to be meaningless without further exposition. This is notably true with respect to winter losses after 1933-34, many of the summer gains after 1935, and the similarities the fluctuations may have with those of other species.

Readers are therefore advised not to strive overhard to draw conclusions from Figure 4 but to keep it in mind for reference as the discussion progresses.

BOBWHITE WINTER LOSSES AND THRESHOLDS OF BASIC SECURITY

DISSOCIATION OF VARIABLES

During the winters of my active participation in the field work at Prairie du Sac, emphasis was placed upon detection and analysis of mortality and

of intercovey adjustments. Fewer details were obtained thereafter, except as related to census figures, until the beginning of the supplementary studies by Professor Leopold's graduate students in 1941-42.

Prior to banding started by Kabat in 1942-43, conclusions arrived at through field observations on winter movements of Prairie du Sac bobwhites were checked mainly with the findings from earlier banding near Madison (*17*) and with those from Stoddard's extensive banding program in southeastern United States (*66*, pp. 167-182). Local movements were commonly followed with satisfaction for coveys living well inside of the traversable boundaries of the area, but the fortunes of border coveys were not always readily ascertained. Abrupt disappearance of such flocks that had been wintering without much loss was considered reasonably good evidence that they had moved out. Sometimes such coveys soon returned and sometimes the courses of those departing were traced. The area's winter egress tended to exceed its ingress, especially at the northeast corner where the birds could almost be depended upon to leave before a winter was over.

When a covey starved, the evidence was characteristic and not likely to be overlooked or misinterpreted by an experienced observer, even when complicated by killing weather and by raiding and scavenging of flesh-eaters (*18*, *27*).

Outright losses due to weather were less easily appraised, as were the consequences of burning, close pasturing, fall plowing, and similar human activities that may often leave large tracts of land nearly foodless and coverless for bobwhites. Circumstantial evidence of declines associated with these factors was acceptable only with reservations.

Voluminous data on food habits of locally resident predatory birds and mammals—particularly horned owls, which proved to be exceptionally good indicators of vulnerability (*26*)—were obtained in the early years of the investigations. Much "sign" of killing or eating of bobwhites was found then and dur-

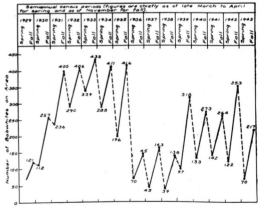

FIG. 4. Gross fluctuations, spring and fall, recorded for bobwhites at Prairie du Sac, Wis., 1929-43.

ing the last four winters. Responses of predators were correlated as much as possible with changes in availability of covey groups at times of hunger crises and of increased intraspecific friction, massing, and wandering (24, 33, 34). At its best, this was the most carefully executed part of the work, but the data are by no means of uniform quality.

Losses from accident, shooting (especially if illegal), pathology, and atypical immaturity, though difficult to record on a quantitative basis, clearly were not responsible for any extreme reductions in the bobwhite populations wintering at Prairie du Sac. Discoveries of feathers and picked bones were considered of very limited analytical significance unless contemporaneous loss trends were otherwise well defined—as by the contents of owl pellets or nests, bunched remains of covey groups at sites of melted snowdrifts, and underweight carcasses or feeble flight powers of members of starving coveys.

EMERGENCY AND NONEMERGENCY LOSSES

Emergency losses for bobwhites are those reflecting weather extremes, drastic though temporary reduction in habitability of environment, and the impacts of many of the special hazards introduced by civilized man. Losses from subhuman predators, collectively, are here considered of emergency types *only* when suffered by bobwhites made vulnerable primarily as a result of an emergency, i.e., birds weakened by hunger, crippled by shot wounds, evicted by snow or ice storms, floods, fires, etc.

Nonemergency wintering losses are broadly those suffered by populations of "normal," mature birds living free from adverse human interference in environment which, favorable or not, retains practically its full winter habitability. In other words, nonemergency losses at Prairie du Sac are of types to be expected in mild, open winters, among bobwhites that have little to contend with except their own selves and their usual kinds of enemies and competitors, hence are about the minimal losses consistent with anything to be called naturalness.

In looking over the following wintering histories the reader may avoid the confusion of possible finedrawn distinctions by keeping a few things strictly in mind. In practice, our chief preliminary task will be to winnow out, by one means or another, the declines occurring: (a) directly through predation; (b) through population adjustments of the birds, themselves; or (c) through predation and adjustments in combination.

This done, when we have reconstructed a picture of what the histories were or probably would have been in the absence of emergencies, we should be better able to search for basic attributes of north-central bobwhite populations.

WINTERING HISTORIES

Bobwhites at Prairie du Sac were known to have been greatly reduced during the winter of 1928-29. Snow was said to have accumulated to a depth of four feet. Gastrow, before the program was organized, observed a single covey, which shrank from 30 to about six birds. The surviving population of the area was calculated on a pro rata basis to be about 22 (33, p. 422), but this figure is no longer thought to be reliable. Lacking area-wide census data, we are entitled to say of the 1928-29 survival only that it may have been as low as at any time in subsequent winters.

Starvation and associated losses were locally severe in southern Wisconsin in 1929-30, the first winter of intensive study (18). However, the Prairie du Sac population happened to have been well fed, and a survival figure of 112 of 121 birds, as of April 1, was arrived at (21, p. 112).

The winter of 1930-31 was sufficiently open to afford relief from food crises to the bobwhites of the southern Wisconsin observational areas. At Prairie du Sac, the reported survival of 236 of 257 birds (21, p. 112) indicates a loss rate differing little from that of 1929-30.

The winter of 1931-32 was similar in open-ness to that of 1930-31 until March. Then, snows brought about widespread deprivation by burying the food left on the ground. The severest emergency losses observed took place in the Wisconsin River bottomlands southwest of Prairie du Sac. Here, 15 birds survived of a population that totalled 90 at the time of a January census and between 83 and 87 in early February (18, p. 28). On the study area east of Prairie du Sac, a 400-bird population was reduced to 290, but the losses ascribed on fair to excellent evidence to starvation were only in the vicinity of 39 (21, pp. 112-113). As the majority of the coveys had been living with slight loss from any cause for weeks prior to the lethal snowfall (18, pp. 25-27), they would in all likelihood have successfully wintered except for the emergency. Adding the 39 to the 290, we get 329, representing probable survival under nonemergency conditions.

Contrasted with the moderately light emergency losses at Prairie du Sac at the end of the 1931-32 winter, the earlier reduction from 400 to 329 comprises a substantial nonemergency loss, which for the first time in the Wisconsin investigations, occurred largely through predation upon physically "normal" bobwhites. (For data on the role of horned owls in this reduction, see 24, pp. 247 and 250, and 34, pp. 834-835, family studies 7 and 9.)

Bobwhites wintering at Prairie du Sac in 1932-33 were mainly favorably situated with respect to weather and food. A population of 406 was reduced through nonemergency losses to 339 by spring (21, p. 112), or to a level comparable with that surviving in 1931-32 up to the spring crisis. As in 1931-32, the preponderance of the mortality specifically traced was through predation upon apparently "normal" birds (19).

Nor were emergency losses considered important at Prairie du Sac in 1933-34, with the possible exception of those attributed indirectly to the debrushing of roadsides by relief labor. The debrushing was indeed drastic, and I estimated at the time and on the ground that 30 to 40 birds (the equivalent of two full coveys) had been evicted without the possibility of their finding suitable quarters elsewhere in the area (*33*, p. 362). This number added to the survival figure of 288 actually recorded for an initial population of 433 would give a nonemergency survival close to that calculated for 1931-32 and to that realized for 1932-33. I now question that the 1933-34 evictions were quite as lethal as then thought; but whether they were or not, it is apparent that the total nonemergency decline brought the population down to around 300 birds.

The winter of 1934-35 was one of weather-linked food shortages. Much habitat was made untenable, and many birds were forced to invade covey ranges already having populations as high as were usually accommodated. A spectacular acceleration of predation—especially by horned owls—followed this massing, but range units not seriously affected by the food shortage wintered about as many birds as in the past, and the predation on these tended to cease with relief from crowding (*23*, pp. 563-565; *24*, pp. 247-248; *33*, pp. 362-364; *34*, pp. 800-801 and p. 849, special phase study 13). A total of about 100 birds either left the area or died from starvation and storm. Despite reduction by spring to 196 birds from an initial 411, some 350 got along fairly well during the first half of the winter before the crisis became acute. Nonemergency losses would therefore appear about as substantial as for similar densities during the preceding three winters. We may see that a survival in excess of 300 birds—possibly 330?—could reasonably have been expected had it not been for the emergency conditions.

In 1935-36, there was a period of deep snow and some cold weather between late December and the middle of January, but, of an initial 416 birds, a fairly well-established population of at least 329 remained by midwinter. This suggests both a nonemergency decline not differing greatly from those of 1931-35 and the prospect of a comparable—300?—spring survival except in the event of lethal emergency. The emergency that did come resulted from several weeks of the severest winter weather shown by Weather Bureau records for the region. Blizzard followed blizzard, with prolonged, intense cold and much snow. One after another of the Prairie du Sac coverts was nearly or quite depopulated of bobwhites. By spring, only 70 of the birds were left alive on the area (*52*, p. 415).

Emergency losses in 1935-36 showed some departures from those observed in earlier years (*27, 52*). Typically, the bobwhites subsist on concentrated foods such as grains and the more nutritious of weed seeds and are unable to endure much fasting. They starve when heavy snowfalls cut off their food supply for a week or more at a stretch. In cold weather, the hunger-weakened are the likeliest to succumb. On rare occasions, sound birds in unsheltered places fall victim to cold, and others are imprisoned to starve or suffocate in hard-packed snowdrifts. In 1935-36, many birds died that should have been in fit enough

condition to have endured any of the weather crises between 1929-30 and 1934-35. Throughout the north-central states as well as at Prairie du Sac, bobwhites not only starved in the deep snow or died from the stress of the blizzards but also seemed to be worn down by the sustained cold.

Ice storms sealed much of the food supply of north-central bobwhites in 1936-37. By early January, a state of real emergency was manifest. The wintering population at Prairie du Sac was cut to 45 from an initial population of 145, a good part of which reduction may be attributed to easily understandable starvation. But the season's temperatures were mild, and, even before the ice storms, a decline of 25 birds had taken place. For the low population density, this was a very high rate of loss during a nonemergency period covering less than half of the winter.

Comparison of this early-season with the all-winter losses of 9 of 121 and 21 of 257 in 1929-30 and 1930-31, respectively, entitles one to ask whether the area would have wintered greatly more birds than it actually did in 1936-37 if the winter had been emergency-free—possibly about 100 instead of the 45? If such an estimate comes anywhere near the truth—and I regard it as the most defendable one I can make—then the situation in 1936-37 was quite different from that in any of the winters 1929-36.

The winter of 1936-37 had the only one of three low to moderate populations so far studied at Prairie du Sac that did not appear to be well accommodated under nonemergency conditions, and all of the five populations beginning winters with 400 birds or more either succeeded in wintering about 300 individuals or showed that they could have done so in the absence of emergencies.

The winter of 1937-38 was also mild, but a food shortage aggravated by heavy snows in late January was attended by extreme losses. The area's spring level of 39 bobwhites remaining of an initial 163 was the lowest for which we have accurate measurement. In contrast with the early, pre-crisis loss of 25 of 145 birds in 1936-37, the 1937-38 population lost only nine before the emergency period—and eight of those were lost in three ranges abandoned in late December and early January. The good survival in the majority of the coveys up to midwinter recalls 1929-31, and it would be easy to believe that, under nonemergency conditions, a much larger population than that actually entering the winter could have gotten through with low losses.

A 1938-39 decline of 41 from a fall level of 138 was recorded. Sudden losses of 13 birds in December and six birds in late winter took place in two border coveys that otherwise lost only two birds, and these 19 were judged simply to have left the area. Even after making allowance for such departure, a mainly nonemergency loss of 22 is still fairly heavy, compared with that for 1929-31 and 1937-38.

The second half of 1939-40 began with a heavy snow and a minimum temperature of 20° F. below zero. Snow and cold worked in deadly combination for several weeks. Only 133 of an initial 318 bob-whites were reported alive on the area by spring, but 38 of those listed in the decline wintered off the area after leaving in January. The effects of a contemporaneous crisis on a 4,200-acre area near Ames, Iowa, were personally studied in detail; 46 birds of a late December population of 231 were definitely known to have died, 37 carcasses having been recovered from three starving coveys, alone *(30)*. Up to midwinter at Prairie du Sac, most coveys maintained their numbers with little loss from any cause. At least 245 and possibly almost 290 were judged, on the basis of the covey records, to have stood a good chance of surviving had it not been for the emergency. In my opinion, the evidence favors the lower of the two figures given above, so perhaps we may suggest a nonemergency survival of about 255.

The decline to 142 from 273 birds in 1940-41 must be said to have been in part associated with unfavorable food conditions in certain covey ranges, although there wasn't much evidence of starvation except in one covey. Field notes described what could be interpreted as a generally marginal habitability of occupied quail land, unrest and local massing of birds. Predation losses were unexpectedly severe, and there was decisive evidence of something more than an ordinary emergency. Even if the starved-out covey had survived, the spring population would have been only 159. It does not seem probable that many more than the 142 actually surviving would have survived a nonemergency winter.

Save for a blizzard in early January, 1941-42 was mild. Predation again was heavy, and only 122 of an initial population of 264 bobwhites were present in spring. In this instance, emergency losses accompanying or following the blizzard had a fundamentally different significance than those characterizing, let us say, 1931-32, when a population had demonstrated its ability to live under nonemergency conditions until spring, finally to show a sharp loss from what almost might be called a freak accident. The February-March population in 1941-42 suffered so much predation during a period of no perceptible emergency that, January blizzard or not, we have no reason to think that the final survival figure would have been much changed if the winter had been entirely mild and snowless. For the density involved, this survival is low, indeed, as was the case in 1940-41.

The exact nature of the 1942-43 losses is badly obscured by undissociated variables. Some birds starved, but it was clear that outright starvation was not the primary medium of reduction. Live specimens handled by Kabat in the course of his banding study were generally in good flesh (unpublished). Heavy predation was noted by Kabat and Gastrow, and much in our interpretation must hinge upon whether that predation was or was not emergency-conditioned. Judging solely from the "paper evidence" made available to me, I would say that the losses look like emergency types, regardless of the lack of a wide-spread hunger crisis. Gastrow's journal refers to eight-foot snowdrifts, and Figure 2

shows a January and February depth of snow surpassed only by the 1935-36 accumulation for those two months. This could very well affect the habitability of many coverts.

Kabat (letter, Sept. 30, 1943) emphasizes night flushes of roosting bobwhites by red foxes, with individual birds alighting in exposed places to become sluggish from chilling and thus easily preyed upon by many predators. The consequences of such disturbances, nevertheless, would be accentuated during periods of cold and crusted snow (as were amply recorded at Prairie du Sac for January and February). It may be logically contended that even the losses of fox-flushed birds were emergency losses—at any rate to the extent that this type of loss would not be expected to occur during the 1930-31 sort of winter. Moreover, the numbers of bobwhites maintaining themselves fairly well until the January storms suggest that upwards of 275 might have survived a nonemergency winter.

Of singular openness until nearly the middle of February, 1943-44 may be classed as a nonemergency winter; nevertheless, of an initial population of 217 bobwhites, only 124 were alive on the area by spring. Predation losses again were heavy, reminding one of the winters of 1940-42.

THE THRESHOLD CONCEPT

As early as 1933, published reference was made to overpopulation of bobwhite environment at Prairie du Sac, but, to a considerable extent, my concept of overpopulation later proved to be erroneous—especially regarding the simple effects of apparent food and cover deficiencies *(18,* pp. 24-27).

At the end of the fourth winter, the low loss rates for 1929-31 and the similarity of the 1931-33 survivals after initially heavy populations had been cut down through nonemergency losses, in combination with like data from other areas, suggested the operation of thresholds of security, then called "carrying capacity" *(21)*. Except in the event of emergencies, populations living below threshold values wintered with slight reduction through predation and self-adjustment. If exceeding thresholds, populations betrayed instability and pronounced vulnerability to predation until again reduced to secure levels.

Thresholds of security, in terms of numbers accommodated, tended to vary with different areas but to be maintained with reasonable constancy for specific areas, winter after winter. The effect of emergencies was to suspend the operation of thresholds—sometimes to the point of making entire populations insecure—but, within limits, a great deal in the nature of environmental change was known to occur without appreciably modifying threshold values. Observed differences in kinds and numbers of predators and "buffers" seemingly were not reflected in changed security of wintering bobwhites.

Data at hand from all north-central sources by the end of the sixth winter chiefly confirmed the trends that were apparent by the end of the fourth and further amplified the threshold concept *(33)*. The importance of obtaining population figures from complete covey ranges and sizable land units was partly recognized. Distinctions were made between "key" habitats and combinations of food and cover and those having no unique significance despite more or less regular use by bobwhites. Such interpretations of threshold phenomena as were advanced did not ignore psychological factors but emphasized the possibilities of a given threshold being primarily a property of the environment.

In retrospect, I think that for many years I was seriously mislead in my attempts to analyze threshold phenomena by the evidences of year-to-year constancy in threshold values shown by the eight territorial subdivisions mapped for the Prairie du Sac area on pp. 394-395 of the six-year report (33). This apparent constancy for the designated land units continued for at least one winter—1935-36—after my personal participation in the field work had ceased. From then on, the constancy seems largely to have been lost for units much smaller than the total area.

It may be said, in short, that former territorial units have been so thoroughly merged and repartitioned since 1936-37 that many of their boundaries, as originally observed, have lost all validity, and the question of whether or not subdivisions of the area display year-to-year constancy in threshold values does not currently appear to be of more than very limited significance. Undoubtedly, the entire Prairie du Sac area shows a greater constancy in threshold values than do its constituent parts.

THRESHOLD CHANGES

From the early winter densities and the calculated nonemergency survivals that are tabulated in Figure 5, it may be judged that about the same basic threshold was operative not only for 1931-35, the last four winters for which data were published, but also very probably for 1935-36, the winter of greatest emergency, and for the low-density winters of 1929-31.

The calculated nonemergency survival of perhaps 100 and the high rate of nonemergency loss for a low-density population in 1936-37 may be construed as reflecting a lowered if not collapsed threshold of security for the area's bobwhites. The possible connection between this and periodic depression of other species in the north-central region will be taken up later.

The nonemergency loss rate for 1937-38 is so comparable with those for 1929-31, when initial populations were likewise below 300, that one might well believe the old threshold to have been again operative.

For 1938-39, the nonemergency rate is higher than would ordinarily be expected under the old high—a 300 to 330-bird—threshold, but it is not utterly excessive for a high threshold. Security, as from predation, is rarely absolute, and loss rates up to 6% per 90 days have been considered nothing unusual for recognizably secure wintering populations (33, p. 375). As will be brought out, however, the high rate for 1938-39 fits in with other evidence pointing to a suspended or depressed threshold.

The 245 to 290 bobwhites calculated to have had a good chance of surviving under nonemergency conditions in 1939-40 could be considered evidence for a threshold restored to about the 1931-36 level. A 290 survival would mean a 9% loss, which would fall on the curve outlined by the 1929-36 and 1937-38 data (lower center, Fig. 5). But, objectively regarded, a nonemergency survival closer to 245 than to 290 seems more in keeping with the field evidence. The selected figure of about 255 gives a loss rate of 20%, which is high for the curve, though still seeming to reflect a substantial basic security.

Survivals for 1940-42 offered the first really clear challenge to previously held ideas of year-to-year constancy in threshold values. Despite the questions as to the degree that the area's threshold was truly suspended or depressed in 1936-37 and the difficulties of appraising the loss rates for 1938-40, the data certainly show that something was different with respect to thresholds in 1940-42.

Complicated though they are by unknowns, the 1942-43 data revealed far too many bobwhites alive by midwinter to support any thesis to the effect that the poor actual survival merely reflected a continuation of the 1940-42 phenomena. In my judgment, the loss rate under nonemergency conditions would have been only a little above the 1929-36 and 1937-38 curve.

Lastly, for 1943-44, the status of the area's threshold apparently had reverted to about that for 1940-42.

From the nonemergency loss rates for 1938-39, 1940-42, and 1943-44, may we suspect the operation of a second basic threshold not much above 100 birds?

Further discussion of threshold phenomena may well be postponed until more evidence has been presented.

BOBWHITE SUMMER GAINS

LIMITATIONS OF SUMMER DATA

It is not implied that the increase of bobwhites on an area from spring to fall is exclusively or even largely the result of breeding of the birds present in the spring or that an individual bobwhite recorded at the time of the last winter or early spring census will necessarily remain on the area to breed. The months from late spring to early fall include, as well as the annual reproductive span of the bobwhite, periods of extensive readjustment in social groups, involving widely separated coverts and areas often to be reckoned in terms of square miles (11; 17; 33, p. 92; 66, pp. 169-182).

The quality of data obtained in warm weather studies of north-central bobwhites must be admitted to be greatly inferior to that for the winter records. Evidences of mortality and other "sign" persisted only briefly; visibility was poor in the dense ground cover; and the birds, old and young, were practically impossible to keep track of on a quantitative scale sufficient to permit bonafide week-to-week population studies. Careful summer work during the first three

Winter survivals and loss rates of wild Galliformes (other than grouse) at Prairie du Sac, Wis., as actually occurring and as calculated for nonemergency conditions. (See text for comments).

Winter		Early winter density	Density by spring	Percent of winter loss, all causes	Calculated Survival under nonemergency conditions	Percent winter loss calculated nonemergency conditions		Early winter density	Calculated survival under nonemergency condition	Percent winter loss calculated for nonemergency conditions
1929-30	Bobwhite, only	121	112	7%	112	7%	Bobwhite and pheasant collectively	121	112	7%
1930-31		257	236	8%	236	8%		257	236	8%
1931-32		400	290	28%	329	18%		400	329	18%
1932-33		406	339	17%	339	17%		406	339	17%
1933-34		433	288	33%	300?	31!%		433	302?	30!%
1934-35		411	196	52%	330?	20?%		413	332?	20?%
1935-36		416	70	83%	300?	28?%		419	303?	28?%
1936-37		145	45	69%	100?	31?%		153	106?	31?%
1937-38		163	39	76%	150?	8?%		172	159?	8?%
1938-39		138	97	30%	116	16%		153	120	22%
1939-40		318	133	53%	255?	20?%		318	258?	19?%
1940-41		273	142	48%	142?	48?%		283	148?	48?%
1941-42		264	122	54%	122	54%		304	152	50%
1942-43		353	70	80%	275+?	22% or less?		383?	283+?	26% or less?
1943-44		217	124	43%	124	43%		231	131	43%

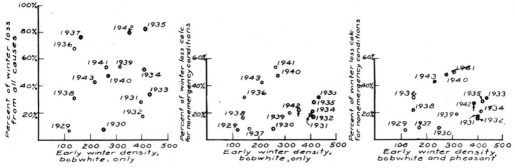

FIG. 5. Rates of loss of wintering bobwhites and ring-necked pheasants in relation to density, Prairie du Sac, Wis., 1929-44.

years in southern Wisconsin yielded scarcely passable data on nesting losses of bobwhites (20); feeding trends of, and preying upon bobwhites by, certain predators, notably horned owls (34); and fragmentary information on mortality and behavior of juvenile bobwhites. After termination of the all-year field work at Prairie du Sac in 1932, the rate of population gain by fall (checked by general observations in others parts of the region) constituted nearly the sole criterion of reproductive success.

Our summer picture is therefore all but lacking in detail, being represented rather literally by not much except the frame. But the spring and fall census data comprising the frame are of good quality and, whatever else may be said of them, afford us something definite enough to grasp. Considering nothing except the census data as they have come to hand, year after year, we are not obliged to make any assumptions whatever as to what the area's bobwhites did or did not do in the intervening months. The data are legitimately usable for analytical purposes to the extent that they show significant trends.

INVERSE RATIOS IN RATES OF GAIN

Summer gains, as shown by numerical differences between spring and fall populations in Figure 4, look highly variable; but, as plotted in percentages in the upper left of Figure 6, they reveal certain patterns.

By the fall of 1932, it had been noted that summer gains tended to be in inverse ratio to spring densities

(21; pp. 124-125), and this inverse relationship was better defined by the fall of 1935 (33, p. 422). At the time of this writing (spring, 1944), the percentages of gain in relation to spring densities for 13 of 14 summers at Prairie du Sac fall pretty well in two curves.

The principal curve, defined by the rates of gain for eight of the 14 summers (1931-35, 1937, 1939, and 1942) is a reverse sigmoid, and, as long as spring to fall rates of gain in bobwhite numbers conform to it, the fall populations should be fairly well predictable from spring densities even before the start of actual breeding.

For such years, we may ordinarily expect Prairie du Sac spring densities of about 40 birds to be followed by fall densities of about 140; spring densities of about 100, by fall densities of about 325; spring densities of about 200, by fall densities of about 400; spring densities approaching 340, by fall densities approaching 440. Stated otherwise, if the fall populations with which we are here concerned were made up only of evenly paired spring residents and their progeny—which, of course, we know is not true—the average numbers of young raised by fall per pair would be about five from a 20-pair spring level; more than four young from a 50-pair level; about two young from a 100-pair level; and not much more than one-half young from the highest observed level of about 170 pairs.

The second curve is too imperfectly defined by the

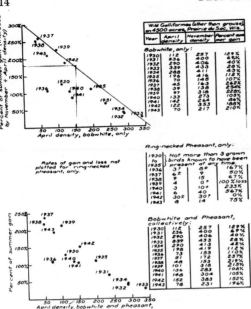

Wild Galliformes (other than grouse) on 4500 acres, Prairie du Sac, Wis.			
Year	April density	November density	Percent incr over gain
Bobwhite, only:			
1930	112	257	129%
1931	236	400	69%
1932	290	406	40%
1933	339	413	28%
1934	288	411	43%
1935	176	416	112%
1936	70	148	107%
1937	45	163	262%
1938	97	338	254%
1939	97	316	228%
1940	133	273	105%
1941	142	264	86%
1942	122	353	188%
1943	70	217	210%
Ring-necked Pheasant, only:			
1930 to 1935	Not more than 3 grown birds known to have been present at any time;		
1936	3±	8±	167%
1937	6±	15	50%
1938	9	0?	67%
1939	4	0?	100% loss
1940	3	10±	233%
1941	6	40	567%
1942	30±	30±	0%
1943	8	14	75%
Bobwhite and Pheasant, collectively:			
1930	112	257	129%
1931	236	400	69%
1932	290	406	28%
1933	339	433	28%
1934	290	419	42%
1935	176	153	112%
1936	73	153	110%
1937	51	172	237%
1938	45	153	219%
1939	101	316	213%
1940	136	283	108%
1941	146	304	105%
1942	152	383	152%
1943	76	231	196%

FIG. 6. Rates of gain of bobwhites and ring-necked pheasants in relation to density, Prairie du Sac, Wis., spring to fall, 1930-43.

gains for 1930, 1938, 1940-41, and 1943 to warrant many comments, particularly as concerns any possible lower asymptote that it may have. So far as it goes, it seems to reflect inverse ratios to spring densities somewhat after the manner of the first curve though on a different level of increase.

Conformation of 13 of the 14 rates of summer gain at Prairie du Sac to one or the other of two density-determined patterns has not been noticeably influenced by emergencies. In southern Iowa, a drought occurred in 1934 that was of such intensity and long duration that it virtually brought about desert conditions in local areas, and this was accompanied by summer declines in bobwhite populations (22). In other parts of the north-central bobwhite range observed during the summer and fall of that year, including southern Wisconsin, the drought seemed less severe than in southern Iowa and more within the toleration limits of the species.

The drought summer of 1936 was one of decidedly low rates of gain in relation to spring densities at Prairie du Sac as well as elsewhere in the region, but, as will be shown, drought would appear to have figured less importantly than what we, for want of understanding and a better term, must still call periodic depression.

THE QUESTION OF PHASE-LINKAGE BETWEEN RATES OF SUMMER GAIN AND NONEMERGENCY WINTER LOSSES

Let us again look at the rates of nonemergency winter losses and rates of summer gains of the Prairie du Sac bobwhites in the lower center of Figure 5 and the upper left of Figure 6.

First, it may be perceived that, *for the density levels involved,* the lowest rate of summer gain—that of 1936—was followed by the highest calculated rate of nonemergency winter loss. Then, three summers —1940, 1941, and 1943—of gains depressed below the main eight-summer curve were followed by distinctly high rates of nonemergency winter loss. A similar thing seemed to have occurred in 1938, but the densities concerned were sufficiently low to make departures from the "normal" gain and loss curves hard to judge.

The slightly elevated (compared with the curve for 1929-36 and 1937-38) rates of nonemergency loss calculated for the winters of 1939-40 and 1942-43 do not follow summers for which rates of gain were depressed below the main curve. They do precede summers of depressed gains.

Of all of the data points falling outside of the two "normal" curves for rates of nonemergency winter loss and summer gain, that representing summer gain for 1930 is the only one standing by itself.

More groundwork will have to be laid before treatment of the present subject may be advantageously concluded in this paper. Meanwhile, the reader may be left with the thought that rates of summer gain and nonemergency winter loss perhaps do run in some sort of phase.

INFLUENCE OF CHANGES IN SO-CALLED LIMITING FACTORS ON THE POPULATION STATUS OF THE PRAIRIE DU SAC BOBWHITES

The "limiting factors" listed in publications on the natural history, ecology, or management of a wild animal usually include extremes of weather; availability of food, cover, and water; various aspects of predation (or of parasitism); and competitive relationships.

Much of the treatment of these factors in the literature is extremely hard to appraise, and a reader may have justifiable doubts as to what is desirably standardized or merely stereotyped, whether an author's statements reflect critical thinking or indoctrination, self-acquired or otherwise. Acknowledgment of some of these deficiencies in my own past writings should not make it less in order for me here to review the question of what really did limit bobwhite populations at Prairie du Sac.

WEATHER AND ENVIRONMENT

There can be no doubt that weather emergencies (notably those manifested by winter starvation) may have influence on population levels of the bobwhite in northern parts of its geographic range. These emergencies, however, show much variability, both in severity and time of onset. Their effects upon local bobwhites do not correspond closely to any natural rhythm that I can detect.

The possible influence of year-to-year variations in food, cover, and water on the Prairie du Sac bobwhites has received a vast amount of attention without yielding much except negative information. Unquestionably, environmental changes have resulted from land-use practices in the course of the 15-year study, and we should not lose sight of the cover deterioration previously mentioned to have been noted by Leopold and Kabat.

For all of the local consequences of roadside debrushing, burning, fall plowing, removal of corn shocks, etc., recorded during investigations of the northern bobwhite, I do not think that the depressed rates of summer gain and the depressed winter thresholds at Prairie du Sac since 1936 may correctly be laid to environmental impoverishment.

The drought of 1936 had its effects on food and ground cover, but, from a short visit made in early winter 1936-37, I would say that conditions did not look appreciably worse than they had in 1931, when rates of summer gain and nonemergency winter loss had followed the patterns defined by the curves for the "normal" years. The 1940-44 data might be construed as reflecting general deterioration of the physical environment were it not for contrary evidence furnished by the summer gain for 1942 and the winter survival up to the beginning of the storm period in January, 1943. And the somewhat depressed rate of summer gain for 1938 and the elevated rate of nonemergency winter loss for 1938-39—coming as they did between the "normal" gains and losses of the summer of 1937, the winter of 1937-38, and the summer of 1939—do not look like early evidence of any permanent change in basic habitability of the Prairie du Sac environment.

Before we here strain to reconcile what is not necessarily reconcilable, let us consider some other things.

PREDATORS AND "BUFFERS"

The reader, checking back on Table 1, will find no figures on predator densities for 1929-30, for the reason that predators were not then studied on all of what was later established as the Prairie du Sac observational area. For the central three square miles, on which most of the work was done in 1929-30, the regularly active predators, "as nearly as could be determined, were one or two horned owls, at least one barred owl, several screech owls, 3 redtails, and a rather sparse population of grey foxes and weasels" (18, p. 22). This suggests that the whole area had, except for fewer foxes and more small owls, a predator population in that winter not differing greatly from that of 1930-31.

Practically any species of raptorial bird and carnivorous mammal listed in this paper may be expected to do more or less preying upon the bobwhite under conditions favoring predation. Of all of these, the horned owl, Cooper's hawk, house cat, and the two foxes were the species to which by far the greater proportion of the actual killing must be charged—to the horned owl possibly more than to all of the others together (24, p. 245). The striped skunk is well known as an egg-eater, and foxes and cats

would be likely predators upon incubating and immature bobwhites. Among the rodents, the ground squirrels, from studies elsewhere (45), may be regarded with suspicion as enemies of eggs and young.

In view of the numerically undetermined, but not noticeably changing, year-to-year status of house cats and ground squirrels, and the irregularity with which Cooper's hawks visited or lived in the area, one problem is to find whether there is a correlation between basic population phenomena of the Prairie du Sac bobwhites and changes in densities of the local horned owls, foxes, and skunks.

Another is the possibility of fluctuations in "buffer" populations influencing the status of the bobwhites. Cottontails and meadow mice were the most important "buffers" showing pronounced changes in numbers.

The lightness of the nonemergency loss rate for the winter of 1929-30 might at first glance look consistent with conventional views on bobwhite-predator-"buffer" relationships. The predator population was rather low, the meadow mouse population was high, and the cottontail population was about "average."

But the 1930-31 rate for a higher bobwhite population was nearly the same as in 1929-30. Though the horned owl and cottontail populations were evidently similar in 1930-31 to what they had been in 1929-30, the meadow mouse "peak" had collapsed months before the winter of 1930-31, and the gray fox population had risen to a moderate level by 1930-31. Cooper's hawks in 1930-31 were unusually persistent in their harassment of bobwhite coveys (18, pp. 22-24).

In 1931-32 and 1932-33, we see similar rates of nonemergency winter losses from similar bobwhite densities. The populations of horned owls and cottontails were also similar, but those of gray foxes and meadow mice were higher in 1932-33 than in 1931-32.

A higher initial density of bobwhites suffered a higher rate of nonemergency loss in 1933-34. The winter was one of fewer horned owls and gray foxes than 1932-33, about the same density of cottontails, and a population of meadow mice apparently approaching another peak.

Nonemergency losses of bobwhites were evidently nearly as great in the winters of 1934-35 and 1935-36 as in 1933-34, from nearly as high initial densities. The horned owl populations for 1934-36 were the largest thus far recorded, being almost twice as high as in 1933-34. Gray fox densities were somewhat higher than in 1933-34; there was no known pronounced change in the cottontails; and meadow mice may have reached a peak in 1934 and fallen off later. A brief visitation by goshawks was not known to have changed any trends in the bobwhite data.

Now we come to the sharply increased loss rate for 1936-37, for a low-density bobwhite population. The horned owl population was lower than in 1934-36, and the adults had no young in nests by late winter and spring. Gray foxes had virtually reached their highest level for the area. Cottontails obviously had declined and meadow mice were probably present in moderate numbers, being on the way to their next abundance peak in 1938.

For all of the nonbreeding of the horned owls, it might be argued that the predator population of 1936-37 was still high enough with respect to available "buffers" to be forced more to seek other prey, such as bobwhites; but, in 1937-38, when cottontails were still low (though meadow mice were nearing another peak), another very high gray fox population and a horned owl population once again "normal" took only about the same relative toll as much smaller numbers of predators having much more available food had taken from the lower bobwhite population of 1929-30.

The winter of 1938-39, when the nonemergency loss rate for bobwhites was rather high for their density level, was one of meadow mouse abundance and "normal" numbers of cottontails and of what might be called an ordinary horned owl population. It was also marked by sharply lowered gray fox densities and by the largest number of red foxes observed on the area in the course of the study up to that time.

Despite lack of certainty as to the numerical value of the nonemergency loss for the winter of 1939-40, my feeling is that it was fairly high for the initial population level of the bobwhites. This, however, is correlated with no differences in numbers of predators or buffers that I can see. It is true that meadow mice had presumably declined following their 1938 peak, but cottontails were again "normal" and about to begin an ascendancy. Horned owls and both species of foxes were less numerous than they had been the previous winter.

The winters of 1940-41 and 1941-42, with their similar, moderate-density populations of bobwhites, showed high rates of nonemergency losses. Cottontails were exceedingly abundant both in 1940-41 and 1941-42, and meadow mice again reached peak numbers in 1942. The 1940-41 population of horned owls was the heaviest of the entire period of study and, in 1941-42, it was still as high as it had been in the previous winters of abundance, 1934-36. Gray fox densities were about the same in 1940-41 as in 1939-40, but this species had almost disappeared from the area by 1941-42. Red fox densities remained about the same in 1940-41 and 1941-42. The predator-prey picture for 1940-41 shows a generally higher level of bobwhite enemies and a lower level of "buffers" than for 1941-42, yet the rate of predation upon similar densities of bobwhites was, if anything, higher in 1941-42.

Regardless of questions remaining unsettled as to how much the severe wintering losses for 1942-43 were emergency or nonemergency in type, we know that there was a great deal of predation upon a fairly heavy bobwhite population. The meadow mouse population was high, as was that of cottontails, though the latter were not as abundant as they had been in the preceding two winters. Horned owls were present in "average" numbers; gray foxes

were barely represented; and red foxes were more abundant than in any previous year of the program. The association of red fox activities emphasized by Kabat might very well represent an intensification of predator pressure of a sort not observed in earlier years. On the other hand, the night flushing by gray foxes described (18, p. 28) for 1931-32—one of the winters of heavy bobwhite densities and high nonemergency survival—would not appear so dissimilar to that of the reds in 1942-43.

The winter of 1943-44 was remarkable for its fox densities. Grays were back about to their 1938-39 level, and the numbers of reds actually killed during the winter surpassed all records for any area of like size of which I know. It seems within reason that upwards of 40 foxes may have ranged over some part of the 4,500-acre tract during particular nights. Horned owls, too, were abundant, as were miscellaneous carnivores. Cottontails and meadow mice had decreased, which, with the above, would appear ideally to fulfill the conventionally accepted requirements for heavy bobwhite losses from predation.

Heavy predation in fact took place, and the possibility is not disproved that so many active and hungry flesh-eaters, by their predacious weight alone, may have forced the wintering population of bobwhites below otherwise operative thresholds of security. Nevertheless, the resemblance of the 1943-44 losses to those of 1940-42 is too great to make wholly convincing any sweeping explanations in terms of increased populations of predators and decreased populations of "buffers."

Rates of gain of the Prairie du Sac bobwhites conformed to the "normal" curve for eight of 14 summers, despite some quite pronounced differences in status of skunks and foxes, as well as of other predators. Of these "conforming" summers, 1937 was one of high skunk densities, and the summers of 1931 to 1935 and that of 1937 evidently had substantial to very high densities of gray foxes. Meadow mouse populations were low about 1935 and 1939 and cottontails were still low in 1937. The predator and "buffer" populations were probably as much out of theoretical balance in 1937 as in any summer of "normal" bobwhite recovery at Prairie du Sac.

In 1936, the year of most severely depressed summer gains for bobwhites, Prairie du Sac skunks were less abundant than in "normal" 1937; gray foxes were at a high level in 1936 but no higher than in 1937; "buffers" admittedly were present in diminished numbers in 1936, but their status by 1937 can hardly be thought to have improved so much as to explain the differences between the bobwhite gains of the two summers.

There is likewise nothing about the known predator and "buffer" situations for 1940 and 1941 to suggest a connection with the depressed gains of the bobwhites. Skunks were abundant in 1941 (about as in "normal" 1937); gray foxes were low in 1940 and still lower in 1941; red foxes were fairly numerous in 1940 and 1941, yet were more abundant in 1942, a year of "normal" bobwhite gain.

If we may judge from the fall and early winter densities, the skunks and the gray and red foxes collectively must have reached, in the summer of 1943, their greatest abundance during the Prairie du Sac studies. At the same time, "buffer" densities were at least fairly low. The depressed rate of gain of the bobwhites for this year may be said to coincide with a state of predator-prey unbalance, but the unbalance does not look any worse than it did for "normal" 1937. Even if the predators did depress the bobwhite gain for 1943, the rate does not fall much short of the "normal" curve. More likely, in my estimation, the rate of gain simply conformed to the second curve, that already outlined by the gains for 1930, 1938, and 1940-41.

The lack of connection between rates of nonemergency winter loss and summer gain and observed differences in populations of predators and "buffers" has seemed to hold pretty well for areas other than Prairie du Sac in the north-central region (29, 33). It denotes much variability in details of mortality, reproduction, and movement of the bobwhites being cancelled out in net effect by the bobwhite's own density factor, in particular. Many of the supposed advantages of favorable breeding seasons or decreased losses from enemies fail to count in the end, as do, in their turn, many of the disadvantages of unfavorable breeding seasons or of increased losses from enemies.

Thus are exemplified the intercompensatory trends in population phenomena of higher vertebrates that have been becoming more and more apparent (31, 32). How widely they apply to the higher vertebrates generally or even to bobwhites over the occupied range of that species is quite undetermined. Less flexibility in bobwhite-predator-"buffer" relationships may exist in southeastern United States, where fewer alternatives of action seem to be open to predators at times of failure of their main staple food, the cotton rat (Sigmodon hispidus) (35).

Known losses of free-living bobwhites from parasites and diseases at Prairie du Sac and elsewhere in the region have been so negligible, and the references to them so scattered in the literature, that the subject may here be dismissed with little additional comment. Let it only be pointed out that ordinary variations in losses through these agencies could be expected to fall into intercompensatory categories and to tend to be counterbalanced, when relatively high, by lowered losses through other agencies, and vice versa. A sweepingly lethal epizootic, if one occurred, would be another thing, logically classifiable as an emergency type of destruction, along with those resulting from hunger crises and storms.

COMPETITIVE GALLIFORMES

It has long been clear that the density factor had a very important role in governing population phenomena of the northern bobwhite. But how broad should be our concept of density factor?—relating to bobwhites alone or to bobwhites plus what else?

The Prairie du Sac bobwhites could hardly have been subject to severely effective interspecific competition from wild animals or the basic patterns in their rates of gain and nonemergency loss would not have been so well maintained. This does not prove, however, that the patterns were totally uninfluenced by interspecific competition.

Among the faunal groups sharing the bobwhite's geographic range in the north-central region, other wild Galliformes might on general grounds be suspected of being the closest competitors. These include the native grouse and certain introduced game birds.

Existing information does not tell us much about interspecific competition in native Galliformes. It seems rather to be expected that such competition might, after a long span of racial adjustment, show less pronounced aspects than competition between native and recently introduced species having overlapping local ranges or territories. We have in native Galliformes, prairie chickens and sharp-tailed grouse (Pediocetes phasianellus) replacing one another in central Wisconsin counties as land is placed under cultivation or reverts, and further possible competitive relations between sharp-tails and ruffed grouse as reversion of former agricultural lands may progress; but the picture is so dominated by plant succession as to make interspecific relations of these grouse seem quite secondary (40).

It should not be thought that significant competition, when occurring, must conform to popular ideas of one species eating the other's food, destroying the other's eggs or young, or driving out the other by attack.

Ruffed grouse were not regarded as real competitors of the bobwhites at Prairie du Sac. During most years, at any rate, the grouse were fairly well segregated in types of habitat rarely visited by bobwhites. During the years of unusually high grouse populations, 1941-44 (see lower Figure 12), the grouse appeared in, and disappeared from, places in which they hadn't been seen for many years; and one would think that, if they had depressive influence on the bobwhites, it would have been at that time. However, the summer of probably the highest actual grouse density, 1942, was one for which the rate of gain of the bobwhite fell very close to the line for the "normal" curve (Figure 6, upper left).

Figure 7 presents largely unpublished data on bobwhite and prairie chicken populations of an area near Hunt City, Ill., furnished through the courtesy of Dr. R. E. Yeatter of the Illinois State Natural History Survey.

Considering the bobwhite data alone, we may see rates of summer gain that are depressed in 1936 and 1937 but otherwise show what looks like typical inverse ratios. Whatever may be the meaning of the irregularities of the prairie chicken gains, if the prairie chickens and bobwhites were mutual depressants, the latter species surely did not find itself at any overwhelming disadvantages as long as its own gains reflected primarily its own densities rather than those of the chickens, even for the summer of 1939

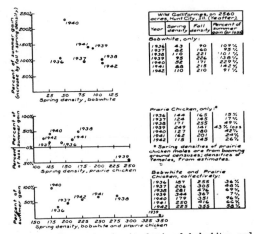

Wild Galliformes, on 2560 acres, Hunt City, Ill. (Yeatter)			
Year	Spring density	Fall density	Percent of summer gain (or loss)
Bobwhite, only:			
1936	43	90	109%
1937	96	160	95%
1938	110	221	101%
1939	95	226	136%
1940	52	171	229%
1941	88	215	142%
1942	110	210	91%
Prairie Chicken, only:*			
1936	144	165	15%
1937	124	145	17%
1938	171	255	49%
1939	249	141	43% loss
1940	127	180	42%
1941	162	201	24%
1942	115	145	26%

* Spring densities of prairie chicken males are from booming ground censuses; densities of females, from estimates.

Bobwhite and Prairie Chicken, collectively:			
1936	187	255	36%
1937	204	305	48%
1938	281	476	69%
1939	344	367	7%
1940	179	351	96%
1941	250	416	66%
1942	225	355	58%

FIG. 7. Rates of gain (or loss) of bobwhites and prairie chickens in relation to density, Hunt City, Ill., spring to fall, 1936-42. Data furnished by R. E. Yeatter.

following the highest spring density of the chickens.

Ignoring the variable introduced by human hunting of bobwhites for sport at Hunt City, we might judge that the survival figures for all except the emergency winters of 1935-36 and 1939-40 would suggest threshold values between 90 and 110, or in vicinity of 100. The two winters of heaviest densities of chickens—1938-39 and 1941-42—were winters of substantial reduction of bobwhites, but the survivals of 95 and 110 birds of the latter species are sufficiently close to the estimated threshold to argue against very much depression resulting from the presence of the chickens.

Preliminary to our consideration of competition between the bobwhite and the introduced ring-necked pheasant at Prairie du Sac, it may be explained that the evidence from elsewhere in the north-central region indicates that wintering bobwhites avoid coverts having many pheasants much as they do places overpopulated with their own kind. Thresholds of security tend to retain their numerical uniformity for given areas but the thresholds apply to mixed populations of bobwhites and pheasants rather than to bobwhites, alone (29, pp. 96-97). The end result of this sort of competition may then be, within limits, the exotic pheasants wintering at the expense of the native bobwhites, essentially bird for bird.

Especially because of the greater mobility and irregularity in habits of the pheasants, the density figures in my possession may often be only approximate or may be accurate only for certain times of the winter, thus necessitating careful estimating for other times. Passably satisfactory data on cold weather pheasant densities are at hand for two long-studied areas besides Prairie du Sac that have mixed bobwhite and pheasant populations (Figures 8 and 9). Unfortunately, for both areas, there are too many undissociated variables in the winter losses of the bobwhites themselves to justify attempts to plot non-

emergency loss rates. The analytical importance of reasonably accurate dissociation is well illustrated by the lower left of Figure 5, in which gross winter loss rates at Prairie du Sac give more of the appearance of a buckshot target than of any meaningful curves.

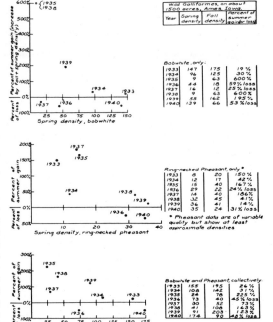

Wild Galliformes, on about 1500 acres, Ames, Iowa			
Year	Spring density	Fall density	Percent of summer gain (or loss)
Bobwhite, only:			
1933	147	175	19%
1934	96	125	30%
1935	9	63	600%
1936	44	18	59% loss
1937	16	12	25% loss
1938	9	63	600%
1939	55	162	195%
1940	139	66	53% loss
Ring-necked Pheasant, only:*			
1933	8	20	150%
1934	12	17	42%
1935	15	40	167%
1936	29	22	24% loss
1937	14	40	186%
1938	32	45	41%
1939	43	52	14%
1940	35	24	31% loss

* Pheasant data are of variable quality but show at least approximate densities.

Bobwhite and Pheasant, collectively:			
1933	155	195	26%
1934	108	142	31%
1935	24	78	225%
1936	73	40	45% loss
1937	30	52	73%
1938	41	108	163%
1939	91	203	123%
1940	174	90	48% loss

FIG. 8. Rates of gain (or loss) of bobwhites and ring-necked pheasants in relation to density, Ames, Iowa, spring to fall, 1933-40. The 1500-acre area has been described in the literature as two areas, but, for purposes of this paper, these may be regarded as a unit.

For Prairie du Sac, the nonemergency losses of bobwhites plus pheasant losses define, in combination, smoother curves than do rates of nonemergency losses of bobwhites, alone (lower right, Figure 5).

In the latter, the nonemergency loss rate beginning in the fall of 1936 still remains high for the total population of bobwhites and pheasants; but its actual numerical value is sufficiently uncertain so that we cannot be absolutely sure that it does not have affinities with the 1938-1940-1941-1943 group, which seems to define a part near the upper asymptote of a sigmoid curve. The curve defined by the nonemergency loss rates beginning in the "normal" falls of 1929-35 and 1937 also would seem to define, for one thing, part of a sigmoid, only in this case a lower asymptote. The rates for 1939 and 1942 give us "off" points, like those for the nonemergency losses of bobwhites, alone.

The rates of summer gain of the combined Galliformes at Ames conform more to definite patterns than do the rates for one component species or the

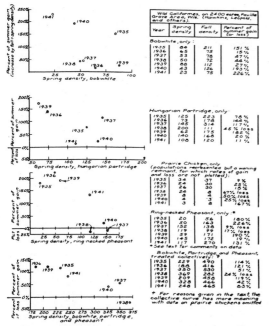

Wild Galliformes, on 2400 acres, Faville Grove Area, Wis. (Hawkins, Leopold, and others).			
Year	Spring density	Fall density	Percent of summer gain (or loss)
Bobwhite, only:			
1935	84	211	151%
1936	65	75	15%
1937	53	78	47%
1938	50	72	44%
1939	88	112	27%
1940	43	126	193%
1941	23	75	226%
Hungarian Partridge, only:			
1935	125	223	78%
1936	73	178	144%
1937	145	314	117%
1938	200	111	45% loss
1939	140	175	182%
1940	140	168	20%
1941	106	120	11%
Prairie Chicken, only (populations represented but a waning remnant, for which rates of gain and loss are not plotted):			
1935	34	37	9%
1936	24	30	25%
1937	24	30	15%
1938	24	8	67% loss
1939	6	3	50% loss
1940	4	3	25% loss
1941	3	8	167%
Ring-necked Pheasant, only:			
1935	20	56	180%
1936	50	163	226%
1937	152	136	9% loss
1938	119	99	17% loss
1939	59	171	190%
1940	145	172	19%
1941	117	270	131%
* See text for comments on data			
Bobwhite, Partridge, and Pheasant, treated collectively: †			
1935	229	490	114%
1936	188	416	121%
1937	350	530	51%
1938	369	282	24% loss
1939	209	466	119%
1940	328	466	42%
1941	246	465	63%
† For reasons given in the text, the collective curve has more meaning with data on prairie chickens omitted			

FIG. 9. Rates of gain (or loss) of bobwhites, Hungarian partridges, prairie chickens, and ring-necked pheasants in relation to density, Faville Grove area, Wis., spring to fall, 1935-41. Data furnished by Department of Wildlife Management, University of Wisconsin.

other (compare lower left, Figure 8, with upper left and left center). The loss for 1936 and the strongly depressed rate for 1937 are thought to have periodic connection. These exceptions notwithstanding, the alignment of the percentages of summer gain (or loss) leaves an impression of agreement with the principle of inverse ratios and of a substantial degree of mutual depression.

The Faville Grove data (Figure 9) were variously arrived at and those for pheasants are considered the weakest, insofar as censuses of these birds were made in January of each year. Professor Leopold, from data available, worked out a rule of thumb method for calculating spring densities, 1937-41, on the basis of a 15% late winter loss plus a gain of 25% of the birds artificially stocked in midwinter. Fall densities were calculated by adding to the January censuses the shooting kills of the preceding falls. Releases of artificially propagated 8-weeks pheasants in August were disregarded. Several discrepancies between the data of Figure 9 and those given by Hawkins (44, pp. 45-49) for the area may be laid to errors missed in proofreading the earlier publication.

The known changes in the waning remnant of prairie chickens at Faville Grove, spring to fall, were slight in terms of numbers of individuals except in 1938 (tabulated data, right center of Figure 9). Moreover, the population was believed to be virtually

without competitive influence on the other resident Galliformes. For these reasons, the curve outlined by the rates of gain or loss of the wild Galliformes, collectively, may be considered more truly informative with the chicken data omitted.

From Figure 9, it is apparent that the curve for the bobwhite-partridge-pheasant combination shows clearer conformity to the principle of inverse ratios than do the curves for any one of these species plotted separately.

Turning again to Figure 6, we see, lower left, that rates of gain for the Prairie du Sac bobwhites and pheasants, collectively, line up better along the main curve, and seemingly along the other one as well, than do the rates for the bobwhite, alone. It would therefore seem that even very low densities of pheasants have their competitive significance to the bobwhite.

THE DENSITY FACTOR

Nearly a quarter-century ago, Pearl and Parker, from experimental data on both vetebrates and invertebrates, stated (64, p. 217): "In general there can be no doubt that this whole matter of influence of density of population, in all senses, upon biological phenomena, deserves a great deal more investigation than it has had. The indications are that it is the most significant element in the biological, as distinguished from the physical, environment of organisms."

To me, the conclusion seems inescapable that much may happen in the ecology of the bobwhite without visibly affecting population trends and that much may be superimposed upon basic natural rhythms, or, shall we say, upon biological undercurrents that are neither easily deflected nor changed in pace. The thresholds of security and the inverse ratios revealed by the Prairie du Sac data—and similarly though less completely and less clearly by data from other north-central areas—may be considered manifestations of such undercurrents.

The observed interchangeability in density equations of the bobwhites and certain of their exotic relatives, on a bird for bird basis, provides some clues as to how density may work as a limiting factor.

If we regard the Prairie du Sac pheasants, in a space-competitive sense, as extra bobwhites, we see accentuated the trends that were already well defined by the bobwhite data, alone.

We may not assume that pheasants, alone, would respond wholly in the same way. Indeed, we may be sure that they would not, for, in other parts of the region, they have displayed vastly more tolerance of crowding than have the bobwhites, sometimes massing locally by hundreds or thousands (29, pp. 96-98). In addition, the pheasants may not be as characteristically sensitive to "undercurrent" influences as the bobwhites. Neither the Ames nor the Faville Grove pheasants (left center, Figure 8, and lower left center, Figure 9) responded during the depressions of 1936 and 1937 in a way comparable to the apparent responses of the bobwhites.

It would seem that the rates of gain and loss of the pheasants, while doubtless not divorced from all rules of order and automatic adjustment, were subject to considerable variation and that the conformation of rates of gain and loss of mixed populations of bobwhites and pheasants to the patterns shown by the Prairie du Sac data (lower right, Figure 5, and lower left, Figure 6) reflected population phenomena that were primarily those of the bobwhites. As the fortunes of the pheasants varied, adjustments in rates of gain and loss of the bobwhites tended to take place accordingly, within the framework set by the bobwhite's own limiting mechanisms.

It would be very easy to get out of our depth philosophically in attempting to consider the role of psychology in self-limiting mechanisms of the bobwhite. In another paper (29, pp. 98-100), I introduced the thought that much of the year-to-year constancy to be noted in threshold values, etc., might be explainable in terms of toleration "traditions" (no anthropomorphic connotations intended) on the part of dominant individuals. All of the evidence of which I know continues to support the "tradition" concept and no evidence impresses me as being contrary to it; however, the subject may here be dropped without further comment for the reason that the "traditions," if truly existent and significant, now would seem to affect merely the local distribution and behavior of the birds rather than the status of the species over large areas. Over the larger areas, populations have a way of following basic patterns, irrespective of many differences that are apparent on a more local scale.

We may ask how much the manifestations of basic patterns of loss and gain that we have before us only hint of McAtee's (55, p. 144) thesis that predation collectively tends to be in proportion to population, later restated by him (56, pp. 125-126) to the effect that the proportion rises and falls progressively with increase or decrease in numbers of the available food organisms?

Granted that the losses classed in this paper as nonemergency in type are not due exclusively to predation, how strictly are they in their totality a function of density?

Predation and other nonemergency losses suffered by bobwhite populations (or by mixed bobwhites and pheasants) in excess of thresholds at Prairie du Sac are characterized by elimination of birds above threshold levels and by relative safety of unhandicapped birds living below thresholds. Nicholson's (61, p. 171) comparison of territory-holding birds with water in an overflowing reservoir is descriptive, with overflow birds like overflow water comprising the part of the whole that is lost.

The threshold type of predation has been considered a modification of the proportional predation type (21). As populations exceed thresholds, they are preyed upon more and more in proportion to numbers, but the populations wintering securely below threshold levels (sometimes with absolutely no loss for months at a time) certainly are not preyed upon proportionally.

Theoretically, one might expect, because of differences in ability of mature birds in regular residence and the not so well established young to take care of themselves, that summer loss rates should be more in proportion to numbers than nonemergency winter loss rates.

Of course, we would be in a better position to inquire into the role of density in determining summer loss rates of bobwhites if we knew more about bobwhite and pheasant densities—of young birds and old—during summer and early fall of each year. In view of the scantiness of existing data, the best we can hope to do is to keep within the limits set by the actual figures for spring and late fall (or early winter) densities and to attempt some computation of densities and loss rates for the intervening periods on the most reliable prorata bases that we can find—without, however, losing sight of the essential unreliability of prorata computations treating with population trends that are marked by much automatic adjustment (31).

The summer of 1931 not only represented, in terms of bobwhite densities, weather, etc., about an "average" summer at Prairie du Sac, but also was the one of most concerted field studies of bobwhite reproduction carried on during the entire program of work in southern Wisconsin and central Iowa.

Data gathered mainly during this summer and mainly from the neighborhood of Prairie du Sac gave the following statistics: bobwhite hens successful in bringing off broods, about 80% (not to be confused with percentage of nests succeeding); young hatched, 13.6 per successful hen; young alive by midsummer, about 70% of those hatched; adults alive by midsummer, about 90% of the spring population. A sex tally of mature bobwhites gave 42.3% hens. Using these values, we get a midsummer density 4.1 times as great as the spring density.

Let us try the latter as a constant in computing midsummer bobwhite densities for different years at Prairie du Sac. We need not really assume, knowing the extreme unlikeliness thereof, that the breeding population for each summer consisted of the identical birds present in spring or even of substantially the same number. Nor need we really assume any year-to-year constancy in adult sex ratios, percentages of hens successful in hatching young, sizes of broods hatched out, mortality rates of adults or young at given seasons, or of other things that we know are actually quite variable. Knowing the great amount of counterbalancing that occurs, we can expect "good" nesting seasons accompanied by little loss of clutches from enemies, weather, accidents, and miscellaneous disturbances to mean early termination of nesting activities and a high "reproductive efficiency." Conversely, during a season of unusual disturbances, we can expect the total number of eggs laid to become greater because of stimulation of increased nesting by abandonment or destruction of clutches, thus resulting in a poorer "reproductive efficiency," though not necessarily in much if any diminution in total numbers of young ultimately hatched during a season (31). And because of the stringing out of the nesting season of southern Wisconsin bobwhites from May to September, we also know that the earlier of a season's young suffer mortality months before the later ones hatch.

Let us nevertheless see what we would get if we ignored these variables and went ahead with our computations.

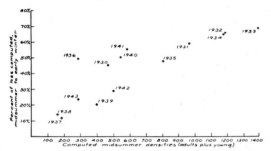

Computation of approximate loss rates of Prairie du Sac, Wis., bobwhites, midsummer to early winter.
(See comments in text.)

Year	Actual spring density	Computed midsummer density*	Actual early winter density	Percent of loss computed, midsummer to early winter
1930	112	467	257	45%
1931	236	966	400	59%
1932	290	1189	406	66%
1933	339	1390	433	69%
1934	288	1181	411	65%
1935	196	804	416	48%
1936	70	287	145	49%
1937	45	185	163	12%
1938	39	160	138	14%
1939	97	398	318	20%
1940	133	545	273	50%
1941	142	582	264	55%
1942	122	500	353	29%
1943	70	287	217	24%

* Spring density multiplied by 4.1, according to rough index derived in text.

There are at least two phases of density response, one shifting into the other and out again, to be traced through Figures 5 and 6. The one that we may call the "normal" or "phase I," defined by the winters of 1929-36 and 1937-38 and the summers of 1931-35, 1937, 1939, and 1942, if not by the winters of 1939-40 and 1942-43, clearly is not due to the same function of density as is the phase of depressed winter thresholds and summer gains that we may call "phase II"—the winters of 1938-39, 1940-42, 1943-44, and the summers of 1930, 1938, 1940-41, and 1943. The summer of 1936 and the following winter, which, at Prairie du Sac, show the lowest rates of summer gain and the highest rates of nonemergency winter loss in relation to density, may conceivably represent a third phase.

* * * * * * *

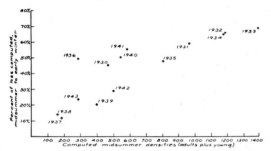

FIG. 10. Computed loss rates of bobwhites in relation to density, Prairie du Sac, Wis., midsummer to early winter, 1930-43. Values obtained through computations should be considered with reservations, as indicated in text.

In Figure 10 are plotted, with reference to known and computed densities at Prairie du Sac, the computed loss rates, midsummer to early winter. The length of the period to which the loss rates apply is intended to be about the same as that used in plotting the winter losses of Figure 5, or approximately four and one half months.

Even while keeping in mind all appropriate reservations, we may be confident that shortcomings of the prorata computations were not the main cause of loss rates in Figure 10 failing to occur in anything resembling straight proportions to density. Be it conceded that the rate for 1936 might reflect something close to proportional losses, the rates for at least 1930 and 1940-41 are patently lower in proportion to numbers involved; those for at least the 1931-35 span, still lower. All of the computed loss rates between 45% and 69% are for computed midsummer densities ranging from 287 to 1,390 birds.

In looking over the rates of nonemergency winter losses and of summer gains plotted in Figures 5 and 6, we may see that the two major curves defined by the data points in each case, density-linked though they are, assuredly are the product of more than density.

LITERATURE CITED

1. **Allee, W. C.** 1931. Animal aggregations. Chicago, ix + 431.
2. 1938. The social life of animals. N. Y. 293.
3. **Back, Roy. N.** 1944. Population fluctuations of the

North Dakota pheasant 1938-1943. North Dakota Outdoors **6(7)**: 8-10.

4. **Baumgartner, F. M.** 1944. Dispersal and survival of game farm bobwhite quail in north-central Oklahoma. Jour. Wildl. Mgt. **8**: 112-118.

5. **Bennett, Logan J., & George O. Hendrickson.** 1938. Censusing the ringneck pheasant in Iowa. Trans. North Amer. Wildl. Conf. **3**: 719-723.

6. **Burt, William Henry.** 1940. Territorial behavior and populations of some small mammals in southern Michigan. Misc. Publ. Mus. Zool., Univ. Mich. **45**: 1-58.

7. **Chapman, Lawrence B.** 1939. Studies of a tree swallow colony (second paper). Bird Banding **10**: 61-72.

8. **Clarke, C. H. Douglas.** 1936. Fluctuations in numbers of ruffed grouse, Bonasa umbellus (Linne), with special reference to Ontario. Biol. Series, Univ. Toronto Studies **41**: 1-118.

9. **Clements, Frederick E., & Victor E. Shelford.** 1939. Bioecology. N. Y. vii + 425.

10. **Davison, Verne E.** 1940. An 8-year census of lesser prairie chickens. Jour. Wildl. Mgt. **4**: 55-62.

11. **Duck, L. G.** 1943. Seasonal movements of bobwhite quail in northwestern Oklahoma. Jour. Wildl. Mgt. **7**: 365-368.

12. **Edminster, Frank C.** 1938. Productivity of the ruffed grouse in New York. Trans. North Amer. Wildl. Conf. **3**: 825-833.

13. **Elton, Charles.** 1942. Voles, mice and lemmings. Problems in population dynamics. Oxford, 1 + 496.

14. **Elton, Charles, E. B. Ford, & John R. Baker.** 1931. The health and parasites of a wild mouse population. Proc. Zool. Soc. London (**1931**): 657-721.

15. **Emlen, John T.** 1940. Sex and age ratios in survival of the California quail. Jour. Wildl. Mgt. **4**: 92-99.

16. **Erickson, Mary M.** 1938. Territory, annual cycle, and numbers in a population of wren-tits (Chamaea fasciata). Univ. Calif. Publ. Zool. **42**: 247-334.

17. **Errington, Paul L.** 1933. Mobility of the northern bobwhite as indicated by banding returns. Bird Banding **4**: 1-7.

18. 1933. The wintering of the Wisconsin bobwhite. Trans. Wis. Acad. Sci., Arts, and Letters **28**: 1-35.

19. 1933. Bobwhite winter survival on an area heavily populated with gray foxes. Iowa State Coll. Jour. Sci. **8**: 127-130.

20. 1933. The nesting and the life equation of the Wisconsin bobwhite. Wilson Bul. **45**: 122-132.

21. 1934. Vulnerability of bobwhite populations to predation. Ecol. **15**: 110-127.

22. 1935. The 1934 drought and southern Iowa bobwhite. Iowa Bird Life **5**: 18-21.

23. 1936. The winter of 1934-35 and Iowa bobwhites. Amer. Midl. Nat. **17**: 554-568.

24. 1937. What is the meaning of predation? Ann. Rept. Smiths. Inst. **1936**: 243-252.

25. 1938. Observations on muskrat damage to corn and other crops in central Iowa. Jour. Agr. Res. **57**: 415-421.

26. 1938. The great horned owl as an indicator of vulnerability in prey populations. Jour. Wildl. Mgt. **2**: 190-205.

27. 1939. The comparative ability of the bobwhite and the ring-necked pheasant to withstand cold and hunger. Wilson Bul. **51**: 22-37.

28. 1939. Reactions of muskrat populations to drought. Ecol. **20**: 168-186.

29. 1941. An eight-winter study of central Iowa bobwhites. Wilson Bul. **53**: 85-102.

30. 1941. Notes on winter-killing of central Iowa bobwhites. Iowa Bird Life **11**: 46-49.

31. 1942. On the analysis of productivity in populations of higher vertebrates. Jour. Wildl. Mgt. **6**: 165-181.

32. 1943. An analysis of mink predation upon muskrats in north-central United States. Iowa Agr. Exp. Sta. Res. Bul. **320**: 797-924.

33. **Errington, Paul L., & F. N. Hamerstrom, Jr.** 1936. The northern bobwhite's winter territory. Iowa Agr. Exp. Sta. Res. Bul. **201**: 301-443.

34. **Errington, Paul L., Frances Hamerstrom, & F. N. Hamerstrom, Jr.** 1940. The great horned owl and its prey in north-central United States. Iowa Agr. Exp. Sta. Res. Bul. **277**: 757-850.

35. **Errington, Paul L., & H. L. Stoddard.** 1938. Modifications in predation theory suggested by ecological studies of the bobwhite quail. Trans. North Amer. Wildl. Conf. **3**: 736-740.

36. **Evans, F. C., & R. Holdenried.** 1943. A population study of the Beechey ground squirrel in central California. Jour. Mammal. **24**: 231-260.

37. **Fisher, Lee William.** 1939. Studies of the eastern ruffed grouse (Bonasa umbellus) in Michigan. Mich. State College Agr. Exp. Sta. Tech. Bul. **166**: 1-46.

38. **Green, R. G., & C. A. Evans.** 1940. Studies on a population cycle of snowshoe hares on the Lake Alexander area. I. Gross annual censuses, 1932-1939. Jour. Wildl. Mgt. **4**: 220-238.

39. 1940. Studies on a population cycle of snowshoe hares on the Lake Alexander area. III. Effects of reproduction and mortality of young hares on the cycle. Jour. Wildl. Mgt. **4**: 347-358.

40. **Hamerstrom, F. N., Jr.** 1939. A study of Wisconsin prairie chicken and sharp-tailed grouse. Wilson Bul. **51**: 105-120.

41. **Hamilton, W. J., Jr.** 1937. The biology of Microtine cycles. Jour. Agr. Res. **54**: 779-790.

42. **Hann, Harry W.** 1937. Life history of the oven-bird in southern Michigan. Wilson Bul. **49**: 145-237.

43. **Haugen, Arnold O.** 1943. Management studies of the cottontail rabbit in southeastern Michigan. Jour. Wildl. Mgt. **7**: 102-119.

44. **Hawkins, Arthur S.** 1940. A wildlife history of Faville Grove, Wisconsin. Trans. Wis. Acad. Sci., Arts and Letters **32**: 29-65.

45. **Horn, E. E.** 1938. Factors in nesting losses of the California valley quail. Trans. North Amer. Wildl. Conf. **3**: 741-746.

46. **Johnson, Fred W.** 1943. Study quail population on Jornada range. New Mexico **21(3)**: 24-25.

47. **Kendeigh, S. C.** 1933. Abundance and conservation of the bobwhite in Ohio. Ohio Jour. Sci. **33**: 1-18.

48. **Kendeigh, S. Charles, & S. Prentiss Baldwin.** 1937. Factors affecting the yearly abundance of birds. Ecol. Monogr. **7**: 91-124.

49. **King, Ralph T.** 1937. Ruffed grouse management. Jour. Forestry **35**: 523-532.

50. **Leopold, Aldo.** 1931. Report on a game survey of the north central states. Sporting Arms and Ammunition Manufacturers' Institute. Madison, Wis. 299.

51. 1933. Game management. N. Y. xxi + 481.

52. 1937. The effect of the winter of 1935-36 on Wisconsin quail. Amer. Midl. Nat. **18**: 408-416.

53. **Leopold, Aldo, & Harry G. Anderson.** 1938. The 1936 cottontail scarcity in Wisconsin. Jour. Mammal. **19**: 110-111.

54. **MacLulich, D. A.** 1937. Fluctuations in the numbers

of the varying hare (Lepus americanus). Biol. Series, Univ. Toronto Studies **43**: 1-136.

55. **McAtee, W. L.** 1932. Effectiveness in nature of the so-called protective adaptations in the animal kingdom, chiefly as illustrated by the food habits of Nearctic birds. Smiths. Misc. Coll. **85** (Publ. 3125): 1-201.

56. 1933. Rejoinder to papers on protective adaptations. Proc. Royal Ent. Soc. London **81**: 113-126.

57. **McClure, H. Elliott.** 1942. Mourning dove production in southwestern Iowa. Auk **59**: 64-75.

58. 1943. Ecology and management of the mourning dove, Zenaidura macroura (Linn.), in Cass County, Iowa. Iowa Agr. Exp. Sta. Res. Bul. **310**: 355-415.

59. **Munger, Thornton T.** 1916. Graphic method of representing and comparing drought intensities. Monthly Weather Rev. **44**: 642-643.

60. **Nice, Margaret Morse.** 1937. Studies in the life history of the song sparrow. I. Trans. Linn. Soc. N. Y., 4. vi + 247.

61. **Nicholson, A. J.** 1933. The balance of animal populations. Jour. Animal Ecol. **2**: 132-178.

62. **Pearl, Raymond.** 1924. Studies in human biology. Baltimore 1-653.

63. 1932. The influence of density of population upon egg production in Drosophila melanogaster. Jour. Exp. Zool. **63**: 57-84.

64. **Pearl, Raymond, & Sylvia L. Parker.** 1922. On the influence of density of population upon the rate of egg production in Drosophila. Proc. Nat. Acad. Sci. **8**: 212-219.

65. **Sanders, Earl.** 1943. Development of a bobwhite management area in southern Iowa. Iowa Agr. Exp. Sta. Res. Bul. **317**: 699-726.

66. **Stoddard, Herbert L., & Others.** 1931. The bobwhite quail; its habits, preservation and increase. N. Y. xxix + 559.

67. **Wade, Douglas E.** 1938. Drouth intensity measurements and the effect of the 1936 drouth on wildlife. Trans. North Amer. Wildl. Conf. **3**: 558-569.

68. **Yeatter, Ralph E.** 1934. The Hungarian partridge in the Great Lakes Region. Univ. Mich. School of Forestry and Conservation Bul. **5**: 1-92.

69. 1943. The prairie chicken in Illinois. Ill. Nat. Hist. Surv. Bul. **22**: 375-416.

70. **Yokum, Charles F.** 1943. The Hungarian partridge Perdix perdix Linn. in the Palouse Region, Washington. Ecol. Monogr. **13**: 167-202.

Reprinted from *Trans. 14th N. Amer. Wildlife Conf.*, 225–231 (1949)

THE ROLE OF INTRASPECIFIC COMPETITION IN GAME MANAGEMENT[1]

DAVID E. DAVIS

The Johns Hopkins School of Hygiene and Public Health, Baltimore, Maryland

INTRODUCTION

Game management consists essentially of manipulating the environment in order to obtain the desired population. The environment, however, consists of a multitude of factors, some favorable, some unfavorable, and still others are favorable under certain conditions and unfavorable under other conditions. The game manager is justifiably puzzled by these complex situations and needs to rely upon some general principles in the hope of predicting the results of environmental manipulation. At present, management is largely empirical and after studying each species separately, alters the environment in some way designed to obtain the desired effect on the population.

In the hope of finding some general principle which will guide the game manager, the role of intraspecific competition will be considered in this paper. It is hoped that the presentation of these results will stimulate others to determine to what extent this hypothesis applies to other species.

COMPETITION

Competition may occur between species (interspecific) which feed on the same plants, require the same nesting sites, etc. Actually such competition is rarely severe because two species seldom have identical requirements. When interspecific competition does occur, it resembles predation in its major effect on populations, at least according to present information.

But competition within the species (intraspecific) may often be severe and have far-reaching effects on the population in terms of pregnancy rates, size of individuals, and survival.

To learn something about the factors limiting populations of animals, an intensive study of the brown rat (*Rattus norvegicus*) has been conducted in Baltimore for about five years. The common rat is a splendid experimental animal because it is abundant (150 per city block), easily trapped, and biologically well-known. A number of experiments have provided information concerning the role of intraspecific competition and the regulation of population.

[1]Contribution from the Rodent Ecology Project of the Department of Parasitology, School of Hygiene and Public Health, The Johns Hopkins University. The work reported in this paper was conducted under grants from the International Health Division of the Rockefeller Foundation and from the U. S. Public Health Service.

1. In order to compare the composition by size categories and the pregnancy rates of stable and of increasing populations, a procedure was developed in city blocks to obtain a sample of rats from a stable population and from an increasing population using a third sample for reference to seasons (Figure 1). Since reproduction is almost the

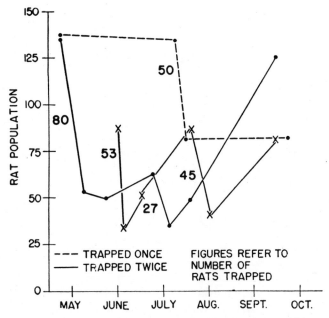

Figure 1. Procedure for obtaining a sample of rats from a population which was stable in May, from the same population when it was increasing in July, and from another population which was stable in July.

sole source of new rats it would be expected that the increasing blocks would show a preponderance of small rats. But Table 1 shows that the weight composition of the populations is essentially the same in stable and in increasing populations. Indeed in some cases the increasing populations have a surprisingly low number of small rats. An explanation of this situation might be that after the population was reduced, more food became available, and the remaining small rats rapidly gained weight, thus increasing the percentage in the higher categories. At any rate, it may be suspected that when the population was reduced and competition thereby reduced, the remaining rats gained weight rapidly.

2. In these same blocks an analysis of the prevalence of pregnancy (Table 1) showed a rise in the increasing blocks in the reproductive season but no rise in the nonreproductive season. It appears that the reduction of the population permitted an increase in reproduction in certain seasons.

TABLE 1. WEIGHT OF RATS IN RELATION TO POPULATION LEVEL

	Percentage in Weight Class (gms.)			Total Rats	Females more than 200 gms.	Per cent Pregnant
	0-199	200-399	400 plus			
1. Stable contrasted with increasing populations						
z. May-July, 1947						
Stable (May)	36	35	29	211	81	34.6
Stable (July)	32	45	23	104	31	19.3
Increasing (July)	28	46	26	133	37	48.6
y. December-February, 1947-48						
Stable (December)	31	33	36	120	38	33.1
Stable (February)	10	43	47	88	32	21.9
Increasing (February)	21	43	36	92	28	17.8
x. April-July, 1948						
Stable (April)	26	38	36	129	44	36.3
Stable (July)	29	40	31	35	13	15.3
Increasing (July)	36	45	19	151	22	40.9
z. September-January, 1948-49						
Stable (September)	22	52	26	121	41	24.4
Stable (January)	37	30	33	111	33	32.0
Increasing (January)	27	43	30	56	22	29.4
2. Increasing populations (Summer, 1946)						
Rapid increase	42	37	21	167	98	10
Slow increase	31	34	35	77	33	18

3. The prevalence of pregnancy is proportional to the size of the rat, the regression equation being: $y = -25.4 + .186x$, where y is the per cent of pregnancies and x is the weight in grams (Davis, unpublished). For example, 23 per cent of rats weighing from 250-274 grams should be pregnant whereas 61 per cent of rats weighing from 450-474 grams should be pregnant. Evidence from domestic rats (King, 1916; Asdell, 1941) indicates that the reproductive rate declines after a maximum in early life and hence large rats in nature should not have the highest pregnancy values. Since it is found that large rats tend to be socially dominant, it seems likely that the dominance relationships permit the large rats to have a high pregnancy rate. This relationship is nicely shown by data for the number of litters.

The rates for different categories of sexual maturity of the females are as follows: For all females with a perforate vagina: 4.79 litters per year. For females with 10-12 visible nipples (medium and large rats): 6.68 litters per year. For rats over 200 millimeters in body length (large rats): 8.40 litters per year.

4. Repeated removal of rats from a population over a long period

of time (14 months) resulted in a gain in weight of the samples (Figure 2). Although there is considerable variation, the trend seems **clear.**

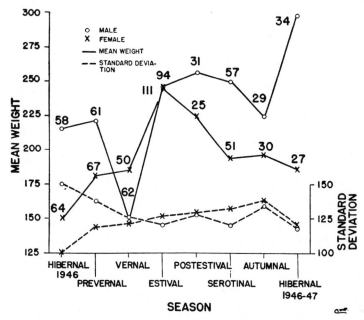

Figure 2. The average weight of rats captured in a homogeneous environment during 14 months. Numbers refer to size of sample.

5. Additional evidence concerning intraspecific competition may be found in the results of several experiments designed to determine the results of introducing alien rats into blocks with stable populations. To our surprise, the population neither increased nor remained stable, but actually decreased. Possibly competition caused a social turmoil and resultant decline.

None of these five lines of evidence is particularly convincing or conclusive by itself, but all five suggest that the removal of some rats makes conditions more suitable for the survivors.

Let us now consider the social organization of rats in search of an explanation. Studies conducted by John B. Calhoun (1949) of our group shows that rats in natural conditions have a social hierarchy just as do rats in cages. The dominant individuals tend to grow faster, and raise more young. It would be logical to expect, therefore, that the removal of some rats, especially the large ones, would permit the remaining rats to grow more rapidly and become pregnant more often. It seems then that the reduction of competition has a favorable effect on the population.

Competition, of course, is only one of the factors limiting populations. These factors may be divided into two types. The first is independent of the density of the species and acts at high or low populations. Environmental factors such as temperature belong in this category. The other type is dependent upon the density of the population and act more severely at high than at low densities. Predators, for example, take a higher proportion of prey at high than at low populations of prey. Competition, also, depends upon the density and is greater at high than at low populations. It should be noted that predation is proportional to the population (N) but competition is proportional to the population squared (N^2). From these theoretical considerations it is at once apparent that intraspecific competition is the factor of major importance in holding the population within the limits of the density-independent factors (environment).

It is now possible to suggest a hypothetical principle which should be investigated to determine its validity. If valid for many species this principle will serve to predict the effect on a population of a change in a specific factor. The hypothesis may be stated thus: To manage a population, alter the environment so that intraspecific competition will change inversely to the desired change in population. Specifically for some species, for example, deer, this means that to increase the deer, food should be increased so that competition will decrease. To decrease the deer, food should be decreased so that competition will increase. Although this example is self-evident, more complex relations are not obvious and may be guided by consideration of their impact upon competition.

SUMMARY

The possibility that intraspecific competition may serve as a guide to management is tested in brown rats living in residential blocks in Baltimore. After a reduction of population the pregnancy rates and size of the survivors increased. Continuous removal of rats resulted in an increase in average weight of the remainder. Probably related to social behavior is the fact that large rats have a higher pregnancy rate than do small rats and also that the introduction of alien rats into blocks resulted in a decline of population. These diverse bits of evidence may be interpreted to indicate that competition is a potent limiting factor. As a hypothesis for use in predicting the effect of specific management practices it is suggested that: To manage a population, alter the environment so that intraspecific competition will change inversely to the desired change in population.

LITERATURE CITED

Asdell, S. A.
  1941. The influence of age and rate of breeding on the ability of the female rat to reproduce and raise young. Cornell Univ. Agr. Exp. Sta. Mem. 238:3-28.
Calhoun, John B.
  1949. Proc. N. Y. Acad. Sci. (in press).
King, H. D.
  1916. The relation of age to fertility in the rat. Anat. Rec., 11:269-287.

DISCUSSION

DR. GRAHAM: Mr. Davis has presented a paper that is extremely stimulating to anyone who is interested in manipulating the environment for the benefit of wild animals. He has gone into an area where he could measure populations considerably more easily than most of us are able to measure our populations in the wild, and he has brought out some points that many people have suspected for many, many years, but have never been able to prove to the satisfaction of most other people.

He not only has shown that competititon between the individuals of a species is perhaps more important than the competition between species, but he has demonstrated rather conclusively the fact that it might conceivably be possible to increase the production of an area by harvesting a heavier crop. This is something that is in line with general observations in many instances, but it certainly definitely is in contradiction to the opinions held by many fieldmen.

DR. THOMAS H. LANGLOIS (Ohio): I cannot help but feel that you have not given adequate treatment to the problem of territorialism. I have seen, too often, cases where an individual thrives and becomes bigger and more able to exert dominance by virtue of having a territorial niche which he claimed as his own, and I wonder if your problem of dominance of any individual may not be the result of those big individuals having territories in which they could be dominant. The ability of a small individual to challenge, or even exert dominance over a large individual, if the little individual has a territorial niche and the big one is a stranger and an intruder, makes me suspect that maybe your problem is complicated somewhat by the number of territorial niches in your territory, rather than, let's say, temperature or some other factor.

One of our men had a problem of rats on his lot on our island at Put-in-Bay, so he started trapping. There were not many rats on that lot, but yet he took 64 rats out by removing one each day for about 64 days. As promptly as a niche was vacated, another rat moved in and took over the territory. Certainly, the welfare of the individual and his ability to survive is associated with his having a territory he feels dominant in. I would like your comment on that.

DR. GRAHAM: Dr. Langlois has raised a question concerning the effect of territorialism upon the size and character of the individuals of a population, and the reactions of the population. I should like to ask Dr. Davis to answer this question.

DR. DAVIS: In our present state of knowledge, I think, at least, of territory as being one of the methods for regulating intraspecific competition. It is a set of ethical rules, let us say; it is the kind of thing that the United Nations is trying to set up among nations; and Dr. Langlois is quite correct, at least to my knowledge, in that a species, individuals which are able to get and maintain a territory, do thrive in that territory.

Another set of ethical rules which helps to reduce the energy expended in intraspecific competition is the problem of social hierarchies. The borderline between social hierarchies and territories is, of course, very hazy. Most species fall into one or the other group, as far as behavior goes; but pigeons, for example, seem, under some circumstances, to be territorial and under other circumstances to have social hierarchies.

In summary, then, I look at these as rules and regulations which reduce the amount of energy the individual has to spend on getting the things that he needs for life.

DR. GRAHAM: Are there any other questions or comments? Is there anyone who has any additional evidence to present in support of the hypothesis proposed?

MR. E. L. CHEATUM (New York): I would like to state that we 'have been investigating the relationship between the reproductive potentials of deer in New York for some years, in relation to the depletion of the food supply in deer range, and, of course, this will be related rather directly to intraspecific competition for the available food supplies. We have found that there is a very striking relationship between depleted, or rather depressed or suppressed ovulation rates in the deer and depleted food supply.

DR. GRAHAM: Are there any other bits of evidence of a similar nature that anyone would like to present?

DR. HENRY S. MOSBY (Virginia): I would like to hear Dr. Davis' hypothesis as to what might occur to depress his population when he introduces animals, such as stocking. He referred to it as having a tendency to depress the population.

DR. GRAHAM: Dr. Mosby has asked what explanation Dr. Davis can give to explain the depression of populations on the introduction of additional animals—the stable population.

DR. DAVIS: I cannot give you an explanation, but I can tell you the types of research we are doing in order to try to understand that; and that, of course, is essentially an outline of what we think might be the explanation. Our best guess, and the thing we are trying to study, is that there is a sufficient (shall I call it?) psychological turmoil developed when we introduce those rats to account for it. Let us assume that we put a five- or six-hundred-gram rat into a colony which, up to that time, was quite happy; everybody knew who was married to whom, and whose children were whose, and so on and so forth. It seems likely to me that there would be a considerable amount of commotion. That 600-gram rat would try to establish himself as the dominant, and I think it is entirely possible that factors of that type would explain the decline.

I would like very much to know if those who have stocked animals in areas have any evidence that this stocking has been followed by a decrease.

DR. GRAHAM: Is there anyone who has such evidence? It is a little hard to get. I think there is perhaps some evidence in connection with fish available along this line. Is there anyone in the audience (I see Dr. Eschmeyer has gone) who has any specific information on this question?

$$22$$

Reprinted from *Ardea*, **39**, 58–71 (1951)

Population Ecology of the Great Tit

H. N. KLUIJVER

PART 5. SECOND BROODS

§ 15. Start of laying.

After the first brood has left the nest, some pairs of Great Tits start a second brood, and very rarely there may even be a third brood. A third brood was never found for certain at O.N.O. At Hoenderloo in 1923, however, after the completion of 2 broods, started on 18th April and 27th May respectively, a third brood was started in the same box on 2nd July. It seems highly probable that these three broods were the product of the same pair, as the nestbox was situated in the centre of a large pine wood in which there was only the one pair of Great Tits breeding. Evidence given later suggests that the occurrence of a third brood may have been connected with the very low breeding density. The clutch-size of the three broods was 4, 8 and 6 eggs respectively. MORBACH (1951) refers to two third broods in Luxembourg.

The egg-laying period is usually over by the end of June, but at O.N.O. laying continued into the first week of July in the years 1912, 1913, 1915, 1917, 1922, 1924, 1929 and 1944, the latest dates being 11th July 1917 and 9th July 1924. Such late broods occurred more often during the earlier than the later years of observation, the decline probably being connected with the increase in population density at O.N.O., as discussed later.

TABLE 22

Interval between date of laying of first egg of first and second broods of the same pair.

Interval (in days)	No. of pairs	Interval (in days)	No. of pairs
10—14	2	35—39	10
15—19	2	40—44	32
20—24	3	45—49	33
25—29	1	50—54	19
30—34	3	55—59	12
		60—64	3
		65—70	1

Many pairs begin a second brood very soon (4-6 days) after the young of the first brood leave the nest. It is unusual for the female to start building a new nest before this time, but very rarely she may even start egg-laying before it (PAULUSSEN 1949). Such an immediate start of the second brood seems to be commoner in Blue and Coal Tits.

The minimum period required for a first brood is 34 days (6 for laying, 12 for incubation, and 16 for feeding the young), but the usual period is 40-43 days (8-12 for laying, 12-13 for incubation and 17-20 for feeding the young). The interval between the laying dates of the first eggs of the first and second broods of 121 ringed pairs is shown in Table 22. Recorded intervals of less than 34 days are all due to failure of the first brood. Most pairs start their second brood 40-49 days after the start of the first brood, but some pairs wait longer, even up to 65 days, thus taking a period of at least 20 days for feeding the young out of the nest and for building the nest of the second brood.

§ 16. Influence of age on percentage of second broods.

A second brood does not follow by any means invariably and the frequency with which it occurs varies with a number of factors. It is not just a characteristic of a particular individual bird, as the same individual may vary in this matter from one year to the next. In what follows, the "percentage of second broods" is the percentage of available pairs which undertook second broods at a particular place in a particular year. Each brood with an initial date 40 or more days after the earliest brood of the year in question is counted as a second brood. The number of available pairs was calculated from the number of first broods (see § 14).

The data in Table 23 show that the first-year females have second broods less frequently, on the average, than do the older birds.

TABLE 23

Percentage of second broods in first-year and older females at O.N.O.

Year	Older females		First-year females	
	Available	2nd broods	Available	2nd broods
1936	23	10	19	0
1937	34	8	29	6
1938	31	6	25	6
1939	19	6	21	4
1940	16	13	17	8
1947	12	3	4	0
1948	17	7	5	3
1949	21	1	11	1
1950	25	12	1	0
Total	198	66(33%)	132	28(21%)

Of the females of unknown age (because unringed) only 21% had second broods, which strongly supports the view that most of these birds are in their first year. (N.B. The age of the male bird has no influence on the percentage of second broods.)

The question must now be raised of whether the smaller percentage of pairs raising second broods among first-year than older individuals might be due to differences in the environmental conditions of the first-year and older females. It has already been shown that first-year females start egg-laying a few days later than the older females, and it will be shown later that the initial date slightly affects the percentage of second broods. The differences concerned are, however, far too small to account for the marked differences in the percentage of second broods, which are set out in Table 23.

It will also be shown later that the percentage of second broods is much influenced by both habitat and population density. If therefore, the first-year and older females were not evenly distributed over the different habitats (and there is some evidence that they are not, see § 33) then age difference in the percentage of second broods might be related to differences in habitat or population density. This, however, is not the case. Table 24 shows clearly that the age of the bird affects the percentage of second broods independently of the habitat.

TABLE 24

Percentage of second broods of first-year and older females in densely populated (mixed) wood and in thinly populated (coniferous) wood.

Year	Mixed wood					Coniferous wood				
	Density per 10 ha	First year		Older females		Density per 10 ha	First year		Older females	
		1st br.	2nd br.	1st br.	2nd br.		1st br.	2nd br.	1st br.	2nd br.
1936	5.6	8	0	15	3	3.4	5	0	8	7
1937	6.7	17	5	23	6	4.8	11	2	12	4
1938	6.1	17	4	20	8	2.7	7	5	13	10
1939	4.6	14	3	12	3	1.6	7	2	4	2
1940	4.9	12	5	11	8	1.9	5	3	6	6
Average	5.9	68	17 (25%)	81	28 (35%)	2.9	35	12 (34%)	43	29 (67%)

It may therefore be accepted that the smaller percentage of second broods among first-year females is definitely correlated with their age per se.

§ 17. Annual and local differences in the percentage of
second broods.

Table 25 shows that at all localities the percentage of pairs with second
broods differs markedly in different years. What is the cause of this?
The proportion of first year females in the population is different in
different years, and as fewer first year than older females have second
broods this is presumably one of the causes of the variability.

Is the percentage of second broods also influenced by meteorological
and other external factors? The average air temperature in May and
June appear to have little significance in this respect. For O.N.O., the
correlation coefficients between air temperature and the proportion of
second broods over a series of 32 years are not more than —0.11 and
—0.18 respectively. I did not calculate such correlations for other
meteorological factors. There is, however, another indication that
weather conditions cannot be important. At any one time the weather
conditions are usually similar throughout Holland. Hence if weather
conditions influence the proportion of second broods, one would expect a high correlation between the percentage of second broods in
different localities in the same year. A statistically significant correlation (+0.68 for a series of 14 years) is found in the annual percentage
of second broods at O.N.O. as compared with Hoenderloo, which are
25 km. apart. There is, however, no such correlation between the annual
percentages at Mastbos and Liesbos, both near Breda and very near
each other. Hence meteorological factors cannot be of great importance
in determining the percentage of second broods.

One might plausibly expect a connection between the percentage of
second broods in any particular year and the date at which the birds
start their first broods. It might be expected that, if the first broods
are started late, the birds complete the feeding of their young so late
that their reproductive organs are already regressing. (For the relation
between start of breeding and the winter and spring temperatures see § 14.)

I therefore calculated the correlation between the percentage of second broods and the average initial date of the first broods over a period
of 37 years (the O.N.O. data for 1912-1943 and 1946-1950). This suggested that there was a tendency for fewer second broods in years
when the first broods were started late, but the correlation coefficient
was only — 0.13 which is not significant. In some years with a very
late start, such as 1917 and 1929, Great Tits also tended to finish the
breeding season very late, and the percentage of second broods seems
not necessarily much smaller in such years than in others.

It will subsequently be shown, that the initial date and the percentage of second broods are nevertheless definitely correlated (see § 18). The causes of the annual differences in the percentage of second broods on O.N.O. will be further discussed in the same paragraph.

Table 25 shows that there are also marked local differences in the percentage of second broods. The percentage of second broods is always higher at Hoenderloo than at O.N.O., and always lower in the Liesbos.

TABLE 25

Annual and local differences in percentage of second broods.

Year	Hoenderloo	Mastbos	O.N.O.	Liesbos
1922	72		30	
23	82	93	33	
24	87	75	93	45
25	85	67	46	44
26	100	64	80	42
27	84		60	34
28	78	100	65	21
29	86	91	33	33
30	61	59	31	43
31	48		16	16
32	56	27	28	22
33	54	78	53	36
34	63	53	27	35
Average for 9 years in common	76	68	51	36

The Liesbos is a pure deciduous wood, O.N.O. is mixed, the Mastbos is a pine wood with much deciduous undergrowth, while Hoenderloo mainly consists of pine wood. There thus appears to be a connection between the vegetation and the percentage of second broods, and in this connection it is striking that at O.N.O. the percentage of second broods tends to be higher in the Pine Wood than in the Mixed Wood, as is clearly shown in Table 26. In 15 out of 19 years there was a higher proportion of second broods in the Pine Wood than in the Mixed Wood, in two years the percentage was about equal, and in only two years was the proportion higher in the Mixed Wood. The averages amount to 62% second broods in the Pine Wood and 40% in the Mixed Wood.

TABLE 26

Percentage of second broods in Mixed Wood and Pine Wood at O.N.O.

Year	% second br.		year	% second br.		year	% second br.	
	mixed	pine		mixed	pine		mixed	pine
1922	28	45	1929	36	35	1935	30	53
23	32	86	30	17	34	36	12	43
24	91	100	31	6	27	37	34	23
25	62	61	32	27	55	38	36	100
26	86	75	33	50	67	39	21	50
27	74	90	34	23	56	40	47	83
28	52	94						

BERNDT and FRIELING (1939), who studied in 1936 nestboxes in a wood near Steckby (Saxony) also stated a higher percentage of second broods in coniferous woods than in broadleaved woods. The same is clear from the data of Table 27, which are discussed in the next paragraph. In this paragraph we will show that the influence of habitat on percentage of second broods works partly by the way of population density.

§ 18. Population density and second broods.

Since 1921 WOLDA, with the help of volunteer observers, organised in many Dutch woods a regular inspection of nestboxes similar to that carried out at O.N.O. This organisation became particularly efficient in the period 1930-1934, and the results for 16 areas where boxes were carefully inspected are given in Table 27 [1]).

It is quite clear from this Table that the frequency of second broods is normally higher in coniferous than in broadleaved woods. The obvious suggestion is that this difference is due to some factor which directly influences fertility e.g. the availability of food supply. This, however, is uncertain and it may be suggested that the frequency of second broods is connected also with the population density of the breeding Tits. For it is in deciduous woods that the Great Tit reaches

---

1) The fact that the data of Table 27 are collected over a period of 5 years does not much increase the variability of the figures in question. Spring climate was rather normal during all these years.

its highest population density, and its density is much lower in pine woods.

In Table 27, the density of breeding pairs (expressed in number per 10 ha) was calculated for each year from the number of first broods found in the nestboxes. This presumes that all those pairs present bred in the boxes, which, though not always quite correct, is certainly close to the truth. For these boxes provide an optimal breeding-site for the Great Tit, which prefers them to most natural breeding holes. Since, in all areas mentioned in Table 27, the boxes were fairly equally distributed, and further a proportion of them were every year left unoccupied, it may be accepted that the calculated breeding densities are close to the true figures.

The density data from the Mastbos and the Liesbos near Breda were not included in Table 27. These woods are respectively 93 ha and 77 ha in area, but the areas where the nestboxes were situated are not exactly known. Further in most years only a proportion and not all of the boxes was inspected regularly during the breeding season and the particular boxes inspected were rather different in different years. Moreover the boxes were spaced rather thinly and unevenly, and in most years all the boxes were occupied. Finally, especially in the Liesbos, there are many old deciduous trees, which doubtless provide natural breeding holes for the Great Tit (e.g. holes of the Great Spotted Woodpecker). As neither the area of the ground occupied by nestboxes, nor the number of pairs of tits, is exactly known, the population density in these areas could not be calculated.

From the data in Table 27, which are graphically represented in Fig. 7 it is clear that the percentage of second broods is higher when the Tit population is less dense. With Great Tit densities of less than 4 pairs per 10 ha there was always a high percentage of second broods, varying between 40% and 100%. At these low densities, the correlation between breeding density and percentage of second broods is not close, but when the density rises above about 4 pairs per 10 ha, such a correlation is immediately apparent. In very dense populations, the percentage of second broods decreases to less than 10%.

Ardea

5

TABLE 27

Population density, fecundity and

Locality and habitat p = park w = wood d = deciduous c = coniferous m = mixed	Area in hectares	1930						1931					
		Density	Clutch first brood	Clutch second brood	% Second broods	Fecundity rate	Production rate	Density	Clutch first brood	Clutch second brood	% Second broods	Fecundity rate	Production rate
Bloemendaal (p, d)	13	—	—	—	—	—	—	—	—	—	—	—	—
Haarlem (p, d)	38	—	—	—	—	—	—	—	—	—	—	—	—
Doorn (p + w, m + c)	30	8.0	8.6	6.8	21	10.0	9.0	7.7	8.5	6.0	9	8.9	7.7
O.N.O., (w, c + m)	129	5.6	8.7	7.6	38	11.6	6.6	5.1	8.6	7.6	16	9.8	5.7
Heemstede (p + w, d)	55	—	—	—	—	—	—	—	—	—	—	—	—
Alkmaar (p, d)	21	—	—	—	—	—	—	6.2	8.5	6.0	16	9.5	8.7
Nunspeet (w, c + m)	84	1.9	10.3	9.1	75	17.1	10.9	2.7	10.2	8.0	70	15.8	13.7
Ugchelen 1 (w, m)	120	1.8	10.4	8.8	50	14.8	8.2	1.0	10.2	7.0	33	12.5	10.3
Ugchelen 2 (w, c)	100	0.9	9.0	7.0	56	12.9	7.4	—	—	—	—	—	—
Driebergen (w, c)	218	1.2	9.8	8.1	76	16.0	10.1	1.3	8.4	8.0	30	10.8	6.7
Hoenderloo (w, c + d)	504	1.4	10.4	8.4	61	15.5	9.5	1.6	9.8	8.4	48	13.8	7.7
Chaam (w, c)	±300	±0.5	10.7	6.9	57	14.6	9.4	±0.6	10.0	7.1	56	14.0	9.9
Appelscha (w, c)	±300	±0.4	10.5	8.4	60	15.5	12.8	±0.5	9.8	7.8	81	16.1	14.6
Texel (w, c)	318	0.4	9.1	8.0	100	17.1	9.1	0.7	10.0	8.5	82	17.0	12.4
Eext (w, c + m)	±300	—	—	—	—	—	—	±0.2	9.9	9.8	57	15.5	9.8
Heerde (w, c + m + d)	±300	±0.3	9.5	8.2	90	16.9	9.7	±0.9	9.7	8.4	43	13.3	10.2
Breda, Liesb. (w, d)	77	?	10.7	6.9	43	13.7	10.0	?	10.8	6.0	16	11.8	10.3
Breda, Mastb. (w, c + m)	93	?	9.6	7.9	59	14.3	7.2	—	—	—	—	—	—

production at 16 areas 1930-1934.

1932						1933						1934					
Density	Clutch first brood	Clutch second brood	% Second broods	Fecundity rate	Production rate	Density	Clutch first brood	Clutch second brood	% Second broods	Fecundity rate	Production rate	Density	Clutch first brood	Clutch second brood	% Second broods	Fecundity rate	Production rate
												19.2	7.2	7.0	4	7.5	6.6
—										—	—	13.9	7.6	5.0	2	7.7	6.0
9.5	7.7	6.0	5	8.0	4.6	12.6	7.2	6.1	14	8.0	5.1	7.7	7.2	5.5	9	7.9	6.6
8.0	8.0	6.0	8	7.6	6.3	6.3	8.0	6.7	31	10.1	6.7	6.6	8.4	6.9	27	10.3	5.6
4.4	9.1	6.7	28	11.0	6.5	4.6	9.0	7.5	53	13.0	5.2	6.5	8.4	5.7	28	10.0	8.0
—						—						4.3	8.8	7.0	22	10.3	6.4
4.8	8.9	8.0	20	10.5	9.2	5.2	8.5	7.0	27	10.4	8.4	2.6	9.2	7.1	41	12.1	9.2
3.1	9.5	8.1	35	12.3	6.0	2.9	9.1	8.0	42	12.5	10.3	1.8	10.4	8.0	50	14.4	6.6
1.1	10.2	7.2	46	13.5	10.1	1.2	10.2	8.8	36	13.4	8.6	1.5	9.5	7.4	60	13.9	8.3
—						0.9	8.8	7.4	100	16.2	14.1	1.3	7.8	8.6	47	11.8	6.3
1.1	8.7	8.5	38	11.9	6.7	0.8	9.1	7.3	65	13.8	7.2	1.3	9.7	7.5	63	14.4	7.6
1.7	10.1	7.5	47	13.7	8.3	1.4	9.3	7.5	54	13.4	6.4	±1.1	8.9	7.9	64	14.0	8.3
±0.4	9.4	8.4	91	17.0	12.6	±0.5	9.9	8.5	100	18.4	14.1	±0.9	9.6	7.8	70	15.1	10.2
—						±0.6	9.9	7.0	74	15.1	12.1	0.9	9.2	8.2	53	14.5	7.2
0.9	8.3	7.3	40	11.2	5.3	1.0	8.7	7.3	53	12.6	9.2	±0.5	11.0	8.6	80	17.9	13.6
±0.3	10.5	9.7	100	20.2	12.9	±0.3	11.0	9.4	88	19.3	15.5	±0.4	9.8	7.8	46	13.4	9.5
—						±0.6	8.7	8.1	47	12.5	10.3	—					
?	10.5	5.6	22	11.7	9.5	?	10.9	7.7	36	13.7	9.8	?	10.6	6.5	35	12.9	11.1
?	10.0	9.0	27	12.4	7.0	?	10.0	8.7	78	16.8	8.1	?	9.7	8.8	53	14.4	8.1

The existence of this inverse correlation does not necessarily mean that there is a direct causal relation between density and the percentage

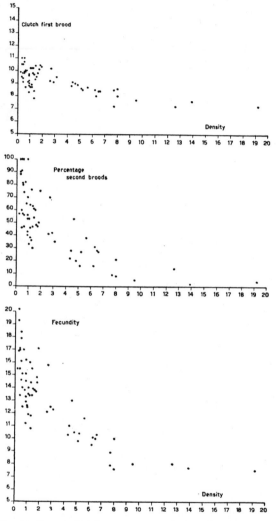

Fig. 7. Correlation between population density (pairs per 10 ha) and clutch size, percentage of second broods and fecundity (average number of eggs per pair). Data from 16 areas 1930-34 derived from Table 27.

of second broods, as both these factors might be independently deter mined by a third factor, e.g. the nature of the habitat. It is clear from Table 27 that both breeding density and the percentage of second broods (and to a smaller extent also clutch-size) are influenced by the nature of the habitat. The influence of the latter can, however, be eliminated by comparing the population density and the percentage of second broods on the same area in successive years. The figures in question are given for O.N.O. in Table 28 and Fig. 8, and the correlation coefficient works out at — 0.60, which is highly significant.

TABLE 28

Percentage of second broods, population density and initial date at O.N.O.

Year	% 2nd broods	Density	Initial date	Year	% 2nd broods	Density	Initial date
1912	45	1.7	30 Ap.	1931	16	5.0	3 May
13	32	3.1	28 Ap.	32	28	4.4	3 May
14	71	3.2	22 Ap.	33	53	4.6	23 Ap.
15	70	2.3	4 May	34	27	6.6	22 Ap.
16	33	4.9	27 Ap.	35	44	4.3	27 Ap.
17	47	1.5	11 May	36	14	6.0	27 Ap.
18	92	3.1	20 Ap.	37	24	6.4	23 Ap.
19	15	5.2	2 May	38	39	5.9	21 Ap.
20	75	2.2	19 Ap.	39	32	3.4	27 Ap.
21	41	2.6	24 Ap.	40	51	3.5	24 Ap.
22	30	2.3	6 May	41	33	7.2	28 Ap.
23	33	3.0	26 Ap.	42	78	1.4	26 Ap.
24	93	2.2	1 May	43	40	5.9	15 Ap.
25	46	7.1	21 Ap.	46	74	3.9	14 Ap.
26	80	3.2	21 Ap.	47	16	5.6	22 Ap.
27	60	4.9	25 Ap.	48	31	5.7	18 Ap.
28	65	3.8	30 Ap.	49	2	8.8	17 Ap.
29	33	3.7	7 May	50	38	6.4	20 Ap.
30	31	5.6	29 Ap.				

The O.N.O. observations cover 37 years, in which time the nature of the woods gradually changed and the Great Tit population increased (see § 9). One part of O.N.O., viz. Part II (Pine Wood), remained effectively unchanged, however, until 1943, and data for the percentage of second broods on this area are available from 1922-1940. At the start of this period, the Scots pines were about 50 years old, and their increase in age to about 70 years did not appreciably alter the appearence of the habitat. In Part II, as on O.N.O. as a whole, a negative correlation exists between population density and the percentage of second broods, the figures amounting to — 0.51, which with 17 degrees of freedom, is highly significant.

In Part I (Mixed Wood) the nature of the habitat changed gradually, causing a rising trend in the population of the Great Tit. Here the correlation coefficient between population density and the percentage of second broods for the years 1922-1940 amounts to — 0.63, which is also highly significant. Owing, however, to the change in the nature of the habitat during this period, this correlation affords less clear evidence for a direct influence of density on the percentage of second broods than in the case of the Pine Wood. In the Mixed Wood, however, it may be noted that the percentage of second broods is very high in years with a low population density, such as 1942 and 1946, while in 1949, when the population density was very high, there were scarcely any second broods. As the habitat changed only slightly during the forties, these fluctuations cannot be due to changes in the nature of the vegetation.

At Hoenderloo, the correlation coefficient between density and the percentage of second broods is — 0.42 (12 d.f.), for Driebergen — 0.26 (10 d.f.), for Mastbos — 0.36 (10 d.f.) and for Liesbos — 0.17 (10 d.f.) [1]. All these correlation coefficients are negative, though none of them are statistically significant. The population densities were, however, rather low, and we stated on p. 65 that the inverse correlation between density and the percentage of second broods is most apparent in the dense populations.

TABLE

Number of breeding pairs and reproduction at Aardenburg

		Few nestboxes				
Year	Pairs	Average clutch size		% of second broods	Fecundity rate	Production rate
		First brood	Second brood			
1926	5	9.8 (±1.7)	8.0 (±0.0)	80	16.2	10.8
1927	6	9.7 (±3.4)	7.8 (±0.8)	83	14.9	12.8
1928	8	10.4 (±1.5)	6.7 (±2.3)	38	13.3	10.5
1929	8	9.6 (±1.5)	8.8 (±2.0)	50	12.5	9.6

1) In Mastbos and Liesbos it was not possible to determine the actual densities. It was, however, possible to calculate the correlation coefficients in question for a particular group of boxes which were inspected every year.

In conclusion, the data from the estate of Aardenburg near Doorn illustrate the above correlation very clearly. Mr. A. A. Tjittes supervised the nesting boxes on this area from 1926 up to 1934. Up to 1929 inclusive, there were few boxes in the wood. In 1930 the number was increased, as a result of which the number of Great Tit pairs increased greatly. From an average of 6 to 8 pairs, it immediately rose to 19-24 pairs.

From Table 29 it can be seen that the increase in the number of pairs is obviously correlated with a decrease in the percentage of second broods. No obvious change took place in the nature of the habitat between 1929 and 1930, in addition to the erection of many more nest-boxes. The figures from Aardenburg therefore make it highly probable that the percentage of second broods is directly influenced by the population density. This does not, of course, mean that population density is the only factor influencing the percentage of second broods. Doubtless other factors are also involved. In particular, the large differences that we found in the percentage of second broods in different localities are due not only to differences in population density, but also to a more direct influence of the habitat. It seems possible that the quantity of food available during the critical part of the nesting period may be important, but this possibility was not studied, as we did not make measurements of food density [1]).

29

near Doorn (standard deviations in brackets).

Year	Pairs	Many nestboxes		% of second broods	Fecundity rate	Production rate
		Average clutch size				
		First brood	Second brood			
1930	24	8.6 (±1.5)	6.8 (±0.8)	21	10.0	9.0
1931	23	8.5 (±1.5)	6.0 (±0.0)	9	8.9	7.7
1932	24	8.0 (±1.4)	6.0 (±0.0)	8	7.6	6.3
1933	19	8.0 (±2.1)	6.7 (±1.2)	31	10.1	6.7
1934	23	7.2 (±1.7)	5.5 (±0.7)	9	7.9	6.6

1) It seems desirable to study further the relation between population density and the percentage of second broods by means of a direct experiment, which is now in preparation. A uniform wood, which appears to provide a suitable habitat except

The influence of population density on the percentage of second broods throws further light on the apparent influence of the initial date (i.e. start of first broods). We saw that with a late initial date there is some tendency for the percentage of second broods to be smaller, though

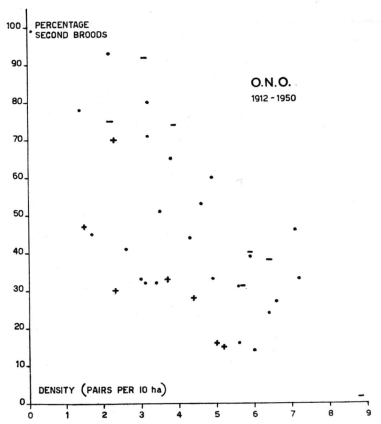

Fig. 8. Correlation between percentage of second broods on O.N.O. and both population density and initial date. Detailed explanation in text p. 71.

for O.N.O. the correlation coefficient amounted to only —0.13, which is unsignificant. We also know that a late initial date is the result of a severe winter and cold spring, and that the latter often causes a reduction in the population. But a low population density tends to increase the percentage of second broods. Hence the influence of a late

for a shortage of nesting sites, will have many nestboxes set up in one part, and only a few nestboxes set up in another part. The number of pairs of Great Tits and the number of second broods, will be studied in both areas over a period of years.

initial date on the percentage of second broods is often obscured by the inverse effect of a low population density. Likewise the effect of population density on the percentage of second broods is somewhat obscured by the above influence of the initial date. This is clear from Fig. 8. Each dot in this figure represents the percentage of second broods together with the population density in a particular year. The dots have been replaced by + marks for the years in which the initial date was more than 5 days later than the average (26 April), and by — marks for the years in which the initial date was more than 5 days earlier than the average. The + marks are for the greater part to be found at the lower side of the group of dots and the — marks at its upper side. Thus Fig. 8 shows clearly that a late initial date tends to decrease the percentage of second broods, wheras an early initial date tends to increase this percentage. The partial correlation coefficients amount to -0.73 between population density and the percentage of second broods, and to -0.54 between the initial date and the percentage of second broods, both values being highly significant. It may therefore be concluded that, both population density and initial date influence the percentage of second broods.

In § 16 it was shown that the proportion of second broods is smaller among first-year than older females. The question therefore arises of whether the higher percentage of second broods in sparse populations might be due to a smaller proportion of first-year individuals. This is not, however, the case. First, the analysis in Table 24 shows clearly that both age and habitat influence the percentage of second broods independently of each other. Further it will be shown in § 33 that on O.N.O. the proportion of first-year individuals tends to be slightly higher, not lower, in the thinly populated Pine Wood than in the densely populated Mixed Wood; this would tend to produce a slightly lower, not higher, percentage of second broods in the Pine Wood. In fact, the influence of habitat or population density (or both) is paramount, and it overrides the possible slight difference due to the age-composition in the two habitats.

23

Reprinted from *J. Mamm.*, **31**(3), 247–259 (1950)

THE ADRENO-PITUITARY SYSTEM AND POPULATION CYCLES IN MAMMALS

By John J. Christian

Since Elton (1924) published his first account of population cycles, there has been a tremendous amount of work done to determine what animals have cyclic fluctuations and what causes these fluctuations to occur with such regularity. Hatt (1930) and Hamilton (1937) both reviewed and discussed vole fluctuations at length. Elton later (1942) reviewed the whole subject in detail. In 1948 Rowan described the ten-year cycle as seen in central Canada with particular reference to snowshoe hares and grouse. Most recently Clarke (1949) briefly summarized the present status of the subject. One thing is abundantly clear in all of these discussions: the cause of cyclic population fluctuations remains unknown. The number of species now known to cycle are legion, in fact the non-cyclic mammal is the exception rather than the rule. The problem of the sudden die-off following a population peak appears to be a major key to the whole problem. Similarly, it has become increasingly clear that infectious diseases or parasites are not the primary factor causing these die-offs. Thus in any explanation of mammalian cycles, we are faced with the problems of accounting for the population-wide sudden death, which has no easily apparent cause, and the fixed cycle length of each species with its attendant population increase during this period.

Since intensive investigations into mammalian cycles began, there have been discoveries in many fields which now appear to be of importance in explaining cyclic phenomena. Selye (1946) elucidated the role of the adreno-pituitary system of the individual in response to stress, and described the diseases resulting from the derailment of this mechanism. The concepts of territoriality and home range have been applied to mammals (Burt, 1940; 1943) and will obviously have a bearing on mammalian populations. The vital statistics of many localized populations have been worked out for varying lengths of time. Lastly, there has been considerable work done on reproduction in mammals, especially with reference to environmental factors influencing the annual reproductive cycle. All of these will be dealt with at greater length later. It now seems that sufficient information is at hand to postulate an explanation of mammalian cycles, although many details will still remain to be worked out.

247

Although it may seem reversed, I am starting at the end of the cycle, because the die-off is its most startling phenomenon as well as one of the major problems to be explained. This crash has been sufficiently described by Elton (1942) to make unnecessary any further description of its general aspects here. Suffice it to say that the crash occurs following a peak in the population, and terminates a cycle of increasing abundance. More detailed accounts of several investigations on crashes are given below.

In 1909 Piper described an outbreak and subsequent die-off of meadow voles (*Microtus montanus*) in Nevada. The peak was reached in the winter of 1907–08, and was followed by the crash from January to March of 1908. All attempts to relate the die-off to a causative organism failed.

Elton, Ford, Baker, and Gardner (1931) intensively investigated a predicted crash in *Apodemus sylvaticus* in late winter and early spring, and were wholly unable to demonstrate any causative organism. No recognizable pathological lesions were demonstrated. These authors did not definitely rule out the possibility of a virus infection, but they believed it extremely unlikely on the basis of negative lesion findings. Their conclusion was "that some obscure biochemical or psychological condition of the mouse population caused the animals to die in captivity (after being brought in from the wild) with the most unusual regularity and rapidity." Animals brought in to the laboratory died within a day or so under identical conditions to those previously kept for three weeks or more. They state that something was at work in the mouse population that was able, either without shock or stress of the mice being trapped and brought into captivity, or with the aid of these factors, to destroy the mice extremely quickly, and before any important lesions had developed. They assume that the mice were weakened by some factor the investigation did not reveal. These authors describe one *Microtus* dying in convulsions, this species having experienced a crash similar to *Apodemus*. No lesions could be found in the brain or elsewhere in this animal.

Findlay and Middleton (1934) definitely ruled out bacteria and viruses as causative agents during a die-off of *Microtus agrestis*. Preceding death in these animals there was a lethargic period followed by a period of convulsive activity. The mice "exhibited head retraction, circular movements, hunching of the back, and frequently paralysis of the hind limbs associated with periodic convulsive movements. Sometimes sudden death occurred during these convulsions, more commonly the voles passed into a comatose condition, followed after a short period by death." Cysts of *Toxoplasma* were found in the brains of 19 out of 34 voles in sufficiently good condition for careful examination. This organism is quite widespread normally, and occurred here in little over half of the animals that died. It hardly seems probable in view of other authors' findings that *Toxoplasma* can be seriously considered as the prime cause of decimation in the voles. The salient fact is that here again are animals dying without demonstrable cause.

Hamilton (1937) studied *Microtus pennsylvanicus* populations in the region of Ithaca, New York, from 1924 to 1936, with intensive work from the spring of

331

1933 to the fall of 1936. Over 4000 animals were examined during this time. A population high was recorded in the winter of 1935–36, and was followed by a period of marked mortality from March to June 1936. Hamilton assumed that disease was the paramount factor in reducing the spring population, and points out that epidemiologically ideal conditions existed for the spread of disease during a population peak. Animals of all ages were affected. The symptoms found in the dying mice were sluggishness, ruffled fur, and lethargy. After incipient drowsiness, a characteristic spasmodic twitching of the neck and shoulders occurred with the animal falling on its side. The animals finally died in convulsions after preliminary extension of the hind legs. These symptoms are remarkably like those described by Findlay and Middleton above. Laboratory examinations of mice dying in convulsions revealed no pathological lesions, nor could a causative organism be demonstrated. Again a filtrable virus could not be definitely ruled out. Brain examinations were completely negative. External parasites were extremely abundant, but Elton (1942) has shown that this is the result of the increased age of the animals which have wintered over from the preceding fall. Once again we have the deaths unexplained.

Green and his co-workers (1938*a*, 1938*b*, 1939), during an intensive study of the periodic die-off in snowshoe hares in Minnesota, were able to demonstrate that a very small number of deaths could be attributed to infectious disease. The majority of the animals exhibited a characteristic syndrome which the authors have termed "shock disease." This syndrome was characterized primarily by fatty degeneration and atrophy of the liver with a coincident striking decrease in liver glycogen and a hypoglycemia preceding death. Petechial or ecchymotic brain hemorrhages, and congestion and hemorrhage of the adrenals, thyroid, and kidneys were frequent findings in a smaller number of animals. The hares characteristically died in convulsive seizures with sudden onset, running movements, hind-leg extension, retraction of the head and neck, and sudden leaps with clonic seizures upon alighting. Other animals were typically lethargic or comatose. Some showed a combination of these two states. The similarity of these symptoms, and those of the mice above, to those of hypoglycemic convulsions is notable. Best and Taylor (1939) give the following description of the symptoms of hypoglycemia produced by any means: "The signs and symptoms vary in the different species. The first signs in the rabbit are hyperexcitability and desire for food. The excitability becomes greater, and mild, and then severe, convulsions are exhibited. The head is retracted and the hind limbs extended in the intervals between convulsive seizures. Coma is frequent. The animals may exhibit rigor mortis immediately after death. The signs in dogs are quite similar. Mice, in some instances, may become comatose without exhibiting convulsions."

In these hares the NPN rose progressively as the symptoms developed with uremia being found in some. These authors (Green *et al.*) emphasized the inability of the animals to withstand any undo exertion or stress which would leave normal animals unaffected. These findings were demonstrated equally in all hares older than seven weeks, and this is of importance, since it shows that factors which induce shock disease exert sufficient influence to produce deaths in both young and old animals in nature. The population studied reached its peak in 1933, and the decline was continuous until 1937, with the greatest die-off being in the fall and winter of 1935. Similar observations were made in the northern United States and central Manitoba. The conclusions reached were that shock disease is primarily responsible for decimations of snowshoe hares, that local conditions were attributable, and that severe upsets in carbohydrate metabolism were the primary effect on the hares.

Throughout these various investigations there is a remarkable consistency in the findings, regardless of the species. With the possible exception of Findlay and Middleton's *Toxoplasma* infections no pathogenic organisms could be demonstrated. The onset of symptoms, terminating in convulsions, shows a similar pattern in every case, and all would appear to be hypoglycemic in nature. All authors pointed to the increased susceptibility to stress. It seems, then, that we are dealing with a phenomenon common to all of the animals, and that Green and Larson (1938a) have adequately described its pathology and physiology. Universally these die-offs follow a population peak, occur primarily in the late winter and early spring, and involve severe metabolic disturbances coincident with a markedly decreased resistance to stress.

In searching for a common cause for these changes we find that Selye's adaptation syndrome (1946, 1947) provides us with an answer to the problem. In other words we are probably dealing with the symptoms of adrenal exhaustion on a population-wide basis. Selye (1946) has already suggested that Green's shock disease in snowshoe hares resembled a disease of adaptation. The adaptation syndrome is the sum of the non-specific physiological and morphological responses to stress other than the specific adaptive reactions (such as serologic reactions to specific antigens). Selye has divided these into (1) shock, (2) counter-shock (these two combined are the "alarm reaction"), (3) resistance, and (4) exhaustion phases. Here we are probably dealing with the exhaustion phase, where the animal is no longer able to react to added stress after long, continued stresses, and the whole mechanism breaks down. Characteristic of the exhaustion phase (Selye, 1947), like those of the alarm reaction, are hypoglycemia, involution of lymphoid tissue, adrenal cortical hypertrophy, decreased liver glycogen, diminished fat content of the adipose tissue, lipid deposition in the liver, increased NPN with a negative nitrogen balance, reversal of the albumin/globulin ratio, decrease in blood chlorides, rise in blood potassium, and a marked decrease of the ascorbic acid, ketosteroid, and plasmal granule content of the adrenal cortex. Frequently during the stage of exhaustion, the adrenal cortex is found to be hemorrhagic. Often the liver only shows signs of cloudy swelling and decrease in size, but may, however, show more intensive atrophy and degenerative changes. These liver changes may be augmented by muscular exercise (Selye, 1939b).

In listing the changes during the exhaustion (or alarm reaction) phase of the adaptation syndrome we have practically repeated the findings in animals during the periodic die-off. Green and Larson (1938a) clearly showed that the convulsions in hares did not occur until the liver glycogen dropped below 0.2 per cent, demonstrating that they were caused by a progressive fall in the glucose reserves. Experimentally they were able to stop or alleviate the convulsions temporarily by intravenous glucose injections, definitely affirming that the convulsions were hypoglycemic in nature. Epinephrine injections were ineffective. It seems more than likely that in most die-offs in mammals we are dealing with a manifestation of the adaptation syndrome with the terminal convulsions precipitated in many cases by some sudden stress, such as fear, captivity, or exertion.

The adrenal response to stress is stimulated by the secretion of adrenocorticotrophic hormone from the anterior pituitary. The mechanism by which the pituitary is stimulated is as yet unknown, but may be brought about by the release of adrenalin (Michael, 1949). As a result of demands on the pituitary of greater life-sustaining importance in response to stress, there is a shift to this function from gonad (Selye, 1939a) and growth (Selye, 1936) stimulating functions; so that growth slows or ceases and the gonads may become atrophic or at least diminished in functional activity. The importance of this factor will be discussed below. Also of significance is that, when the adaptation syndrome is elicited to one stress, the animal becomes more susceptible to other stresses (Selye, 1947). Fasting (Selye, 1946) and muscular exertion (Selye, 1939b) markedly increase the susceptibility of mammals to alarming stimuli, and may augment the changes which have already taken place. This is especially true of the hypoglycemic phase, and it is probably these factors which determine the final outcome in the form of hypoglycemic convulsions.

The adaptation syndrome may be elicited by a wide variety of alarming stimuli: environmental changes, pregnancy (Sayers and Sayers, 1949), toxic chemicals, bacterial infections (Selye, 1946), fear, rage, emotional upsets (Selye, 1946; Pincus, 1949; Michael, 1949), cold, muscular exertion, and a great many others (Selye, 1946). Assuming that the die-off is due to a manifestation of the adaptation syndrome, we must then investigate the stresses which are at work to produce this effect on a population-wide basis in mammals.

In all of the studies so far recorded, the die-off has been in the winter or early spring following a population peak the preceding autumn. Hence we have an extremely high population coming into a period of relative food scarcity, cold, and other severe climatic stresses with their attendant sequelae. Along with the attainment of high population levels, animals are progressively crowded into less and less favorable environments from an originally highly favorable nucleus (Hatt, 1930; Elton, 1942; Burt, 1949). This means less favorable foods, less good cover, and often unfavorable water supply, and other factors of submarginal living. In all probability every population peak in the wild exceeds the normal carrying capacity of the environment for that and other species, in particular during the succeeding winter. Under these conditions we would expect the following stresses: (1) food scarcity, (2) lack of proper cover, (3) increased muscular exertion resulting from longer food forage trips, (4) fights with other individuals (Calhoun, 1949), (5) increased exposure to cold from longer forage trips and inadequate cover, (6) fighting resulting from territorial encroachment (Burt, 1940; Calhoun, 1949), (7) utilization of inadequate foods, (8) increased exposure to predators due to lack of cover as well as migration of predators into areas of abundant food supply in the form of a peak population, and (9) nutritional deficiencies. In summary nearly all emotional, nutritional, exertional, and climatic stresses are augmented and operable upon the entire population. It is of interest to note that this would impose similar stresses upon animals of all ages, agreeing well with Green and Larson's finding in hares (1938*a* and 1938*b*) in which all animals over seven weeks of age were involved. Apparently parental care ceases at about seven weeks (Aldous, 1937).

We now have a population under highly stressed conditions and undoubtedly taxing its adreno-pituitary system to the maximum. At this time (i.e., in early winter) the day length begins to increase and stimulate the pituitary in most mammals to secrete gonadotrophic hormones with resultant gonad development prior to the breeding season. That daily light stimulates the gonads via the pituitary has been shown for the snowshoe hare (Lyman, 1943), *Microtus agrestis* (Baker and Ransom, 1932), ferrets (Bissonnette, 1932, 1935; Allanson, 1932), raccoons (Bissonnette and Csech, 1937), white-footed mouse (Whitaker, 1940), mink (Hansson, 1947), and the list will probably be extended to many more. In some mammals (e.g. the ground squirrel) light has no effect upon the gonad development, but a period of low temperature is the factor which stimulates the pituitary to gonadotrophic activity (Wells and Zalesky, 1940); while in others light is the primary factor, but low temperatures may augment the reproductive activity (Whitaker, 1940). In any event in all mammals with a spring breeding season we have a factor placing increased demands on the pituitary at a time when it is already maximally stimulated to produce adrenocorticotrophic hormone in response to stress. That the gonadotrophic function is stimulated in spite of adrenal demands is evident from the fact that the breeding season with production of young does ensue. The argument might be offered that in some mammals the males are primarily affected by light, but this may be answered by pregnancy demands in the female (Sayers and Sayers, 1949; Hatfield, 1935), and growth demands in the young (Selye and Collip, 1936). We have already seen that the pituitary cannot simultaneously respond maximally to increased demands of all types, but that one type must succumb to another. If, then, the pituitary is under maximum demand as a result of the already-listed stresses, and a direct stimulus (in the form of light or low temperature) causes stimulation of gonadotrophic activity, it stands to reason that this latter must occur at the expense of the adrenotrophic activity. Hence at a time when stresses are maximal, we suddenly have an increased pituitary load, and the animals begin to succumb with symptoms of exhaustion of the adreno-pituitary mechanism, and the periodic decimation is in full swing. As one may see from the preceding discussion, all of the animals succumbing at more or less the same time is a function of the high population with its inherent stresses.

The time it takes an animal to succumb depends upon the severity and duration of the applied stress. Thus, in a population which reaches a peak in a brief time, the accompanying stresses would be more acutely developed and their application more intense, resulting in a more sudden and pronounced population death. When a population develops more slowly, the stresses increase at a slower rate, and are applied more gradually, which would tend to attenuate the final collapse (Sayers and Sayers, 1949).

A word may be said here on the many discussions of disease as a cause of the periodic die-off. That epidemics are common during some population peaks has been shown and the subject reviewed by Elton (1931, 1942). As an example, rabies has been suggested as a population control in the arctic Canidae (Cowan, 1949). The status of rabies in the wild and its relation to population decimations has been reviewed by Gier (1948). One would expect epidemics to break out during a population peak due to decreased resistance and crowding. We have already seen that increased resistance to one stress leads to decreased resistance to others, and that resistance to all stresses is diminished during the exhaustion stage of adaptation. We can understand why infectious disease, although secondary, is not uncommon in many die-offs, and why it has been investigated as a possible primary cause. Parasities were similarly suspected, and Elton (1942) showed them not to be a cause of decimations other than as a secondary factor.

We now have a working hypothesis for the die-off terminating a cycle. Exhaustion of the adreno-pituitary system resulting from increased stresses inherent in a high population, especially in winter, plus the late winter demands of the reproductive system, due to increased light or other factors, precipitates population-wide death with the symptoms of adrenal insufficiency and hypoglycemic convulsions. The only mammal known not to have a cycle is the beaver (Clarke, 1949), and it seems likely that, throughout the recorded history of the animal, trapping pressure has been sufficient to keep the population level below the highly stressed condition. Similarly, muskrats are cyclic throughout their range except where water levels are controlled. In these areas there is usually a controlled high level of trapping pressure to keep the population from peaking to stress levels. Even these, however, become subject to crashes when extremely adverse climatic conditions lower the carrying capacity of the environment, and all the stresses of a peak population succeed with their sequelae. Such a condition probably occurred throughout the muskrat producing areas of northwestern Pennsylvania and Ohio in the fall and early winter of 1946–47. Several different pathogens were isolated from the dead muskrats at this time, primarily enteric organisms, which were in all likelihood secondary to exhaustion of the adaptive mechanism. This explanation of the die-off accounts for crashes in discontinuous large areas, as well as in different species, being out of phase, since only in one homogeneous large area will the population pressures be necessarily applied simultaneously. Species with similar habitats, size, and food requirements would be expected to be in phase, and such is usually the case (Elton, 1942).

Up to this point the adaptation syndrome has sufficed to explain the observed phenomena, but this still leaves two aspects of mammalian cycles unexplained: (1) the fixed cycle length for each general size of mammal, and (2) the differences in degree of cycling between the sub-arctic and more temperate and tropical climates. First let me say that what has been said and what will follow applies primarily to rodents, since the predators merely reflect the cycles of their food supply. For example, the foxes of Ungava closely follow the vole and lemming cycles upon which they depend for food (Elton, 1942).

Voles, mice, and similar small mammals have an approximately four year cycle (McCleod, 1933; Hamilton, 1937; Elton, 1942; Clarke, 1949). Larger mammals, such as hares, rabbits, and muskrats, have about a ten year cycle (Elton, 1942; Rowan, 1948; Clarke, 1949; Green *et al.*, 1938a, 1939; Chitty, 1946, 1948), while the still larger mammals have a still longer cycle (Clarke, 1949). The first fact which strikes one is the apparent relationship between cycle length and body size, but this in itself is not the answer. It seems likely that the short period of sexual immaturity, the short gestation period, and the large litter sizes of the smaller mammals are the prime factors in the short cycle. These factors lead to a lower reproductive potential in the medium-sized animals, for example, snowshoe hares require at least a year to reach sexual maturity and have a gestation period of 38 days (Aldous, 1937) as opposed to 25 days and 21 days, respectively, for the meadow vole (*Microtus*) (Hamilton, 1941). The larger animals, such as caribou, probably have an even longer cycle correlated with a gestation period of 7–8 months, and reach sexual maturity at $1\frac{1}{2}$ years (Asdell, 1946). In other words, it seems highly probable that the cycle length is a function of the reproductive potential of the given animal. Those with the highest reproductive capacity build most rapidly to a population peak in excess of the carrying capacity of the environment. It is notable that the most prolific animals of their respective size groups (e.g. meadow voles and snowshoe hares) are the most prominently cyclic. Starting with a population low, we have a period of relatively heavy predator pressure on a minimal population (Elton, 1942). At this time the excess of births over the death rate will be very small, but must exist for a population increase to occur, and it is this excess with which we are primarily interested. Taking only this excess, it is understandable that this segment of the population will increase exponentially in relation to its reproductive potential. In the highly prolific *Microtus* with a short period of immaturity and short gestation period, an average litter size of 5.5 (Asdell, 1946), and capable of having up to 17 litters annually (Bailey, 1924), it can be seen that rapid increases to enormous levels are possible, and do occur. An additional factor, pointed out by Hamilton (1937), is the progressively increasing chances of immediate post-parturient mating with a resulting increased number of litters per year as the population rises and contacts between the sexes increase. A further well-known fact tending to accelerate the rate of reproduction is the increase in average litter size for about the first half-dozen litters. In an increasing population the average age of adults is similarly increasing, which means that more maximum-sized litters are born, as well as more females reaching reproductive maturity. Since the potential reproductive capacity of any species, as well as the death rate, is relatively fixed, the cycle length will be related directly to the time it takes the population excess over the death-rate to peak to a point beyond the carrying capacity of the environment, and hence highly stressed conditions. It can be seen that in this fashion the lower the reproductive capacity of the species, the longer will be its cycle. Environment in this discussion is used in its broadest sense, including predators, interspecific relations, and all other factors. The existence of annual cycles has not been forgotten, but, since we are speaking in terms relative to the environment, they will not affect the over-all picture. This explanation carries with it a considerable flexibility if this consideration is kept in mind, i.e. all terms in relation to the environment. Exceptionally severe or favorable environmental factors could possibly shorten or lengthen the cycle a year or so, depending upon whether operative early or late in the cycle. If extremely unfavorable conditions were brought to bear on a population at its ebb, it is conceivable that peak attainment could be delayed a year. Conversely, if this occurred when the population was high, but not at its peak, a relative peak would be caused with a resultant die-off. This same mechanism would tend to put isolated populations in phase, when not too widely separated, as in the southern mountain blocks of Norway (Hatt, 1930). These populations occur in a sufficiently small area to be subject to the same major climatic upsets, and one extremely severe upset of this kind would be sufficient to put the populations in phase for a considerable period of time. Over such a large area as Canada with its divergent climates one would not expect to find populations in phase from one extreme to the other, and such is the case (Clarke, 1949; Chitty, 1943, 1946).

One next might well ask why some population peaks assume plague proportions whereas the majority do not. On a purely theoretical basis it would seem that the maximum carrying capacity of the environment would lie much closer to the peak than any other point of the cycle. If this is so, then highly favorable conditions, particularly with respect to food, for the final season of the cycle would allow an enormously greater peak than normal. This seems to be associated usually with abrupt land-use changes which result in a greatly increased food supply, improved cover conditions, or a relative decrease in predators in relation to the sudden increase in carrying capacity of the area, or probably all of these combined. Under these circumstances, the attainment of highly stressed conditions would be delayed until both prey and predator populations have achieved a balance with the newly increased carrying capacity. The immigration of predator populations usually lags somewhat behind the development of high rodent populations with the result that for a short while the rodent peak would be limited only by the available food supply and living space.

The evidence concerning litter size during the course of a cycle is contradictory. On the basis of the proposed theory a population, as soon as it reaches a highly stressed condition, would be expected to experience a decreased fecundity with decreased litter size. When the stresses are relieved by a die-off, the birth-rate would again be expected to increase, slowly at first, and then more rapidly as predator pressure and its stresses become relatively less important. This becomes evident from the preceding discussion concerning the ability of the pituitary to secrete large quantities of gonadotrophic hormone in the face of demands for the life-sustaining adrenocorticotrophic hormone. Fasting has been shown to result in decreased reproductive performance (Slonaker and Card, 1923*a,b,c*), probably acting primarily through the pituitary-adrenal system, as previously discussed. Aldous (1937) states that there was a gradual decline in the young per litter of snowshoe hares starting before the population decline, and continuing with it. This is understandable on the basis of the population obtaining stressed conditions prior to the actual occurrence of death. Elton (1924) describes large litters in hares during the period of increase, and smaller during the decline. Piper (1928) found no signs of breeding at the peak of an outbreak of house mice. Hamilton (1937) similarly describes a decrease in the size of *Microtus* litters during the decline. Green, Larson, and Bell (1939), however, found no significant difference in the litter sizes between the peak year and the height of the decline. Elton, Ford, Baker, and Gardner (1931) also state that the litter size in *Apodemus* decreases during the decline. That the birthrate decreases due to decreased litter size or sometimes total reproductive failure shortly preceding, during, or shortly after the population crash would be expected; but much more conclusive experimental work must be done on this point. Likewise, if Hamilton's theory (1937) of increasing chances of mating with increasing population is correct, and there is no reason to suppose that it is not, one would expect fewer litters during the early period of the cycle, and gradually increasing until the population decline.

We have, as yet, not explained why the cycles are so much more pronounced in the far north and become progressively less pronounced with decreasing latitude. This aspect of the problem is still without any experimental evidence from the field. The change from north to south is gradual; and in looking for constantly changing climatic factors which might cause this effect, temperature, day-length, and light intensity appear as three obvious factors which vary regularly and progressively with latitude. All three of these may be contributing factors if we assume, lacking conclusive evidence from the field, that the reproductive capacity in the north is greater than in the south for the same species. Whitaker (1940) has shown in a series of experiments on the white-footed mouse (*Peromyscus leucopus*) that increasing daylight stimulates the gonads to gametogenesis. However, light of low intensity caused a continuous breeding condition, regardless of day length, while more intense light caused the activity of the gonads to follow the increases and decreases in day length. He also demonstrated that low temperatures were capable of keeping an animal in a state of continuous reproductive activity with a light intensity intermediate to the above two experiments regardless of the length of day. Wells and Zalesky (1940) showed that low temperatures were necessary for continuous spermatogenesis in the male ground squirrel (*Citellus tridecemlineatus*) as opposed to the normally seasonal behavior. Low temperatures only slightly hastened the onset of spermatogenesis from an aspermatic condition in these animals. These studies provide an experimental basis from which we might postulate an increasing reproductive rate the farther north one progresses. Conversely, in areas where summer temperatures are high, there is a suppression of breeding activity in mice during July and August (Hamilton, 1941; Howard, 1949). This is further supported by Laurie's paper (1946) on house mice in different environments. The mice living in cold stores ($-21°$ to $-10°$C.) had significantly larger litters than those in warmer environments, but the number of litters was unchanged. In the same paper she reviews some of the experimental evidence for suppression of sexual activity by high temperatures (midsummer or tropical). We then have experimentally conditions which would indicate that intensive field work on one species (simultaneously carried out in a subarctic and a temperate habitat) would reveal a greater productive capacity per female per year in the far north where low temperatures, lower light intensities, and prolonged days would maintain a high level of reproductive activity. This would be opposed further south by a fall in the breeding rate during the hot summer months with a generally lower reproductive potential.

In addition to these facts, and correlated with them, it is interesting to note that the animals with the most pronounced cycles have the highest reproductive potentials in their size groups, as well as being the most daylight-active. These include meadow voles (*Microtus*), snowshoe hares (*Lepus*), lemmings, and house mice.

SUMMARY

Exhaustion of the adreno-pituitary system subsequent to the stresses inherent in a high population level, severe climatic conditions, and the demands of the spring breeding season have been proposed as the cause of cyclic declines in mammals.

The length of the cycle is probably a function of the reproductive potential of the species involved, and this potential is applicable to that portion of the population in excess of the death rate for the development of peaks. The terminating factor is the attainment of a population above the carrying capacity of the environment.

The possibility has been suggested that longer daylight, decreased light intensity, and lower temperatures during the breeding season in the north maintain a higher reproductive potential in turn leading to more pronounced cyclic changes than in more temperate climates. Field data on this matter is needed.

LITERATURE CITED

ALDOUS, C. M. 1937. Notes on the life history of the snowshoe hare. Jour. Mamm., 18: 46–56.

ALLANSON, M. 1932. The reproductive processes of certain Mammals. III. The reproductive cycle of the male ferret. Proc. Roy. Soc., B, 110: 295–312.

ASDELL, S. A. 1946. Patterns of mammalian reproduction. Comstock. Ithaca. 1–437.

BAILEY, V. 1924. Breeding, feeding, and other life habits of meadow mice (*Microtus*). Jour. Agric. Res., 27: 523–535.

BAKER, J. R., AND R. M. RANSOM. 1932. Factors affecting the breeding of the field mouse (*Microtus agrestis*). I. Light. Proc. Roy. Soc. B, 110: 313–322.

BEST, C. H., AND N. B. TAYLOR. 1939. The physiological basis of medical practice. Williams and Wilkins, Baltimore. Pp. 942–943.

BISSONNETTE, T. H. 1932. Modification of mammalian sexual cycles; reactions of ferrets (*Putorius vulgaris*) of both sexes to electric light added after dark in November and December. Proc. Roy. Soc. B, 110: 322–336.

————. 1935. The avenue of reception of sexually stimulating light in ferrets. Anat. Rec., 64. Suppl., 1: 89.

BISSONNETTE, T. H., AND A. G. CSECH. 1937. Modification of mammalian sexual cycles. VII. Fertile matings of raccoons in December instead of February etc. Proc. Roy. Soc., B, 123: 246–254.

BURT, W. H. 1940. Territorial behavior and populations of some small mammals in southern Michigan. Misc. Publ. Mus. Zool. Univ. Mich., No. 45: 1–58.

————. 1943. Territoriality and home range concepts as applied to mammals. Jour. Mamm., 24: 346–352.

————. 1949. Territoriality. Jour. Mamm., 30: 25–27.

CALHOUN, J. B. 1949. A method for self-control of population growth among mammals living in the wild. Science, 109: 333–335.

CHITTY, H. 1946. The snowshoe rabbit enquiry, 1942–43. Canad. Field-Naturalist, 60: 67–70.

————. 1948. The snowshoe rabbit enquiry, 1943–46. Jour. Animal. Ecol., 17: 39–44.

CLARKE, C. H. D. 1949. Fluctuations in populations. Jour. Mamm., 30: 21–27.

COWAN, I. Mc. 1949. Rabies as a possible population control of Arctic Canidae. Jour. Mamm., 30: 396–398.

ELTON, C. 1924. Periodic fluctuations in the numbers of animals, their causes and effects. Brit. Jour. Exp. Biol., 2: 119–163.

————. 1931. The Study of epidemic diseases among wild animals. Jour. Hygiene (Cambridge), 31: 435–456.

————. 1942. Voles, mice, and lemmings. Oxford, Clarendon Press. Pp. 1–496.

ELTON, C., E. B. FORD, J. R. BAKER, AND A. D. GARDNER. 1931. Health and parasites of a wild mouse population. Proc. Zool. Soc. Lond., 2: 657–721.

FINDLAY, G. M., AND A. D. MIDDLETON. 1934. Epidemic disease among voles (*Microtus*) with special reference to *Toxoplasma*. Jour. Animal Ecol., 3: 150–160.

GIER, H. T. 1948. Rabies in the wild. Jour. Wildl. Mgt., 12: 142–152.

GREEN, R. G., AND C. L. LARSON. 1938*a*. A description of shock disease in the snowshoe hare. Amer. Jour. Hygiene, 28: 190–212.

————. 1938*b*. Shock disease and the snowshoe hare cycle. Science, 87: 298–299.

GREEN, R. G., C. L. LARSON, AND J. F. BELL. 1939. Shock disease as the cause of the periodic decimation of snowshoe hares. Amer. Jour. Hygiene, Sect. B, 30: 83–102.

HAMILTON, W. J., JR. 1937. The biology of microtine cycles. Jour. Agric. Res., 54: 779–790.

————. 1941. Reproduction of the field mouse (*Microtus pennsylvanicus*). Cornell Univ. Agric. Exp. Sta., Mem. 237.

HANSSON, A. 1947. The physiology of reproduction in mink (*Mustela vison*) with special reference to delayed implantation. Acta. Zool., 28: 1–136.

HATFIELD, D. 1935. A natural history study of *Microtus californicus*. Jour. Mamm., 16: 261–270.

HATT, R. T. 1930. The Biology of the voles of New York. Roosevelt Wildl. Bull., 5 (4): 513–623.

HOWARD, W. E. 1949. Dispersal, amount of inbreeding, and longevity in a local population of prairie deermice on the George Reserve, southern Michigan. Contrib. Lab. Vert. Biol., Univ. Mich., No. 43: 1–50.

LAURIE, E. M. O. 1946. The reproduction of the house-mouse (*Mus musculus*) living in different environments. Proc. Roy. Soc., B, 133: 248–281.

LYMAN, C. P. 1943. Control of coat color in the varying hare, *Lepus americanus* Erxleben. Bull. Mus. Comp. Zool., Harvard. Univ., 93: 393–461.

McCLEOD, J. A. 1933. A parasitological survey of the Genus Citellus in Manitoba. Carad. Jour. Res., 9: 108–127.

MICHAEL, S. T. 1949. Adaptation to brief stress. Yale Jour. Biol. and Med., 22: 71–92.

PINCUS, G. 1949. Adrenal cortex function in stress. Ann. N. Y. Acad. Sci., 50, art. 6: 635–645.

PIPER, S. E. 1909. Mouse plagues, their control and prevention. Yearbook, U. S. Dept. Agric. 301–310.

ROWAN, W. 1948. The ten year cycle. Univ. Alta. Ext., Alta. Pp. 1–15.

SAYERS, G., AND M. A. 1949. The pituitary-adrenal system. Ann. N. Y. Acad. Sci., 50, art. 6: 522–539.

SELYE, H., AND J. B. COLLIP. 1936. Fundamental factors in the interpretation of stimuli influencing the endocrine glands. Endocrin., 20: 667–672.

SELYE, H. 1939a. The effect of adaptation to various damaging agents on the female sex organs in the rat. Endocrin., 25: 615–624.

———. 1939b. Effect of muscular exercise on the fat content of the liver. Anat. Rec., 73: 391–400.

———. 1946. The general adaptation syndrome and the diseases of adaptation. Jour. Clin. Endocrin., 6: 117–230.

———. 1947. Textbook of endocrinology. Acta. Endocrin., Montreal. Univ. Pp. 837–867 (The general adaptation syndrome and the diseases of adaptation).

SLONAKER, J. R., AND T. A. CARD. 1923a. Effect of restricted diet. II. On pubescence and menopause. Amer. Jour. Physiol., 64: 35–43.

———. 1923b. Effects of restricted diet. III. On the numbers of litters and young born. Amer. Jour. Physiol., 64: 167–180.

———. 1923c. Effects of restricted diet. IV. On the age of greatest productivity. Amer. Jour. Physiol., 64: 203–209.

WELLS, L. J., AND M. ZALESKY. 1940. Effects of low environmental temperature on the reproductive organs of male mammals with annual aspermia. Amer. Jour. Anat., 66: 429–448.

WHITAKER, W. L. 1940. Some effects of artificial illumination on reproduction in the white-footed mouse, *Peromyscus leucopus noveboracensis*. Jour. Exp. Zool., 83: 33–60.

Wyeth Institute, 900 Broad Street, Philadelphia 30, Penna. Received January 9, 1950.

V
Applied Behavior

Editor's Comments on Papers 24 Through 27

The evolutionary trend of emphasis from about 1910, when behavior was not considered as an ecological factor, to the situation in 1960, when behavior was accepted as an integral part of the ecology of a species, has allowed the application of basic knowledge to the solution of practical problems.

The papers reprinted thus far have represented "basic biology," which can be considered to be biological knowledge obtained for its own sake. But such knowledge may be applied, of course, and some of the papers (e.g., Myers and Poole) were a mixture of basic and applied approaches. The application of basic knowledge to human problems requires time. Indeed, the lag may be one or two decades. This lag results from several features of human behavior. First, some time is required for the simple communication of the ideas; second, the persons who have the opportunities to apply the ideas must *accept* them, which usually occurs only after experimentation on their part; and, third, the user has to be convinced. Behavior has perhaps been difficult to apply because it is not as readily measured as are anatomical and physiological traits.

A first example is the use of information on social organization in poultry production (Guhl and Warren, 1946). Although Masure and Allee (1934) described the peck order of chickens, the application of these behavioral principles to the ecological conditions of poultry production has hardly started. The paper reproduced here indicated the potential for improving poultry production. It happens that the knowledge of social rank has not been very useful to the poultry industry. One reason has been that genetic selection has been the favorite tool for improvement of production, and as long as gains were achieved, there was little interest in behavior. Furthermore, management practices emphasized labor-saving devices such as mechanical feeders and big brooder houses. Poultry kept in flocks of several thousand birds of one sex, one color, and one age do not develop a rank organization. The article presented here deals with fertility, which had little application because the producers wanted infertile eggs or used artificial insemination for breeding genetic lines. Nevertheless, although the knowledge of behavior did not have great use, it did have value in promoting an understanding of poultry production.

Another is the complex relation of flock organization and food supply of pigeons (Murton et al., 1966). A careful observational study of the social behavior in flocks led the way to understanding the feeding habits and the amount of food required to maintain a bird through the winter. The observations on behavior are excellent. Application of this knowledge of behavior is already showing results. The essence of the research (several long papers) is that pigeons shot in the fall would either die or displace another pigeon by the time damage is done to the clover crop in late winter. Thus shooting in the fall is a waste of time and benefits only the sellers of gunshot. However, shooting in late winter, especially in certain mild winters, removes the survivors and actually reduces the number of pigeons and the damage. Thus knowledge of behavior may have definite applied value.

Game management is now applying behavioral principles (Buechner, 1960). Innumerable papers on game species in North America and Africa could be cited that date from 1960. These consider the life history of the species (kob, elk, cats, elephants), with great emphasis on social organization and reproduction, thereby setting the stage for management. Indeed enough is known about many species (deer, ducks, quail, trout, etc.) that application of principles could rapidly produce increased harvest. But unfortunately "people management" is the greatest problem, and it is difficult to demonstrate behavioral principles in part because of their relation to ecological features. The problem of doe seasons is a spectacular example. As populations increase, the low-ranking does produce fewer fawns and grow less. Harvest of these does would benefit the herd and reduce damage to the habitat. But it is easier for the sportsman to see overbrowsing conditions than low rank, and thus campaigns to feed the deer are popular. A parallel situation occurs with respect to disease. Does that have low resistance because of crowding are the stimulus for schemes to use vaccines, antihilminthics, etc. Unfortunately, the effects of behavior on the ecological conditions are not simple and thus are difficult to apply.

Another applied topic is the behavior and ecology of primates, studied now by anthropologists who obviously are hoping to better understand the origin of man. A few anecdotal papers appeared about 1935, but the real impetus came when Emlen and Schaller (1960) showed that the gorilla could be profitably and safely studied. The reputation of the gorilla for ferocity prevented, for several decades, adequate research on its behavior, which was necessary to permit comparisons with other apes and to develop plans for conservation. I well remember a meeting about 1954 of a committee concerned with providing funds for research on primates. Emlen had presented a proposal that was eminently meritorious, but several members were opposed to awarding funds because they felt that Emlen was exposing himself to harm and that nothing could be learned by observation. Fortunately, knowledge of behavior (aggression, territory, rank, etc.) was adequate to convince these people that the gorilla could successfully be studied, and it was.

Reference

Buechner, Helmut K. 1960. The Bighorn Sheep in the United States, its past, present, and future. *Wildlife Monog.* 4: 1–174.

24

Reprinted with permission from *Poultry Sci.*, **25**(5), 460–463, 467–472 (1946)

Number of Offspring Sired by Cockerels Related to Social Dominance in Chickens[1]

A. M. GUHL AND D. C. WARREN

Kansas Agricultural Experiment Station

(Received for publication April 9, 1946)

THE social organization in flocks of chickens described by Schjelderup-Ebbe (1922, 1935) has during more recent years received much experimental treatment by Allee and his co-workers, and has been discussed in summaries by Allee (1942, 1943, 1945) and by Collias (1944). The social hierarchy, or peck-order, is based on dominance-subordination relationships among the members of the flock. Cocks, however, normally do not peck the hens but form a social order among themselves. Evidences have been presented that high social rank may offer biological advantages for birds at the upper levels of the social order as contrasted with those at the lowest levels. In flocks of hens, those at the top half of the peck-order lay more eggs than those at the bottom half (Sanctuary, 1932). Dominant birds have a greater freedom of the pen and freer access to food (Masure and Allee, 1934). In small flocks of hens containing several cocks, Guhl, Collias and Allee (1945) found that the superior male mated at a higher frequency than his inferiors, the sexual activity of which he suppressed more or less completely.

In the present study the primary interest was to determine whether the dominant males also leave more offspring than their subordinate cocks. The sire of

each chick was made evident by the use of males which differed from each other, or from the hens, by some hereditary character.

THE BIRDS AND THEIR TREATMENT

The study was conducted with two flocks during their breeding season. In the initial test the males selected were of breeds which differed in body weight and color. These variations may have introduced complicating factors. With the second flock the procedure was somewhat more refined by the use of males which possessed less marked differences, all being the same color and of similar size.

1944 flock. The cockerels, a Rhode Island Red, a Barred Plymouth Rock and a White Leghorn were placed in a pen containing 30 Rhode Island Red pullets. All these birds were one year old and in their first breeding season. The pen was uncrowded as there were 8.4 square feet per bird, males plus females. After the second day (cf., Moore and Byerly, 1942) the eggs obtained were held for incubation. Eggs were collected for 40 consecutive days. Sample observations on sexual behavior were made intermittently on 22 days, and the experiment ran from April 11 to May 22, inclusive. The chicks sired by the Rhode Island Red had red down, those sired by the Barred Rock black down, and those sired by the White Leghorn white down. The dominance order of

---

[1] Contribution No. 204 Department of Zoology and No. 165 Department of Poultry Husbandry.

Editor's Note: Tables 4, 5, and 6 have been omitted, owing to limitations of space.

the males was in the order mentioned and the males are listed accordingly in the tables.

This flock was permitted to range during the last three weeks of the experiment in a yard 100 feet by 60 feet.

1945 flock. A White Plymouth Rock[2] male known to be homozygous for the dominant white factor, a White Plymouth Rock male known to carry recessive white, and a Wyandotte male which was homozygous for rose comb were introduced into a pen containing 36 Barred Plymouth Rock pullets. All the birds, except the Wyandotte male, which was older, were one year old. Space relations averaged 5.6 square feet per bird. Egg collections for incubation began on the third day and continued for 37 consecutive days. Observations on mating activities were made on 46 days between May·10 and July 27, inclusive. The chicks sired by the *dominant white* male had white down, those sired by the *recessive white* had black down, and those sired by the *rose comb* developed a rose comb. These males are listed in the tables in their order of social dominance beginning with the individual of highest rank.

Procedure. The males in each flock were selected from among several of their breed or phenotype on the basis of the results of a fertility test made by· the artificial insemination of their semina into several pullets other than those used in these flocks. For the 1945 flock this test served also for the determination of the homozygosity of the characters used to trace the sire of the chicks. With the assurance that the spermatozoa of each male were functional, a further test was made of the competitive fertilizing capacity of the three males selected for each flock. Equal volumes of semen from each of the three

[2] This bird was obtained from Dr. R. G. Jaap of the Oklahoma Agricultural Experiment Station.

males of each flock were mixed and artificially inseminated into several hens of the same breed as those to which each set of males were later exposed. Of the 56 chicks obtained in 1944, 9 were sired by the Rhode Island Red, 19 by the Barred Plymouth Rock, and 28 by the White Leghorn; the 1945 test yielded 60 chicks of which 39 were sired by the *dominant white*, 10 by the *recessive white* and 11 had rose combs. The results indicated that the sperm of each male were capable of fertilizing eggs when in physiological and numerical competition with those from the other males, a situation that would be typical with mixed semina in the reproductive tract of females of this polygamous species.

These flocks were under observation at frequent intervals. During these periods records were made of pecks delivered and received to determine dominance relationships, and data on sexual activity were obtained. These observations were made late in the afternoons at which time matings are most frequent (Heuser, 1916; Philips, 1919; Upp, 1928; Skard, 1937; Parker, McKenzie and Kempster, 1940). The length of the observation periods varied, consequently the numerical values given as means for the two flocks in the tables are not comparable as rates of sexual activity. The types of behavior recorded have been previously defined and described (Guhl, Collias and Allee, 1945). When a cockerel mounted a pullet his behavior was recorded as a mating whether or not the mating pattern was completed, and when it appeared that the vents met the mating was noted as a completed or successful mating.

OBSERVATIONS

A summary of the data obtained is presented in Table 1. In both flocks the male at the intermediate level of the male peck-

TABLE 1.—*Summary of the basic data*

Peck-order	Courtings	Total matings	Completed matings	Eggs fertilized	Viable chicks
		1944 flock			
R. I. Red	219	59	51	236	201
Barred Rock	250	76	35	165	139
White Leghorn	73	8	3	56	54
Total	542	143	89	457	394
		1945 flock			
Dominant white	710	175	112	267	221
Recessive white	2184	244	54	129	120
Rose comb	71	8	0	0	0
Total	2965	427	166	396	341

order was seen to court and attempt to mate the most and the inferior cockerel the fewest number of times. When only those matings which appeared to be completed are considered the socially dominant male in each flock was most success-

TABLE 2.—*Summary of the interferences with mating among the cocks with regard to social position*

Social positions	1944	1945
1 on 2*	65	273
1 on 3	17	8
2 on 3	2	5
2 on 1	21	140
3 on 2	0	1
3 on 1	0	1
Total	105	428

* Attacks made by a male at one social level on that of another rank.

ful. Nearly all of the courtings and matings of the inferior White Leghorn occurred when the flock was permitted to range out-of-doors, where close contacts between the males were reduced. The *rose comb* male was inhibited from mating successfully although he courted occasionally throughout this phase of the experiment.

Acts of treading frequently stimulated the cockerels to attack the male attempt-

ing to mate. The interferences with mating are summarized in Table 2 (see also, Guhl, Collias and Allee, 1945), from which it is evident that the dominant males made the most attacks and the suppressed inferiors interfered less frequently or not at all when their superiors mounted the pullets. Although these attacks frequently resulted in incomplete matings there were incidences in which the treading male was able to complete the mating pattern.

In general, the number of eggs fertilized by each male shows some relationship to the number of observed completed matings (Table 1). In each flock the top ranking male fertilized the most eggs and the lowest ranking male the fewest or none. The number of viable chicks sired by each male is much in line with the number of eggs he fertilized.

A summary of data on incubation is given in Table 3. The 1945 flock had a higher percentage of infertile eggs than the 1944 flock. Before observations were begun, both flocks were culled by the usual methods followed by poultrymen. In the 1944 flock 5 pullets had a fertility of 50 percent or less, as against 13 pullets in the 1945 flock.

Experience in studies of mating behavior has indicated that the degree to which a dominant cock may suppress or continue to suppress the sexual behavior of his inferiors may vary, as does concomittantly the time required to develop inhibitions on the part of the inferiors (cf., Guhl, Collias and Allee, 1945). An

TABLE 3.—*Incubation record*

	1944		1945	
	No.	%	No.	%
Infertile eggs	211	31.5	309	43.8
Embryo mortality	63	9.4	55	7.8
Viable chicks	394	59.0	341	48.3
Total eggs	668	99.9	705	99.9

examination of the daily performance of the males is therefore essential to a critical analysis of the data. A comparison of the day-by-day records of the three males in each flock was made by paired daily records, using the method of "Student" for paired comparisons. The results are given in Table 4 for daily courtings, total matings whether attempted or successful, completed matings, eggs fertilized and viable chicks. The daily mean for each male is given in the table, as well as the mean daily differences between the males as compared pairs, and the statistical probability or *P*-value of the difference.

The male of the beta levels in the social order courted more and attempted to tread more (Table 4) than did the alpha male. These differences were statistically significant during 1945 but not during 1944. In each flock the male at the bottom level of the hierarchy showed significantly less sexual activity than either of his superiors. In all of the comparisons the dominant male of the pair was more successful in mating, with statistically signifi-

cant differences in all comparisons save that between the R. I. Red and Barred Rock and that tends to approach significance. The dominant member of the pairs also fertilized more eggs and had more viable chicks per day than his paired inferior, differences that show statistical significance in all of the comparisons.

Hens are known to mate at different rates with each of several cocks introduced singly into their pens (Philips, 1919; Upp, 1928; Skard, 1937; Guhl, Collias and Allee, 1945). Some indications of nonrandom mating were noted in both flocks. As an attempt to determine the relative influence of "preferential mating" on the differences observed, another set of paired comparisons were made. In making this analysis (Table 5) the total performance of each male with each female was determined. The males were then compared by their paired records with each female. As in the previous analysis, the males at the second level in the peck-orders courted and attempted to mate more frequently than the males outranking them, but the differences lack statistical significance in the 1944 flock. Statistically the cocks at the omega levels displayed significantly less sexual activity than either of their superiors. Dominance is associated with a higher average of successful matings, number of eggs fertilized and viable chicks sired per hen, each difference being statistically significant except the comparison between the R. I. Red and the Barred Rock.

The indications from both the preceding analyses are that the R. I. Red failed to suppress the Barred Rock to any marked degree, due probably to a similarity in their relative aggressiveness. However, some factors associated with space relationships may have also contributed to the lack of significant differences between the performance of these two males. The

1944 flock was observed under two sets of conditions. During the earlier part of the experiment this flock was confined to the pen and later an outdoor range was made available.

Table 6 presents an analysis of the day-by-day performance of the cockerels in the 1944 flock when they were in a situation comparable with that of the 1945 flock (*i.e.*, they were confined to the pen), and also when they had a yard containing well developed vegetation in which to range. Under the former conditions the Barred Rock courted and attempted to mate significantly more than his superior, the R. I. Red. The latter male, however, tended to be more successful although the greater number of chicks he sired as compared with the Barred Rock is barely significant statistically. The lack of statistically significant differences between these two males in the number of successful matings and eggs fertilized gives some validity to the impression obtained by observation that the R. I. Red displayed insufficient aggressiveness to develop any pronounced suppression of the Barred Rock. The results of the other two sets of comparisons are similar to those in the previous analysis (Table 4).

During the time when the flock was permitted to range some marked changes occurred. In contrast with the previous period, and with the analysis in Table 4, the alpha male courted and attempted to mate more often than the beta male. The comparisons of courting and attempts to mate which involve the alpha male all lack statistical significance. Since this R.I. Red fertilized more eggs and sired more chicks (differences that are statistically significant) than did either of his two inferiors, it appears that the data concerning behavior obtained during the eight observations of this period, were insufficient to consider very seriously from a statistical viewpoint. The R. I. Red attacked the Barred Rock with a marked intensity during these later observations. This increase in domination by the R. I. Red became evident during the first of these three weeks by a gradual decrease in the sexual activity of the Barred Rock and by an obvious increase in the number of scabs and wounds on the latter's comb resulting from hard pecks delivered by the alpha bird. The Leghorn typically remained in the pen near some feeding pullets when his superiors were outdoors, and he hurried into the yard when the other two males entered the pen. Under these conditions the Leghorn display sufficient sexual activity to make the differences between his and the Barred Rock's performance (which decreased during this time) lack statistical significance. The Leghorn's pronounced avoidance of his superiors under the greater spatial relations resulted in a dilution of the domination to which he was subjected. Contrariwise, the Barred Rock and the R. I. Red were frequently in close proximity; apparently the Barred Rock had not yet learned to keep at a distance from his superior. Such contrasting phenomena are not unusual in studies of behavior. The point of interest here is that the changes in the frequency of sexual activity and number of offspring on the part of the Barred Rock and the Leghorn, although in opposite directions, can be related to changes in the intensity of domination to which they were exposed. The alpha R. I. Red, however, maintained his ability to sire significantly more chicks than either of his subordinate males.

All but one of the Leghorn's 54 chicks (Table 1) were sired during the last three weeks, a fact which indicates that the suppression of his sexual behavior did not destroy his fertilizing capacity. Because the *rose comb* male, of the 1945 flock, had no progeny his semen was tested, during the sixth week of the experiment, by artificial insemination into several hens which then produced 25 fertile eggs among the 49 laid. In these flocks sexual suppression did not incapacitate the males from producing functional spermatozoa.

TABLE 7.—*Summary of observations made on the 1945 flock when the cockerels were introduced singly and successively into the pen of the pullets*

	Dominant white	Recessive white	Rose comb	Total
Courtings	351	384	0	735
Matings	38	41	0	79

At the close of the 1945 observations the flock was subdivided to test the males singly and in the absence of competition between them for mates. The same sex ratio was maintained by placing a third of the pullets in each of three similar pens, with a floor area of 3.2 square feet per bird. The cocks were placed into a pen by themselves at some distance from the pullets. During the late afternoons each male was introduced into one of the pens containing pullets and his sexual behavior was observed. Under these conditions a 15 minute exposure yielded data more rapidly than previously, as the birds appeared to be in sexual readiness. The males were rotated among the three pens until each was exposed for six equal periods to each subflock of females. The number of courtings and matings observed are summarized in Table 7. The mating pattern was completed in all but one instance, in which the female struggled. The differences between the number of courtings and matings by the alpha (*dominant white*) and the beta (*recessive white*) males lack statistical significance when analyzed by paired days or by paired hens. This test shows that the

marked differences between the performance of these males when in competition was not a result of inherent differences in sexual behavior. The complete lack of sexual activity on the part of the omega (*rose comb*) male appeared to be a continuation of the conditioning experienced in the larger pen, but this explanation cannot be considered as conclusive because he developed a severe lameness which continued during and beyond this final set of observations.

DISCUSSION

Among the males used in the flocks described, there was a marked relationship of their position in the dominance order to the number of their successful matings, eggs fertilized and viable chicks sired. In the 1945 flock the results of the analyses were somewhat diagrammatic, but in the 1944 flock the differences between two of the males lacked statistical significance in several of these comparisons. In an effort to evaluate the behavior of these six males their sexual behavior may be compared, in a general way, with those studied by Guhl, Collias and Allee (1945). In two heterosexual flocks described in some detail by the latter, the alpha males courted and mated at a higher rate than any of their inferiors; but the rates of the other males did not vary uniformly in a direct relation to their social levels. These differences from the trend found in the present report indicate that further work is needed before it will be possible to decide which pattern is most frequently exhibited.

There are a number of factors which may influence the frequency at which cocks mate. Knowledge of some of these has been gained through general experience of poultrymen, and are commonly used by them as a rule of thumb although they have not, to our knowledge, been tested by quantitative experimentation. Nevertheless, these merit some attention. Among the factors that influence the rate at which males mate are (1) the sex ratio in the flock, (2) breed differences, (3) space relationships, (4) accessibility of males to females, (5) antipathies and tolerations, (6) individual differences in libido, and (7) conditioning effects or learned patterns of behavior. One of the authors has some unpublished data, which need further study, suggesting other factors pertaining to relative sexual readiness and certain psychological situations.

The sex ratio remained virtually constant in the flocks (one pullet in the 1945 flock died about two weeks before the close of observations). Since the omega males were more or less suppressed in their sexual behavior one should probably consider the constancy of the effective sex ratio, i.e., the number of males that are sexually active and successful in mating. The Leghorn male was notably suppressed early in the experiment but less so when this flock ranged in a yard. The suppression of the rose comb male was relatively constant throughout the observations on the 1945 flock. Philips (1919) found that Leghorns mated much more often than did heavier breeds. Breed differences may have been a factor in the 1944 results, but in the 1945 experiment all males were at least of similar size and color.

Space relationships, as expressed in the number of square feet of floor space per bird in the pen, influence the relative frequency at which individuals may meet in the pen. In the present study the number of birds used was below the estimated capacity of the pens. In experiments on mating behavior of males a consideration of space might also include the unit of area per male. In the 1944 experiment the pen had 70 square feet per male. The 96 square feet per male in 1945 remained constant as no outdoor range was used. Some evidence was presented (Table 6) in the 1944 observations that space relations may alter a cock's relative success in mating. Associated with the relative crowding is the accessibility of males for females for mating and of males for males with regard to interference with mating as influenced by the furnishings of the pen or the vegetation on the range. When Heuser (1916) used a labyrinth in the breeding pen there was an appreciable decrease in the number of interferences with mating among the males. In the 1945 pen there was a clear "hen's-eye-view" of the entire floor and the arrangement of the equipment was not altered during the period of observations. The 1944 pen was similar in furnishings, but the outdoor range contained much vegetation.

Less obvious factors which influence the frequency of mating by males have received some attention by Guhl, Collias and Allee (1945). One of their alpha males displayed a marked antipathy towards one male and markedly suppressed his sexual activity. At the same time he was notably tolerant of another male which was permitted to intercept the mating pattern of the dominant male to the advantage of the interloper. In our flocks, among the 1944 males there was some indication of antipathy but it did not persist; in the second flock no obvious incidences of either antipathy or toleration occurred among the males. Guhl et al. (1945) tested the males singly and successively with the hens prior to their admission to the hens as a group. By analytical methods they found differences between some of the cocks which suggested variations in libido. No such test was made with our 1944 males, and that made with the 1945 males (Table 7) followed the basic observations and therefore did not rule out any conditioning which may have occurred previously. Under these conditions the observed differences in mating behavior between the *dominant white* and the *recessive white* males when they were tested singly with the same females were not statistically significant.

The interrelationships among the factors considered above form a more or less interacting complex or syndrome which makes it rather difficult to devise an experiment in which all but one factor, such as social dominance (position in the peck-order) are under complete control. From the foregoing discussion one may conclude that the 1945 experiment approaches such a controlled situation for the study of the relation of relative domination or subordination to success in mating.

Under the 1945 conditions one would expect the male which was most successful in mating to fertilize more eggs and sire more chicks than his less successful penmates. There are, however, some factors other than social dominance which may influence this relationship or the results of a statistical analysis, such as, (1) individual differences among the males in semen production, (2) non-random mating (3) the frequency at which a given female may mate with any male, and (4) infertility which may result from the physiological state of the female.

Individual cocks vary as to the amount of semen produced. Burrows and Titus (1939) found no significant relationship between semen production and display of sexual activity. The males of both sets used in these 1944 and 1945 tests were therefore selected as good producers of semen. Further, since fertilization may depend upon the concentration and density of spermatozoa (Munro, 1938) equal volumes of semen from each of the three males used, in each flock, were mixed and tested by artificial insemination. The results of these relatively simple tests showed that the sperm from each cockerel were capable of fertilizing eggs when in numerical and physiological competition with those of the other two males in his flock.

Selective or non-random mating, if pronounced, might influence the results of a statistical analysis of eggs fertilized by several males in competition for mates. Within the time limits of the observations and for both flocks, the alpha males failed to mate with a total of 8 pullets although the beta males mated with these females. The beta males failed to mate with 4 pullets which were trod by the alpha males. All the males in the 1944 flock failed to mate with one pullet. When both flocks are considered together the males mated infrequently with a total of 17 females, and 6 of these had a fertility of 50 percent or less. These evidences of non-random mating, which Lamoreux (1940) lists as one of the factors contributing to infertility, were not given any special treatment in the analysis. Nevertheless, despite the influence of non-random mating on the results of the "hen-by-hen" analysis (Table 5), except for the comparison of the R. I. Red and Barred Rock, the differences between the males in number of eggs fertilized and chicks sired were statistically significant, with the dominant of the pair being more successful.

The number of times a hen is mated and the time elapsed since the last mating may also contribute to infertility (Lamoreux, 1940). These relationships have their bases, in part, in the fact that the competitive efficiency of sperm may be of relatively short duration (Crew, 1926; Warren and Kilpatrick, 1929; Nalbandov and Card, 1943). It has been shown above (see also Guhl, Collias and Allee, 1945) that domination among the males decreases the frequency at which inferiors may mate and which may thereby impair, on a physiological level, their chances to sire offspring. The analysis of eggs fertilized and chicks sired when made by "day-by-day" comparisons did not give any special consideration to the rates at which each female was mated. Nevertheless, the differences between the males (Table 4) are statistically significant in all comparisons save that of eggs fertilized by the R. I. Red and Barred Rock and this P-value approaches significance.

There is evidence that some of the females which mated frequently tended to lay infertile eggs. In the 1944 flock 9 pullets which mated at rates that were above the flock average, had a fertility of 50 percent or less and 4 of these were poor producers. There were 6 such individuals in 1945 with a low fertility. No adjustments were made in the analysis for these cases of infertility which appear to be related to the physiological state of these females.

Finally, after some consideration of factors, other than social dominance, which may influence the number of chicks a male may sire when in social competition with other males, and some of which were known to be in operation in these flocks, it appears evident that the number of offspring sired by each of the cockerels was related to his relative dominance. One has no way of knowing whether the males used were representative of the population of the more common breeds, although they were selected from groups of males which were taken at random from a large colony. Nor can one, at present, evaluate the relative importance of social dominance as compared with some of the other factors if the latter varied greatly from the conditions under which these flocks were observed. The data here presented show that normal and healthy males do, when in competition, vary greatly in the number of offspring which they produce.

REFERENCES

Allee, W. C., 1942. Social dominance and subordination among vertebrates. Biol. Symposia. 8: 139–162.

———, 1943. Where angels fear to trend; a contribution from general sociology to human ethics. Science. 97: 517–525.

———, 1945. Human conflict and co-operation; the biological background. Chap. XX, pp. 321–367, of Approaches to national unity; fifth symposium of the Conference on Science, Philosophy and Religion, ed. L. Bryson, L. Finkelstein, and R. M. MacIver, New York, Harper & Bros., pp. 1037.

Burrows, W. H. and Titus, H. W., 1939. Some observations on the semen production of the male domestic fowl. Poultry Sci. 18: 8–10.

Collias, N. E., 1944. Aggressive behavior among vertebrate animals. Physiol. Zool. 17: 83–123.

Crew, F. A. E., 1926. On fertility in the domestic fowl. Proc. Roy. Soc., Edinburgh. 46: 230–238.

Guhl, A. M., Collias, N. E. and Allee, W. C., 1945. Mating behavior and the social hierarchy in small flocks of White Leghorns. Physiol. Zool., 18: 365–390.

Heuser, G. F., 1916. A study of the mating behavior of the domestic fowl. Thesis, Master of Science of Agriculture degree, Graduate School of Cornell University. (Unpublished.)

Lamoreux, W. F., 1940. The influence of intensity of egg production upon fertility in the domestic fowl. Jour. Agric. Res. 61: 191–206.

Masure, R. H. and Allee, W. C., 1934. The social order in flocks of the common chicken and the pigeon. Auk. 51: 306–327.

Moore, O. K. and Byerly, T. C., 1942. Relation of time of insemination to percent of fertility. Poultry Sci. 21: 253–255.

Munro, S. S., 1938. The effect of dilution and density on the fertilizing capacity of fowl sperm suspension. Canadian Jour. Res. 16: 281–299.

Nalbandov, A. and Card, L. E., 1943. Effect of stale sperm on fertility and hatchability of chicken eggs. Poultry Sci. 22: 218–226.

Parker, J. E., McKenzie, F. F. and Kempster, H. L., 1940. Observations on the sexual behavior of New Hampshire males. Poultry Sci. 19: 191–197.

Philips, A. G., 1919. Preferential mating in fowls. Poultry Husbandy Jour. 5: 28–32.

Sanctuary, W. C., 1932. A study in avian behavior to determine the nature and persistency of the order of dominance in the domestic fowl and to relate these to certain physiological reactions.

Thesis for M.S. degree, Massachusetts State College, Amherst. (Unpublished.)

Schjelderup-Ebbe, T., 1922. Beiträge zur social-psychologie des Haushuhns. Zeitschr. f. Psychol. 88: 225–252.

————, 1935. Social behavior in birds. Chap. XX in Murchison's Handbook of social psychology, pp. 947–972. Worcester, Mass.: Clark University Press.

Skard, Ase G., 1937. Studies in the psychology of needs; observations and experiments on the sexual needs of hens. Acta psychologica, 2: 175–232.

Upp, C. W., 1928. Preferential mating in fowls. Poultry Sci. 7: 225–232.

Warren, D. C. and Kilpatrick, L., 1929. Fertilization in the domestic fowl. Poultry Sci. 8: 237–256.

25

Reprinted from *Auk*, **51**, 306–313, 327 (July 1934)

THE SOCIAL ORDER IN FLOCKS OF THE COMMON CHICKEN AND THE PIGEON.[1]

BY RALPH H. MASURE AND W. C. ALLEE.

THE problems centering about the organization of flocks of birds are by no means solved despite the attention they have received. A portion of the pertinent literature to date has been summarized by Allee (1931, 1934) both with regard to the observed facts and their general social implications; hence no general literature survey will be undertaken now. It will suffice to call attention to certain work which is directly antecedent to ours. Schjelderup-Ebbe (1922) has analyzed the organization of flocks of Domestic Chickens and of both wild and tame Ducks (1923). In all of these he found a more or less definite organization revealed by the way in which the birds reacted in contact situations. He recognized a so-called peck-order in which the animal highest in the order pecks and is not pecked in return while that at the extreme bottom of the order is pecked without pecking in return. Throughout the entire order, any individual with the peck-right over another remains steadily dominant over it until by a combat their positions are reversed.

More recently (1931) Schjelderup-Ebbe has extended his observations to include a large number of different sorts of birds both in nature and in various kinds of confinement; he finds that when two birds of one species are together, one is despot and the other is subservient. Schjelderup-Ebbe believes that this sort of despotism is one of the fundamental principles of biology.

The older bird of a flock is usually despot because her matured body gives her strength which the young, partially developed birds lack; even after the latter attain their full size and strength, if of the same sex, the older individual maintains her despotic rights. Between the sexes, the larger males are usually despots over the females. When the two sexes are alike in size and strength and the

[1] The work upon which this report is based has been supported in part by a grant from the Rockefeller Foundation to aid investigations in the biological sciences at the University of Chicago. We are indebted to Dr. L. V. Domm for permission to study the group organization of Chickens in certain of the pens under his control and for two post-mortem examinations.

male possesses ornamentation, he is despot; otherwise either may be despot. Often males put on and lose their despotic rights with the assumption and loss of breeding plumage.

We have repeated Schjelderup-Ebbe's observations first on Brown Leghorn Chickens, both males and females but with sexes separate; then with flocks of Pigeons, first with sexes segregated, then when mated, and again with the sexes together. The observations were the result of joint planning and while the senior author was in very close touch with the work throughout, the actual observations were all made and the preliminary summary prepared by the junior author.

Two flocks of Brown Leghorn Chickens were observed in their Whitman Laboratory quarters during January, February and March, 1932. They were housed in pens 14.5 feet long, 5 feet wide and 8 feet high. These pens were in a heated house that was kept between 60° and 70° F. They were each provided with raised roosts which occupied a corner space of 3 by 4 feet. The floor below the roosts was protected from droppings. The cement floor was covered with straw litter which was changed weekly. On a few warm days, the Chickens were given access to separate outside runs which were about 5 feet long and were floored by wire mesh about a foot from the ground to prevent attacks by rats.

All the Chickens were about 10 months old when observations were begun. They had been raised together in the Whitman pens since hatching. One pen contained 11 cockerels and the other at the start contained 26 pullets. Toward the end of January, 12 days after the observations were begun, half of the pullets were removed and on February 27 a Barred Plymouth Rock rooster was added. The flock of cockerels was disturbed only by the experiences of one bird. On February 29, RB was removed because his eye was injured in fighting; he was returned to the flock on March 14 when his eye was completely healed and sight was normal; he was again removed five days later in a much battered condition.

All these Chickens were fed twice daily with mixed grain and in addition the cockerels were given a small amount of grain when the observations were being made, since relatively little pecking was done in the absence of food. In addition to the grain, the pullets were provided with a hopper of mash which remained in the pen constantly. The cockerels received no mash since by omitting it most of their fighting was avoided.

The majority of the observations on contact reactions were made in the afternoon. The observer sat quietly in an adjoining pen, note pad on knee, and was separated from the Chickens under observation only by ordinary chicken-wire netting. When two animals came into contact with each other and one was pecked and retreated, this was considered to be the subservient member of that particular contact pair. Many times two Chickens would meet without showing any signs of pecking or being pecked; such neutral encounters were not considered in deciding on the peck-order. Often such contacts would start a battle, more often with the cockerels than with the pullets; pecking would be mutual. In these cases the Chicken that gave in and retreated was considered to be subservient to the other. The majority of the pecking contacts were observed over and over. Colored celluloid leg bands furnished a ready means of individual identification by the observer; the birds themselves appeared to recognize other individuals by means that were not always apparent to the human observer.

SOCIAL ORDER AMONG THE PULLETS.

The social order in the flock of 26 pullets had not been determined when, twelve days after the observations began, the flock was reduced to 13. The definite status of any one hen had not been established but the evidence at hand allows a fair approximation of the birds in the upper half of the flock and extreme lower part as follows:

1. W	7. GY	13. WW
2. B	8. BY_2	
3. RW	9. G	22. Y
4. GY_2	10. RY	23. BB
5. RY_2	11. RG	24(?)BY
6. RR	12. BG_2	25. M
		26. A

Although the observations were insufficient to establish the contact reactions between any one bird and all of its associates, the data at hand show that there was no absolute despot. One bird, GG, was not observed to receive pecks from another individual but it pecked only A and YY during these observations. YY ranked below the 13 just given and A ranked at the bottom of the indicated order in the larger flock and at the very bottom of the better tested order of the reduced flock: hence it appears that GG cannot be assigned a place in the upper half of the peck-order.

In all the contacts observed in this crowded pen, there was only one reversal. W was seen to peck RY_2 once and the reverse was observed once also. This indicates a strong and stable social organization among the pullets. Pullets of inferior position always gave way to their superiors at the food hopper or water dish. Superiors were often observed pecking food from the bills of their inferiors. M, A, and BB, individuals low in the peck order, spent much of their time on the roost where there were fewer birds and when they did venture onto the floor, they were alert and avoided many pecks by darting away from an approaching hen.

After the flock was reduced to 13 individuals, it was observed for 60 days and the complete order obtained as shown in Table I.

TABLE I. SHOWING THE STATUS OF EACH INDIVIDUAL IN A FLOCK OF 13 BROWN LEGHORN PULLETS ON THE BASIS OF THE PECK-ORDER.

RW pecks all 12: A, BG, BB, M, Y, YY, BG_2, GR, R, GY, RY, RR.
RR pecks 11   : A, BG, BB, M, Y, YY, BG_2, GR, R, GY, RY.
RY pecks 10   : A BG BB M, Y, YY, BG_2, GR, R GY.
GY pecks 9   : A, BG, BB, M, Y, YY, BG_2, GR, R.
R   pecks 8   : A, BG, BB, M, Y, YY, BG_2, GR.
GR pecks 7   : A, BG, BB, M, Y, YY, BG_2.
BG_2 pecks 6   : A, BG, BB, M, Y, YY.

YY pecks 4   : A, BG, BB, M.
M   pecks 4   : A, BG, BB,   Y.
Y   pecks 4   : A, BG, BB,   YY.

BB pecks 2   : A, BG.
BG pecks 1   : A.
A   pecks 0

It is apparent from the data of Table I that RW is the despot of the flock and that down to YY there is a straight line order; then YY, M and Y form a triangle order below which the straight line order continues to A, the lowest member which pecks none. As in the larger flock, superiors often ate food from the bills of their inferiors without resistance from the latter. When a hen was sitting on even one or two eggs, it took much more pecking before she would relinquish her position to the superior hen; even so, no reversals in peck order for this or other reasons were observed in the reduced flock. RR, second in the social order, was much given to pecking her inferiors, more so than RW, the ranking hen. The impression of a social order gained in the observations on the large flock is definitely strengthened. The reduction in flock size did not change any of the observed peck-rights. R, which stood fifth in the social order of the reduced flock, had not been observed in contact with any of this group before the reduction took place; it had been seen to peck G and BR and to be pecked by W and RY_2.

Social Order among the Cockerels.

Practically the complete order for the flock of eleven cockerels was determined during the 70 days of observation. The findings are summarized in Table II, which is built on the same plan as Table I, and so allows ready comparison with the social order obtaining among the pullets. In this table the characters in italic indicate that these peck-rights were not settled; those in heavy faced type indicate that there was one reversal of the peck-order observed in each of these cases.

Table II. Showing the Social Organization of a Flock of 11 Brown Leghorn Cockerels on the Basis of their Peck-order.

BW pecks 9 : W, BY, G, RY, B, BG, Y, R, GY.
BR pecks 8 : W, BY, G, RY,   BG, Y, **R,**   BW.
GY pecks 8 : **W,** BY, G, RY, B, **BG,** Y,   BR.
R   pecks 7 : W, *BY,* **G,** RY, B, BG,   GY.
Y   pecks 6 : W, BY, G, RY,   BG, R.
GB pecks 5 : W, BY, G, RY, B.
B   pecks 4 : W,   G, RY,   Y.
RY pecks 3 : W, BY, G.
G   pecks 2 : W, *BY.*
BY pecks 2 : W,   B.
W   pecks 0.

In this order there are six triangle situations as follows:

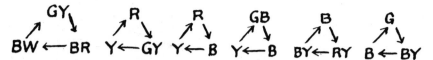

These triangle situations run almost through the whole gamut of the social order. Those high or low in the peck-order are least involved and those in the middle of the order are most concerned. BW, the highest ranking cockerel is involved in one triangle; W, the lowest, in none; B, one of those near mid-rank is involved in four and Y, in three.

The social organization of this flock of cockerels was not as stable as was that of the pullets. After 70 days of observation, peck-rights were still unsettled as indicated in Table II by the letters in italic. They were those of BY–R and BY–G. BY was observed to peck G on six occasions and the reverse was observed on eight occasions. BY was likewise observed to peck R on six occasions and the reverse was true for the same number of contacts. The place of these cockerels must therefore be assigned tentatively and they are placed in their logical positions with reference to other relations which seemed to have been settled.

Four other cases of reversal were observed and are indicated in Table II by heavy faced type. In these the reversal occurred but once in each case. It is of course possible that there may have been an error in observation but the likelihood is no greater here than with the pullets and we are inclined to think that these reversals indicate a less stable and fixed peck-order in the cockerels as compared with the pullets.

On account of a fight which incapacitated BR, other consequences of which will be given later, this individual was removed before the peck-order between him and B had been observed. The other relationships indicate that B would probably have been pecked by BR, but one cannot be certain of this without direct observation.

The cockerels were much more given to fighting than were the pullets. Many times two of them would face each other and start to fight even though the peck-right records showed that the relations between the two were fairly definitely settled. Usually such combats were interrupted by another individual walking between the two combatants, but if the battle was not interrupted, the cockerel which was usually pecked, finally retreated, but not, perhaps, until he had put up quite a fight. The individuals in the lower positions of the social order have a difficult life; they are continually being pecked by their superiors. When food was given, W, which stood lowest in the order, never ventured near those cockerels that were eating but waited until they had left the food and then ate what remained. W also spent most of his time on the roost where the others, if present, seemed to be more tolerant of their inferiors. As a rule very little pecking took place among the cockerels on the roost even though they were somewhat crowded at times. When W did venture on the floor of the pen, he was constantly running to avoid contacts with the other members of the flock. BW, which was first in the order, seemed more inclined to peck his inferiors than did any other cockerel; about half of the observed pecks were delivered by this individual.

The order among cockerels differs from that found among the pullets in the type of organization. That of the pullets is fairly regular while that of the cockerels is built about triangular relationships. For a time there was no despot among the males and in fact BW, the cockerel standing highest, was pecked by BR who stood just below him in rank. On February 29, this latter individual and Y started to fight, as they had many times before, but during this battle BR received a hard peck in the eye which closed the left eye and he retreated. He was then removed to another pen and two weeks later on March 14 he was replaced with the group of cockerels which he had almost dominated. At this time his eye was entirely healed and his sight was apparently normal; he was found to have lost his position in the peck-order completely. In fact he now stood lowest of all and was even pecked by W, which had not been observed to peck a fellow cockerel before. One or all of the following factors may have entered into his loss of status: he had lost an encounter; he had sustained a severe injury and he had been absent from the flock for 14 days. Whatever the reason, he was persecuted so badly that on March 19 his injuries prevented him from standing and he was removed permanently to save his life.

During the five days BR was in the pen after his recovery, he avoided contacts with the others as much as possible and spent much of his time partially hidden under a low shelf upon which the water dish was kept. Not once was he observed to attempt to assert his former high position. After BR was removed no change was observed in the pecking. The absence of BR made BW despot. This change in the relations of BW did not seem to affect his behavior towards the other members of the flock.

*   *   *   *   *   *   *

REFERENCES.

ALLEE, W. C.
  1931. Animal Aggregations: A study in general sociology. Chicago. University of Chicago Press. 431 pp.
  1934. Recent studies in mass physiology. Biol. Rev. *8:* 1–48.
CRAIG, W.
  1908. The voices of pigeons regarded as a means of social control. Amer. Jour. Sociol. *14:* 86–100.
FISCHEL, W.
  1927. Beiträge zur Sociologie des Haushuhns. Biol. Zentrbl. *47:* 678–695.
KATZ, D. UND A. TOLL
  1923. Die Messung von Charakter- und Begabungsunterschieden bei Tieren. (Versuch mit Hühnern). Zeitschr. f. Psychol. u. Physiol. d. Sinnesorg. Abt. I, *93:* 287–311.
KROH, O.
  1927. Weitere Beiträge zur Psychologie des Haushuhns. Zeitschr. f. Psychol. *103:* 203–227.
RABAUD, E.
  1931. Les origines de la société. Essai sur les sociétés animales. Renaissance du livre. Paris, 1931.
SCHJELDERUP-EBBE T.
  1922. Beiträge zur Sozialpsychologie des Haushuhns. Zeitschr. f. Psychol. *88:* 225–52.
  1923. Das Leben der Wildente (Anas bosehas) in der Zeit der Paarung. Psychol. Forsch. *3:* 12–18.
  1931. Die Despotie im sozialen Leben der Vögel. s. 77–137 in Arbeiten zur Biologischen Grundlegung der Soziologie. Leipzig.
  1931a. Soziale Eigentümlichkeiten bei Hühnern. Kwart. psychol. *2:* 206–212.
TAYLOR, W. S.
  1932. The gregariousness of pigeons. J. Comp. Psychol. *13:* 127–131.
WHITMAN, CHARLES O.
  1919. The behavior of pigeons. (Posthumus works, Vol. III, edited by Harvey A. Carr). Carnegie Institution of Washington, Publ. No. 257.

26

Reprinted from *J. Appl. Ecol.*, **3**, 86–91, 93 (May 1966)

The Relationships Between Wood-Pigeons and Their Clover Food Supply and the Mechanism of Population Control

R. K. MURTON, A. J. ISAACSON, and
N. J. WESTWOOD

6. A DISCUSSION OF THE MECHANISM OF POPULATION CONTROL

When clover stocks were lowest and wood-pigeon numbers had declined to a minimum level in parallel with changes in the food supply much clover apparently remained available. In 1964 in February when the lowest clover stocks recorded in any year were experienced it can be calculated (knowing the average clover density per square foot, the acreage available and how much one pigeon eats per day) that there was enough clover left per 100 ac to last one pigeon for 2403 days or a population of thirty-three birds for 73 days. The birds seemed to have left themselves a very considerable safety margin, but it is possible that the area with clover densities sufficient to maintain an adequate feeding rate may have been limiting. Unfortunately our sampling methods were not detailed enough in 1960 and 1961 to evaluate the clover content in individual fields, although this could be done in 1962 and 1964 (in 1963 the fields were snowbound). Thus in 1964, 63% of all the clover acreage in February supported leaf densities below $100/ft^2$ and 43% below $50/ft^2$ (total acreage was 745). On pastures the average leaf density was only $85/ft^2$ (72 ft^2 samples). It was noted above that at leaf densities below $50/ft^2$ the pecking rate showed evidence of a decline that would make the collection of sufficient food impossible (Fig. 7). Considering these arguments, and allowing for errors in field observation, it might be that the birds fed much closer to the limits of their food supply than was at first apparent. Why should total population size be so clearly correlated with food stocks in different years if there did exist an unexploited reserve in each year?

(a) A hypothesis for the mechanism of population control

In Section 3(c) there is evidence that clover leaf survival was proportionately greater at low leaf densities during the first part of the winter (November to February) but that no such relationship occurred with clover densities of over about 150 leaves/ft^2 (Fig. 4). It was suggested that the absence of pigeon feeding at low clover densities could account for this result. Because more leaves were recognizable as having been eaten by pigeons where more leaves were available (Table 3), it must follow that either more birds fed or that each bird fed for a longer time where more clover was available. Table 4 shows that in January pigeons were avoiding fields with low clover densities, but not apparently in February (see below), while Fig. 5 suggests that the birds also sometimes distribute themselves according to food stocks in February.

Birds (animals) have presumably evolved some mechanism enabling them to relate searching effort to the amount of food found and certainly they can recognize a situation where no food exists and respond by searching elsewhere. Unfortunately it is not known in quantitative terms how long a pigeon requires to recognize and react to an unfavourable feeding situation. However, it is feasible that this could be a continuous process with the bird at all times being able to relate searching effort to food returns until a point was reached at which a very high searching rate necessitated flight. Conversely, it is possible that the feeding rate in turn affects what food is eaten and the possession of characteristic searching motor responses could determine within limits what food was selected by a particular species. In other words, a running bird would tend to find some objects more easily than others, whereas walking might enable it to locate those objects that were previously missed. The food eaten would then depend on what was available in a given environment at a suitable density to provide a feeding stimulus for a bird searching at a characteristic rate. Innate and learned food preferences could primarily determine what was eaten but it is perhaps undesirable to think of food collection only in terms of the animal searching specifically for what it requires. If this were so an animal could evolve the mechanism to respond to a threshold point below which the number of food items found per unit of searching time would be inadequate.

We now assume that pigeons which left from the front of the flock as was shown to happen in Section 4(a) did so because they were prevented from obtaining sufficient food under the existing conditions. Birds which left from the rear may have obtained sufficient food and have been ready to return to roost or to rest in trees near the feeding grounds. However, whereas few birds which left the front of the flock tried to settle again (11%) a much higher proportion originating from the rear and middle did so (25–26%). Similarly it seems likely that pigeons arriving at the flock settle outside until their social position has been established; it would be expected that only a very dominant bird would choose to alight in the centre of a flock and so face immediate competition from established feeders. It therefore seems highly probable that front birds may leave a flock to search elsewhere if their intake of food is insufficient. But because the gregarious habit is so fixed this behaviour would result in the birds trying to join another feeding flock. Provided that surplus feeding grounds existed with high enough food densities it would be expected that neither flock distribution nor flock size would be related to clover density except that low clover densities should be avoided. A decrease in clover density should result in a loss of birds from some sites and the establishment of flocks at lower densities so that the birds might be distributed according to density as sometimes seems to happen. Further clover reductions could result in flocks being forced to feed at low clover densities and the birds' distribution might no longer match food availability, as in February in Table 4. But in such an event mortality through the starvation of some birds would be expected or alternatively emigration motivated by food shortage. If a flock was forced to feed at very low clover densities then all the birds should show evidence for a loss of condition and the same would happen if the flock exploited a less nutritive food source like *Brassica* leaves as happened in 1963. Summarized in Table 4 are data for three flocks inhabiting fields 244, 107 and 13A in February which were much bigger than would be expected on the basis of food availability. While the flock on field 244 could be explained by the limited availability of clover fields in that part of the study area it should nevertheless have shown signs of mortality. Unfortunately no data are available for these fields but some of the figures in Table 12 and Fig. 9 which refer to similar situations demonstrate that starvation could well have been occurring. It now seems highly feasible that in places like the study area most of the flocks could go through the winter on fields where clover stocks were never limiting and as a result little or no mortality would occur. But here and there one flock might be found where nearly all the mortality was taking place. The net result would be that total numbers would be related to total food at the worst season but that a detailed examination within the population would show little evidence of loss. Thus in Table 4 the density of pigeons remained the same over the winter at high clover densities but increased at low densities as surplus birds or flocks were forced to use these areas as the total supply of food decreased. While at first apparently vindicating the contention that the pigeons were living close to their food supply it now seems more likely that we were witnessing the first stage of population regulation in relation to the food supply.

One important question remains. Why did not those individuals forced to experience poor feeding conditions leave the flock and feed comfortably nearby? Why even instead of being found at the front of the flock did not these birds follow up in the rear keeping near enough to satisfy their gregarious impulses but far enough away to avoid conflict situations? In fact, it might be thought that the rear of the flock would be the obvious place because rear birds pass over ground already searched by van birds. We have no explanation for the mechanics involved in this ancillary problem and while the general question might be answered on the basis of group-selection the next section explains why this is not really tenable, at least on present knowledge.

(b) *The function of gregarious behaviour*

It is easy to demonstrate that a species behaving as a flock so that competition is introduced at all stages can more effectively exploit the food supply and leave more survivors than a species that feeds in a solitary way and avoids competition until food stocks reach a critical level. Elsewhere a diagram is presented to illustrate this in a simplified way (Murton 1965b), but the basic idea is that if a population of fifty birds has to live on a food supply of 200 items for a limited period and that to survive each bird needs five items, then clearly there is enough food to ensure that forty birds survive. A survival rate approaching this value might be achieved if competition occurred at all times so that some birds were excluded from feeding. Without competition each bird might eat four food items and then the population would have to become extinct. The highest survival rate must be correlated with the highest level of competition that can be achieved. But although a bigger population could survive with gregarious behaviour the advantages would reside only with the surviving animals and another mechanism would be needed to explain why some birds did not break the rules.

Elsewhere (Murton & Isaacson 1962) it was argued that gregarious behaviour would facilitate food finding because only those individuals that had located worthwhile feeding grounds would stay and so could then attract others. The evolution and maintenance of the habit could be explained satisfactorily on this basis without advocating any other function although additional advantages might follow once gregarious behaviour became established. Defence against predators is also often advocated as a function of gregarious behaviour but it seems unlikely to be an important factor in the case of the wood-pigeon. There are no important predators of feeding wood-pigeons in Britain today while the goshawk (*Accipiter gentilis*) and fox (*Vulpes vulpes*) seem the only ones likely to have been important at any time. Goshawks are mainly woodland species and hunt to some extent like the sparrowhawk (*A. nisus*) by dashing round a corner and catching a bird by surprise. It has yet to be shown that gregarious behaviour would reduce the risk from such attack and even if it did it might result in fewer predators—not more pigeons. Thus in general it seems that vertebrate predators are limited by prey density and not prey density by their enemies, a situation that would need to exist to justify the predation defence theories (Errington 1946; Jenkins, Watson & Miller 1964). Hinde (1952) found that flocks of tits tended to disperse at the end of the day when food collection became less imperative but before the birds returned to their roosting holes. A predator could easily adapt to such laxity if the habit really was a predator defence. In the case of the wood-pigeon it would be desirable to discover selective pressures that are currently operative rather than rely on claims that flocking is a relic from days when predation was more important. If predation ceased to be important its selective pressure would immediately be lost and if no other considerations were involved any anti-predator mechanisms would likely be lost at the same time. It seems more reasonable in the wood-pigeon (and perhaps many other species) that gregarious behaviour is directly concerned in some way with feeding or the finding of food, even though anti-predator defences can be found.

The results of this study seem in many ways to support the ideas recently put forward by Wynne-Edwards (1962) that the close adjustment between animal populations and their physical requirements, particularly food, is a manifestation of a self-regulatory process. Wynne-Edwards has argued that population homeostasis has evolved to prevent animals from over-exploiting their resources which implies that natural selection has favoured a mechanism that enables a population to anticipate mortality and make its own adjustments to numbers or density. The regulation of the animal's reproductive rate in anticipation of possible population replacement requirements is but one aspect of his thesis and is not pertinent to the present discussion, although, of course, winter mortality is a facet of all population loss. Particularly relevant here is his theme that 'unchecked competition for food must be successfully averted, otherwise "over-fishing" would be impossible to escape; this could only result in lasting detriment to the predators and the risk, if they persisted in it, that the prey might be exterminated altogether'. Wynne-Edwards argues that if vast overcrowding occurs when a good food supply exists the animals could over-exploit the resources. Nonetheless this did seem to happen when wood-pigeons were suddenly forced by heavy snow falls to feed on a limited *Brassica* supply in the 1962–63 winter, a situation that could not be anticipated by the population. He further argues that an optimum population density is one that could be supported throughout the period of limited food supplies and to achieve this a limit would need to be set on population-density from the beginning. Exactly how this could be achieved is uncertain because pigeon numbers varied considerably at the beginning of each winter relative to clover stocks because numbers had been adjusted to grain supplies independently of clover availability. The population has ranged between 25 and 174/100 ac in December and there seems no reason why much higher densities could not be attained. There is no evidence that the mechanism of population control would fail providing intensive competition could occur between birds. But regulation would occur if pigeons fed in a solitary way as a point would eventually be reached at which competition could no longer be avoided and single birds like flocks could actively or passively become distributed and their numbers adjusted according to food density. The only difference would be that numbers would fall to a lower level and population fluctuations from the breeding season to the period of winter minima would become more extreme. Again the 1962–63 winter was an example of what does sometimes happen. Even so, there can be no doubt about the advantages of introducing early social competition, and our views on this aspect are in accord with those of Wynne-Edwards. The crucial question, however, is whether the flock behaviour of the wood-pigeon (and for that matter other species) has evolved primarily as a homeostatic mechanism—which Wynne-Edwards would claim to be the case—or whether the mechanism of population control operates through a social behaviour that has evolved as a result of other selective pressures. The environment must determine population size, otherwise there would be no consistent variation in bird density in different areas or seasons; but in the first instance the adjustment would involve an intrinsic and active adaptation by the animal, in the second case a passive response to extrinsic environmental pressure.

Wynne-Edwards (1962) maintains that natural selection may favour the group at the expense of the individual. In a very restricted sense this may be true because the gene pool of the Mendelian population sets limits on the genetical diversity of the individual genotypes which must be limited by the variation available from the gene pool. If this in turn contains many factors for social behaviour this might modify the evolutionary process. Haldane (1934) showed many years ago that natural selection can only operate on those individual genotypes which have been entrusted with temporarily carrying the variability available from the gene pool. Recently Wynne-Edwards (1963) has maintained that genetical evidence for group selection does exist and Smith (1964) has given the mathematical requirements for group or kin selection. However, Smith shows that the conditions that have to be satisfied are so special and severe that it seems extremely unlikely that they could operate under field conditions for most, if not all, bird species.

Because at present there appears to be no satisfactory genetical basis for group selection of the sort discussed above, social behaviour ought first to be considered in terms of its advantages or disadvantages for the individual. Gregarious behaviour in the wood-pigeon might evolve because those individuals reacting positively to other feeding birds would have much better chances of locating suitable food sources and as a result have higher survival chances. Social behaviour would then facilitate a hierarchy system which could ensure earlier competition between individuals and the exclusion of surplus birds or flocks to other feeding grounds. This could also lead to a higher survival rate relative to food as discussed above and the process could reinforce social behaviour, but it need not be the cause. The mechanism of population control through social behaviour could, therefore, be fortuitous and be explainable on the grounds of individual selection. At present this view is preferred in the absence of a satisfactory genetical theory.

* * * * * * *

REFERENCES

Arnold, G. W. (1964). Factors within plant associations affecting the behaviour and performance of grazing animals. In: *Grazing in terrestrial and marine environments*. Fourth symposium of the British Ecological Society (Ed. by D. J. Crisp). Oxford.

Errington, P. L. (1946). Predation and vertebrate populations. *Q. Rev. Biol.* **21**, 144–77, 221–45.

Haldane, J. B. S. (1934). *The causes of evolution*, 2nd edn. London.

Hinde, R. A. (1952). The behaviour of the great tit (*Parus major*) and some other related species. *Behaviour*, Suppl. No. 2, 1–201.

Hughes, R. E., Milner, C. & Dale, J. (1964). Selectivity in grazing. In: *Grazing in terrestrial and marine environments*. Fourth symposium of the British Ecological Society (Ed. by D. J. Crisp). Oxford.

Jenkins, D., Watson, A. & Miller, G. R. (1964). Predation and red grouse populations. *J. appl. Ecol.* **1**, 183–95.

Kluyver, H. N. (1957). Roosting habits, sexual dominance and survival in the great tit. *Cold Spring Harb. Symp. quant. Biol.* **22**, 281–5.

Lockie, J. D. (1956). Winter fighting in feeding flocks of rooks, jackdaws and carrion crows. *Bird Study*, **3**, 180–90.

Masure, R. H. & Allee, W. C. (1934). The social order in flocks of the common chicken and the pigeon. *Auk*, **51**, 306–27.

Murton, R. K. (1965a). Natural and artificial population control in the wood-pigeon. *Ann. appl. Biol.* **55**, 177–92.

Murton, R. K. (1965b). *The wood-pigeon*. London.

Murton, R. K. & Isaacson, A. J. (1962). The functional basis of some behaviour in the wood-pigeon *Columba palumbus*. *Ibis*, **104**, 503–21.

Murton, R. K. & Isaacson, A. J. (1964). Productivity and egg predation in the wood-pigeon. *Ardea*, **52**, 30–47.

Murton, R. K., Isaacson, A. J. & Westwood, N. J. (1963a). The feeding ecology of the wood-pigeon. *Br. Birds*, **56**, 345–75.

Murton, R. K., Isaacson, A. J. & Westwood, N. J. (1963b). The use of baits treated with alpha-chloralose to catch wood-pigeons. *Ann. appl. Biol.* **52**, 271–93.

Murton, R. K., Westwood, N. J. & Isaacson, A. J. (1964a). A preliminary investigation of the factors regulating population size in the wood-pigeon. *Ibis*, **106**, 482–507.

Murton, R. K., Westwood, N. J. & Isaacson, A. J. (1964b). The feeding habits of the wood-pigeon *Columba palumbus*, Stock dove *C. oenas* and turtle dove *Streptopelia turtur*. *Ibis*, **106**, 174–88.

Smith, J. Maynard (1964). Group selection and kin selection. *Nature, Lond.* **201**, 1145–7.

Thomas, A. S. (1960). Changes in vegetation since the advent of myxomatosis. *J. Ecol.* **48**, 287–306.

Wynne-Edwards, V. C. (1962). *Animal dispersion in relation to social behaviour*. Edinburgh.

Wynne-Edwards, V. C. (1963). Intergroup selection in the evolution of social systems. *Nature, Lond.* **200**, 623–6.

(*Received* 22 *July* 1965)

27

Reprinted from *Zoologica*, **45**(4), 41–52 (1960)

Distribution and Status of the Mountain Gorilla
(*Gorilla gorilla beringei*)—1959[1]

JOHN T. EMLEN, JR., & GEORGE B. SCHALLER

Department of Zoology, University of Wisconsin, Madison, Wis.

(Text-figures 1-5)

THIS report describes the results of a survey of the distribution of the Mountain or Eastern Gorilla (*Gorilla gorilla beringei*) as defined by Coolidge (1929) throughout the extent of its geographic range in the Kivu and Oriental Provinces of the Belgian Congo and the Kigezi District of Uganda. No attempt is made to review the literature or to summarize the data from museum specimens. The objective is to construct as complete a picture as possible of the distribution and status of these animals as of 1959. To this end historical records are included only as they shed light on the current situation.

The survey in its over-all objectives included comparisons of behavior and adaptations in various parts of the range and was organized as the first stage of a systematic study of the life history and ecology of the Mountain Gorilla. These latter aspects of the study are currently being pursued and will be reported at a later date.

The survey extended from early March to the end of July, 1959. Approximately four weeks were spent in the Birunga Volcano area, two in the Kayonza (Impenetrable) Forest of Uganda, two in the Lubero region west of Lake Edward, two in the Lake Kivu-Mt. Kahuzi country, two in the southern region bordering Lake Tanganyika and three in the broad equatorial forest west of the Rift Escarpment. Time did not permit a visit to the northern region which is largely inaccessible by road. Most of the survey work was done on foot in the first two regions and by car in the others.

The study was conducted under the sponsorship of the New York Zoological Society with funds partly from the Society and partly from a grant by the National Science Foundation. Full cooperation, including the provision of housing facilities and guides and the granting of permission to visit restricted areas, was tendered by the Institute of National Parks of the Belgian Congo. Helpful assistance was extended by the Institute for Scientific Research in Central Africa (I. R. S. A. C.) and the Uganda Game Department. Technical cooperation was offered by these organizations, by staff members of the Medical School of Makerere College, the Uganda Agricultural Research Station at Kawanda, the Virus Research Laboratory at Entebbe and the National Institute for Agronomic Research in the Congo (I. N. E. A. C.).

The list of persons whose assistance should be acknowledged is too long to be included in its entirety. A partial list is included as an appendix to this report.

PROCEDURE AND METHODS

The data for this report are based, in the last analysis, on current observations in each area as recorded personally or as communicated directly to us by residents and visitors. The procedure followed may be considered as falling into three steps: (1) general reconnaissance of the whole region, (2) local reconnaissance and (3) on-the-spot checking of details.

The general reconnaissance was accomplished by examining reports and by interviewing government officials, prospectors, travelling missionaries and others. Published reports of early expeditions and surveys were also examined and a considerable correspondence was

---

[1] A supplement containing regional maps and tables summarizing the raw data and a list of authorities and references is available on request from the senior author.

41

established with experienced residents in all parts of the region. The information so assembled was useful in obtaining a preliminary picture of the outlines of distribution and a basis for intensive local work.

Local reconnaissance, the second step in the survey procedure, was accomplished largely through interviews with local administrators and regional native chiefs. European settlers were also visited as encountered or were deliberately sought out when reported to have pertinent experience or information. Their reports were often quite specific and detailed, but more often required further checking at the local level. Many of these people were able to call in natives from various parts of their regions to supplement, confirm or refute their reports.

Local checking of details was done personally on foot or by interviewing native residents. More than 400 local natives and native groups were interviewed. Workmen in the small road maintenance crews scattered every few kilometers along the Congo forest roads were generally well acquainted with conditions for several miles around their semi-permanent camps and proved to be one of our best sources of information. The elders of the small roadside villages also provided much helpful information. We usually stopped our car at a central point in a village and within a few minutes were surrounded by a few to 20 or more men eager to pool their knowledge of gorilla distribution and gorilla history in the immediate vicinity. Most of the reports were negative and these were, of course, as important as positive reports for charting the details of gorilla distribution. We repeatedly checked the accuracy of positive reports by challenging the informants to show us the evidence. Without exception they were able to do so.

Personal observations of gorillas and gorilla spoor were made whenever possible. In the Lubero and Uganda border regions personal observations provided the bulk of the data. In the other regions we personally checked at sample points and particularly where we had reason to question a reported occurrence. A foot transect was also used to cover a broad, roadless and relatively unknown area of equatorial forest in the western region. Except when we were led to known sites by locally experienced natives, the procedure was to follow trails as available into the heart of the area, then strike out in a straight line with a compass, looking for all beds, droppings, trails and other traces of gorilla activity and obtaining a crude quantitative index of the incidence of such traces per hour or mile of transect.

Data were recorded on the best large-scale maps available for each region (1:50,000, 1:200,000 or 1:500,000) and subsequently transcribed to a standard base (1:500,000, C. N. Ki, 1958) from which the distribution maps (Text-figs. 1 and 3) accompanying this report were prepared. A sample larger scale map is presented in Text-fig. 2.

GEOGRAPHICAL DISTRIBUTION

The range of the Mountain Gorilla extends from the equator south to about 4° 20′ south latitude, a distance of about 300 miles, and from longitude 26° 30′ east to the west escarpment of the Rift Valley at about 29° east longitude, with a small extension into mountain forests of the Valley floor north of Lake Kivu, an east-west distance of slightly more than 200 miles (Text-fig. 1). It is in the form of a triangle with its points near Lubutu on the northwest, Lubero on the northeast and Fizi on the south and covers an area of about 35,000 square miles. The boundaries are quite distinct along the east and southwest fronts, and no records of stragglers were encountered beyond these limits in fairly intensive surveys. We unfortunately have very little detailed information on the north boundary, however, and there may well be additional pockets in the Maiko Valley and westward toward Ponthierville.

Within the borders of this range, most of the gorillas are concentrated in about 60 more or less isolated tracts of 10 to 120 square miles each (solid black areas in Text-fig. 1), separated from each other by distances of 2 to 30 miles. Thus the area occupied by substantial populations is only about 1,900 square miles, or 5½ per cent. of the total area of the range.

A block of about 6,500 square miles in the western region between the Lowa and the Lugula Rivers (hatched areas in Text-fig. 1) apparently contains a relatively sparse but uniform population of wandering troops and individuals. The vast stretches of forest between and around these populated areas are essentially unoccupied except for stragglers—single adult or sub-adult males or occasionally small groups which leave the troops to live solitary lives as much as 20 miles from the nearest colony. Such animals are encountered at irregular intervals and in a variety of situations, sometimes establishing themselves for considerable periods of time near farms or native villages.

NUMERICAL STATUS

Although we have very inadequate data for estimating the numbers of gorillas present within the range described above, there is a legitimate

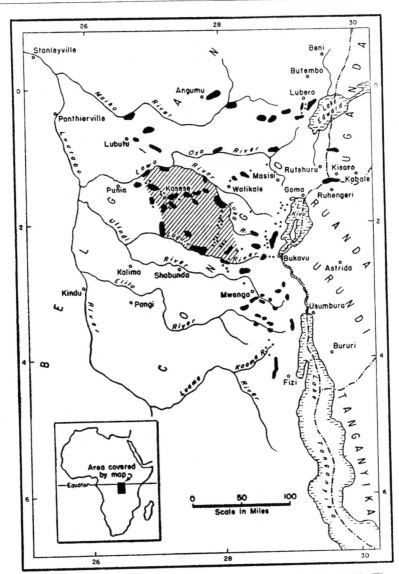

TEXT-FIG. 1. Geographic distribution of the Mountain Gorilla—1959. The sixty black areas indicate the location and approximate shape of sixty "gorilla areas" ranging in size from about 10 to about 100 square miles each. The small dots indicate records of stragglers outside of these areas. The hatching marks a central region of fairly continuous but sparse population. (Detailed maps and tables presenting a summary of the raw data on which the map is constructed may be had from the senior author on request. A sample map of this type is presented in Text-fig. 2).

demand for quantitative information on this rare and localized animal which should not be disregarded. We therefore submit the following data with the warning that the conclusions are extremely tenuous.

Intensive surveys of local gorilla populations were conducted on five plots in three areas in the northeastern part of the range. It was hoped that these surveys would provide indications of population densities in these areas. We found, however, that, because of the wandering habits of the gorilla, a survey conducted on a plot of 5 to 10 square miles over a period of a few weeks may be quite atypical of even the local situation. On one 10-square-mile plot, for instance, the population was known to have nearly quadru-

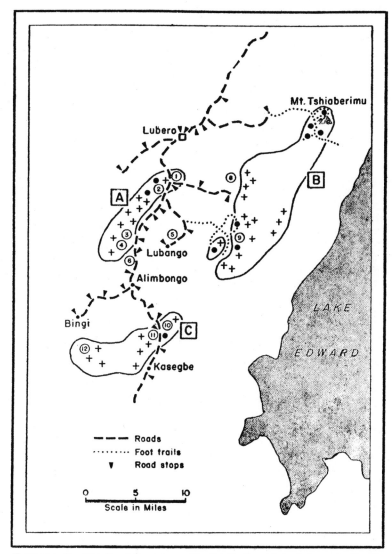

TEXT-FIG. 2. Details of the geographic distribution of the Mountain Gorilla in the Lubero Region (a sample of the regional maps from which the general map in Text-fig. 1 is constructed). Encircled numbers indicate the locations of specific records reported to us; large circular spots indicate the location of personal observations of gorillas or gorilla spoor; plus signs show where gorillas were reported to be present but without specific observations recorded. Dashed lines indicate routes travelled by car; dotted lines are those traversed on foot. Short bars at the side of the routes mark the places where local natives were interviewed.

pled through immigration shortly after our twenty-day survey was completed; on another, six days of intensive work over four of the six square miles revealed no fresh sign; then two troops were discovered together at the western end of the plot. Taking all factors into consideration we estimate that population densities in these areas generally averaged somewhere between one and five animals per square mile. Higher levels were encountered, but these were thought to represent temporary concentrations.

Calculations based on predation statistics provided another independent source of information on population size in the Lubero region where gorillas, coming close to villages or raiding native gardens, are commonly hunted with spears. Data from the three hospitals serving this region showed an average of five cases of gorilla injury treated per year, all apparently the victims of counter-attacking gorillas. Hospital attendants and others estimated that this indicated an actual incidence of at least 10 injuries (half not reporting for treatment) per year. They and two patients who had sustained leg injuries also agreed that the number of raids organized against gorillas each year was considerably greater than the number of injuries, and that most of these raids were pressed until the offending gorilla was killed. Animals which inflicted injury, in particular, were marked for further raids of attrition and rarely escaped with their lives. Although these calculations are very crude, the figures suggest a slaughter of at least 20 and perhaps more than twice that number of gorillas in this 110-square-mile area every year. The large number of fairly fresh gorilla skulls (14) which we saw in the villages is further evidence of such a slaughter. While we still know very little about the population dynamics of these slow-reproducing animals, the population level required for survival and maintenance under this type of predation pressure must be of the same order of magnitude as that indicated in the plot surveys.

Comparisons of population level in different parts of the range were essayed by indices based on the number of bed sites, droppings or feeding areas encountered per mile or per hour of trailless foot transect. Unfortunately conditions of vegetation varied so greatly that direct comparisons of figures are of little value. For instance, beds were more easily seen in bamboo forests where the animals generally built 8 to 10 feet off the ground than they were in the broad-leaf forest where most of them bedded on the surface. Tallies of droppings were also difficult to equate because of marked differences in the density of concealing ground cover. Thus, the relative abundance of gorillas in different parts of the range could only be crudely estimated by a subjective appraisal of all available evidence. Such evidence suggests that the densities in the lowland areas to the west are, in general, similar to those on the more intensively studied survey plots in the northeast sector.

If we accept the rough figures and estimates presented above, the total population of Mountain Gorillas probably lies somewhere between 3,000 and 15,000 individuals.

ECOLOGICAL DISTRIBUTION

Although the range of the Mountain Gorilla is small, it possesses considerable ecological diversity. Gorillas are found at altitudes ranging from 1,600 feet on the Lowa River above Kima to over 13,000 feet near the summit of Mt. Muhavura in the Birunga Volcanoes. Frosts occur frequently on the higher slopes of the volcanoes while temperatures are decidedly tropical in the areas west and south of Walikale. The topography is notably rugged along the eastern front from Lubero to Fizi but becomes quite flat in many parts of the western region. Gorillas seemed to favor steep slopes and were characteristically associated with the hillier sections where there was a choice. In the precipitous mountains of the Rift gorillas often frequented slopes very difficult for us to negotiate in following them.

Vegetation is of major importance in gorilla distribution. The animal's range lies wholly within the forest zone though extending to its very edge as it borders on the grasslands or plains to the east and the savannah-woodlands to the south. Keay (1959) recognizes two major forest types in this area, the boundaries of which are shown, together with the gorilla distribution data, in Text-fig. 3. Of the 60 gorilla concentration areas plotted on this map, 25 are located in forests classified by Keay as belonging to the Montane Forest Communities, and 35 in the Equatorial Forest or "Evergreen Forest of Low and Middle Altitudes." There are no obvious indications of preference for one or the other of these two types in our data except in the southern region where range extension is greatest in the Montane Forest zone.

I. Montane Forest Communities

Montane Forests can be divided into a number of subtypes with irregular and interdigitating distributions (Keay, *op. cit.;* Lebrun, 1935; Robyns, 1948), four of which are utilized extensively by gorillas. Each of these, in turn, has distinctive seral stages of regeneration. The pertinent sub-types of the "Montane Forest Communities" are listed and described below together with comments on their utilization by gorillas:

(a) *Bamboo Forests.*—Pure stands of bamboo such as those found in the Volcanoes are characterized by a dense though shallow and rather translucent canopy 20 to 25 feet above the ground, roofing an essentially open subcanopy and supported by an abundance of irregularly clustered subvertical stems. The shrub and ground strata are sparse and low or completely absent (Text-fig. 4a). Bamboo forest is a prominent type at altitudes of 7,000 to 9,000 ft. in the Lubero and Birunga Volcano regions and locally elsewhere, forming pure stands or mixtures with various low broad-leafed trees. It supplies food and shelter for gorillas but is util-

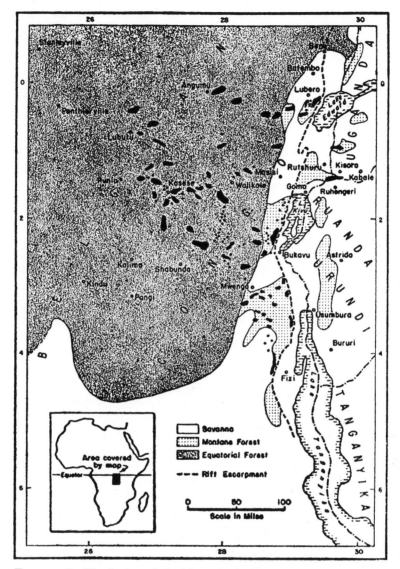

TEXT-FIG. 3. Distribution of the Mountain Gorilla with respect to vegetation types. The indicated boundaries of the Equatorial Forest and the "Montane Forest Communities" are taken in substance from the new UNESCO map (Keay, 1959), refined in some details on the basis of large-scale local maps and personal observation.

ized extensively only in mixed stands and along borders.

(b) *Montane Scrub Forest.*—The low forest above the bamboo zone in the eastern part of the Volcanoes is a mosaic of rather dense 20-, 15- and 10-foot leafy canopies interspersed with open patches of three-foot sedges and herbs (Text-fig. 4b). Scrub forests of this general physiognomy are found irregularly distributed at altitudes of about 10,000 feet along the Rift ridge

and on the Birunga Volcanoes. Gorillas occur sporadically in these forests wherever tall herbs are prominent in the understory.

(c) *Hagenia Woodland.*—In its best form in the central part of the Birunga Volcanoes, this type presents a rather dense but incomplete canopy 40 to 50 feet in height supported by heavy, sub-horizontal, epiphyte-covered branches. The subcanopy is quite open, especially under pure stands. Sufficient light penetrates to suport a

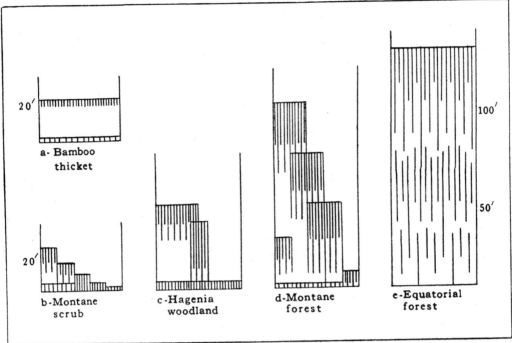

TEXT-FIG. 4. Schematic diagrams depicting the height, cover and stratification characteristics of the five principal vegetation types utilized by Mountain Gorillas. Heights are in feet; foliage density is indicated by intensity of shading. In general, the best gorilla habitats are those which have the rankest herb stratum. The diagrams depict conditions in mature stands. For verbal description and gorilla utilization see text.

luxurious and almost continuous stratum of herbs about five feet tall (Text-fig. 4c). This type is largely confined to the Birunga Volcanoes, but the two dominant tree species also occur in mixtures with other Montane evergreens to produce stands intermediate between this and type *d*. It is excellent gorilla habitat.

(d) *Montane Evergreen Forest.*—Apparently mature stands of this type along the ridge of the Rift Escarpment as at Mulenge have an incomplete upper canopy 80 to 120 feet high and several indistinctly defined lower canopies. Epiphytes are quite numerous. The herb-shrub stratum is sparse and low where the tree canopy is continuous but quite dense under breaks in the canopy (Text-fig. 4d). This type seems to be best represented in the Mwenga-Fizi region but is also found in the Kahuzi, Lubero and Kayonza Forest areas.[2] It tends to be quite irregular on rough topography with highest canopies on the gentler slopes and a low, more continuous canopy on the ridges. It is rather inferior as gorilla

habitat except along edges and where regenerating.

The three gorilla troops whose activities and movements we followed in the Birunga Volcano area all wandered back and forth between several forest subtypes. Because of the patchy nature of forest distribution in these areas it seems unlikely that any gorilla troops are confined to a single type. In the less varied forests of the Rift rim and the Kayonza forest, however, many animals must live entirely within the Montane Evergreen Forest type, its seral stages and the bordering areas of elephant grass, bracken and tree fern.

II. Equatorial Forest

Mature equatorial forests are said to be quite uniform in physiognomy throughout their broad range west of the Rift Escarpment. Very little mature forest remains, however, and the various seral stages of regeneration form an intricate and irregular mosaic throughout the region. Selected stages in this regeneration process are described below with comments on their utilization by gorillas. The diagrams in Text-fig. 5 are merely subjective evaluations of conditions and are not based on specific data.

---

[2]These tall forests are at times transitional with equatorial forests and some have been classified as such. At higher altitudes they are shorter and with more epiphytes and fewer lianas.

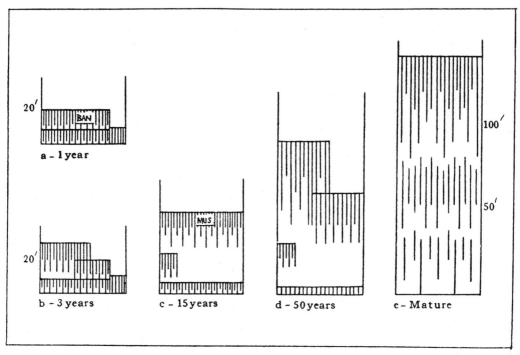

TEXT-FIG. 5. Schematic diagrams depicting the physiognomic characteristics of five seral stages in the recovery of equatorial forest following a clearing for banana cultivation. Times given are very approximate.

Newly Cleared Field.—Equatorial forests are characteristically cleared for cultivation in a crude and incomplete manner with simple hand tools and fire. A field is generally irregular in shape and a few acres in extent, with a scattering of hardwood trees and "Spirit" trees left uncut. Stumps six to eight feet high are left standing, and the ground remains strewn with half-burned fallen logs. Plantings of manioc or rice and bananas are made almost immediately in shallowly turned soil. Gorillas are often seen at the edges of newly cleared fields and may briefly wander into them.

(a) *At One Year.*—Banana plants reach their full height of about 20 feet in a year and are harvested at about that time. They form a rather dense but irregularly broken canopy down to the ten-foot level. Herbs and vines form a substratum from this point to the ground (Text-fig. 5a). Gorillas visit such stands for food (banana stems and various other herbs) but retire to bigger timber for resting and sleeping.

(b) *At Three Years.*—After a few years the bananas which remain and the new ones which have sprung from the rootstocks of the originals are gradually choked out by second growth trees, vines and shrubs. The whole makes a dense

tangled mass, but a partially open subcanopy is usually present under the taller trees (Text-fig. 5b). This is excellent gorilla feeding habitat.

(c) *At Fifteen Years.*—The second growth Musanga trees are now 40 to 50 feet tall and produce a rather thin canopy of big leaves over an open subcanopy.[3] Young specimens of more durable species of trees form an incomplete second story of 10 to 20 per cent. The herb stratum is luxurious and about five to six feet tall (Text-fig. 5c). Gorilla utilization is now declining, but much food remains and the dominant Musangas provide both food and favored bedding situations.

(d) *At Fifty Years.* In many parts of the Kivu Province, the demands for agricultural land force a recutting of regenerating forests every 10 or 20 years. In more inaccessible areas away from settlements and roads, however, older forests are found which show signs of immaturity presumably related to land clearing activities in early times. These forests may have upper canopies 100 or more feet high, but incomplete and uneven. There is also a large num-

---

[3]According to Ross (1954), the life span of this tree is about 15 to 20 years.

ber of younger trees pushing up through the subcanopy, and, importantly, a definite lower stratum of herbs and ferns (Text-fig. 5d). Gorillas find considerable food in this herb stratum.

(e) *Mature Equatorial Forest.* — A mature equatorial forest in the central Kivu Province has a nearly complete canopy 120 to 180 feet high and a rather even, thinly foliaged subcanopy, sometimes vaguely divisible into several strata, extending down to the ground. Herbaceous and ground vegetation is very sparse or absent (Text-fig. 5e). The mature forest, presumably because of the scarcity of food materials near the ground level, is sparsely inhabited by gorillas.

Gorilla distribution in the equatorial forest region is determined to a considerable extent by the amount and distribution of patches of regenerating forest. The optimum situation is provided by a mosaic of old fields in various stages of regeneration.

DISTRIBUTIONAL DYNAMICS

An examination of the maps in Text-figs. 1 & 3 shows that many of the larger population units are located flush against prominent physical barriers such as rivers, escarpments and forest boundaries. The Lugulu River north of Shabunda, the Luka and Lowa Rivers west of Walikale and the eastern forest boundary from Fizi to Lubero are examples. The impression suggested by this situation is that there has been a centrifugal expansion from one or more centers, a piling up of animals against certain major barriers, and perhaps a partial evacuation of large tracts near the centers.

Range Boundaries.—The factors limiting dispersal at the north, east, south and west boundaries each present special problems. A problem of the northern region is the bridging or circumventing of the Oso-Lowa River barrier. We incline to the theory that the region above this barrier was invaded from the south, but regardless of the direction of movement the problem of a crossing remains to be explained. The gorilla is notoriously shy of water[4] and a direct fording or rafting of this formidable river system west of the vicinity of Masisi is unlikely under conditions resembling those which prevail today. Either the penetration was made during an ancient interpluvial period of low water or was accomplished near the headwaters. This second possibility presents other difficulties associated with the crossing of expanses of what is now open grassland, but is supported by the presence

———
[4]In driving animals into capture nets C. Cordier could not force gorillas to cross streams more than 15 or 20 feet wide and two or three feet deep.

of isolated colonies on forested "islands" in the Rift valley floor. It suggests passage during a pluvial period with expanded forest boundaries.

The eastern boundary of the range is nicely demarcated by the Rift Escarpment along much of its length (Text-fig. 3). The absence of gorillas in the broad tongues of montane grassland which extend up over the rim north and south of Lake Kivu, however, indicate that the actual effective barrier is not the escarpment itself, but the forest-grassland boundary. As already noted, two populations have penetrated this barrier and established themselves on montane-forest "islands" in the predominantly grass-covered valley floor. The crossing to the eastern volcano area was probably by way of irregular forest bridges on Mts. Nyamuragira and Nyiragongo, presumably in pluvial times. The population in the Kayonza Forest to the north probably arrived there by way of the volcanoes. The Semliki Plains apparently have constituted an effective barrier to dispersal into the Ruwenzori forests east of Lubero.

The southward extension of gorillas into the Mwenga-Fizi region probably occurred near the headwaters of the Lugulu and Ulindi Rivers. It is interesting and perhaps significant that this southernmost arm of gorilla range lies entirely within the montane forest zone. The fragmented nature of the populations in this region and the small size of the units may be related to the highly irregular topography and vegetation.

We have no explanations or worthy speculations on the limiting factors of the western boundary from the Lugulu River to the Maiko. Altitudes become progressively lower, prevailing temperatures higher and agricultural operations more intensive, but none of these conditions seem adequate to explain the abrupt discontinuation of gorilla populations short of the Lualaba River.

The occurrence of a separate population of gorillas (*G. g. gorilla*) in the Cameroons and Gabon 1,000 miles to the west is evidence for a former occupation of the intervening regions. Vague and inconclusive records suggest that gorillas may have occurred north of the great bend of the Congo within the past few hundred years (Coolidge, 1936), and physiographic features argue strongly that the connection which provided the assumed channel for gene exchange must have been located in this region and around the headwaters of the various large rivers of the right bank. This region now supports a savannah vegetation unsuitable for gorilla habitation, and we conclude that the bridging populations must have lived here during one of the pluvial periods of the Pleistocene when the forest boundary lay farther north than at present.

Local Distribution.—As already noted, gorillas are not randomly distributed through the forest but are aggregated for the most part into small, well-defined population units on concentration areas comprising only about 5½ per cent. of the total range. In some cases the boundaries of these areas coincide with natural boundaries which appear to have operated as barriers to dispersal, but in most cases the forest continues unchanged, and we have been unable to find a simple explanation for the discontinuity.

Social behavior acts as an aggregating factor at least at the local level. Gorillas tend to live in troops of about 5 to 20 individuals and to restrict their activities for extended periods to home ranges of familiar terrain (unpubl. data). From a few to several dozen such troops combine to make a population unit, apparently selecting contiguous home ranges rather than dispersing themselves randomly through the forest. The nature of these social interactions and their role in the derivation of clumped distribution patterns is now being investigated.

The actual geographic pattern of distribution of concentration areas is undoubtedly determined by a combination of physical and environmental factors operating through time. Population units probably drift or are pushed about as local conditions change. Evidence for movements of this sort was encountered in a number of reports of recent invasions and of evacuations or extirpations. More often, however, our informants told of local stability over periods of several decades. Written records, furthermore, suggest a general stability of distribution in the Lubero and Birunga Volcano areas for 30 to 40 years.

As noted above, physical barriers have apparently influenced and in part determined the present location of many of the concentration areas. Of the 60 areas plotted on our map, 17 have one side bordering on a large river, and 13 front on the forest-grassland boundary. Without historical data we can only guess as to how this situation developed, but one is tempted to visualize the population units slowly drifting over the map like wandering amoebas until their progress is physically blocked by a barrier. In those cases where man has moved the location of a barrier as by pushing back the forest boundary, the retraction of the population unit has been as immediate as it was inevitable.

The forces or influences responsible for the drifting of population unts within the outer boundaries set by physical barriers can be visualized, at least in part, by examining the effects of man-made disturbances at the local level. The most important of these disturbances during the last century has been the repetitive clearing of the forest in small patches to create a shifting kaleidoscopic pattern of successional stages of forest regeneration. Gorillas have responded by adjusting their local foraging activities to meet the temporarily favorable conditions created in the recently cultivated fields. There is also evidence of the expansion or movement of gorilla population units into developing agricultural or mining regions. The large concentrations of gorilla areas around Kasese and south of Mwenga correlate with a heavy influx of native families into these mining areas in recent times. A road engineer east of Kasese found gorillas moving in around the workers' villages and garden plots while construction work was still in progress, and several mining engineers told of gorillas appearing around new camp sites within a year or two of their establishment. Such incompletely documented anecdotes may represent returns after initial displacement rather than actual invasions, but in either event they suggest a certain fluidity of movement.

The long-standing government policy of resettling natives from the deep forests on relatively narrow strips along the roadways probably accounts in part for the present concentration of gorillas along these human thoroughfares and their relative scarcity in the hinterlands. The animals have apparently moved into these disturbance areas and remained despite persistent harassment. A long-established mining community surrounded by the villages and shambas of native workers may thus form a sort of nucleus for a gorilla population unit as at Utu, Kima or Lulingo.

Further evidence for the role of rotational forest clearing in gorilla movements is found in historical records of evacuation by gorillas following the emigration of human populations. The first consequence of an exodus of human beings may be an improved environment freed of direct human disturbance. Thus the village of Mangombe near Kima still had a large local population of gorillas three years after it was abandoned as a mining camp and allowed to grow up to a jungle of shrubs and vines. A former resident reported that "the gorillas from the surrounding shambas moved into the village as the people moved out." Abandoned mining centers like those near Lutunguru and Alimbongo apparently had good gorilla populations during and immediately following their period of activity and have subsequently lost them. It must be said, however, that the records here are very fragmentary and quite controversial.

The old and long-abandoned Arab slave route[5] from Angumu to Walikale resulted in vegetational disturbances which can still be detected by foresters and which apparently drew gorillas in from the surrounding deep forests. Reports reaching government surveyors told of gorillas persisting in this area for many decades, but current reports indicate that the animals have now largely disappeared.

As with most studies of distributional dynamics, our direct observations are limited to the effects of man-made disturbances operating over a very limited period of time. Our documentation of the environmental disturbances and distributional responses of prehistory are much less precise, but the evidence for frequent and major disturbances is overwhelming, and we can assume that responses in the past were basically similar to those we observe today. Gorilla distribution has thus been a dynamic phenomenon from time immemorial and doubtless will continue to be.

What the observed tendencies and trends indicate for the fututre of the Mountain Gorilla varies in different parts of the range. Peripherally and in the areas of good arable soils as in the Volcanoes, the Kayonza Forest and the Lubero district, range boundaries are rapidly being pushed back by the conversion of land to stabilized agriculture. The continued existence of gorillas in these regions will depend on the maintenance of reserves set aside for the preservation of gorilla habitat, *i.e.* large continuous expanses of suitable forest. The rigid protection against all types of human disturbance in the Albert Park gorilla sanctuary has been extremely desirable from many points of view, but experience in other areas indicates that complete control of trespass is not always necessary. Excessive disturbance must be avoided, but above all, the destruction of habitat by forest clearing must be stopped.

In the equatorial forest zone west of the Rift Escarpment the conditions and needs are totally different. Here the boundaries of the gorilla range are little affected by pressures of land utilization except around the large population centers. Conditions of soil and climate enforce and will probably continue to enforce a type of agriculture which is beneficial rather than destructive to gorilla habitat. Protection of the forest against this rotational system of agriculture would actually be undesirable. Gorilla conservation here must take the form of restrictions on

---
[5]This route led south of Lake Kivu into Tanganyika and was probably at its peak of activity during the 1880's.

killing. **Present killing rates are apparently not serious, but the introduction of even small numbers of firearms would be disastrous regardless of legal restrictions on their use.**

Above all it must be remembered that the total geographic range of the Mountain Gorilla is extremely small, and the history has shown that animals so situated are highly vulnerable.

SUMMARY

The total geographic range of the Mountain Gorilla in 1959 covered an area of about 35,000 square miles south of the equator and west of the Albertine Rift. Within this range the animals were mostly aggregated into about 60 small areas of concentration covering roughly 1,900 square miles of forest habitat. Population densities within these areas probably ran between one and five animals per square mile.

Gorilla populations were found through several climatic zones and in various types of forest. They were most numerous where there was an abundance of herbaceous undergrowth as in fairly open stands or in patches of secondary forests in early stages of regeneration.

The location of concentration areas along rivers and forest edges suggests that population units, though relatively stable and localized over periods of decades, may drift within boundaries prescribed by physical barriers. Evidence for such drifting is provided by observations on distributional responses of populations to man-made changes in the forest habitat.

The future of the Mountain Gorilla depends on the protection of tracts of forest habitat in the eastern areas and on the control of firearms in the equatorial forest areas west of the Rift.

ACKNOWLEDGMENTS

For general direction and guidance in planning and conducting the expedition we wish to thank, in particular, Dr. Harold Coolidge of the National Academy of Science and Dr. Fairfield Osborn of the New York Zoological Society. For critically reading the manuscript we thank Dr. James Chapin of the American Museum of Natural History.

For technical advice and logistic assistance we are grateful to W. M. Baumgartel, hotel keeper at Kisoro, U.; R. Bouckhaert, Conservateur, P. N. A., at Rumangabo, C. B.; Dr. C. M. Chenery, Agricultural Researcher at Kawanda, U.; C. Cordier, animal collector and naturalist at Kabunga, C. B.; C. Cornet d'Elzius, Conservateur, P. N. A., at Ruindi, C. B.; E. Fangoudis, white hunter, Kabale, U.; Prof. A. Galloway, Makerere College Medical School, Kampala, U.; Rev. P. Hurlburt, Missionary, Kitsambiro, C. B.; B.

Kinloch, Dir. Uganda Game Dept., Kampala, U.; L. Mertens, Sec. Prov. Kivu, Bukavu, C. B.; M. Micha, Conservateur en Chef des Parcs, P. N. A., Rumangabo, C. B.; Dr. U. Rahm, Mammalogist, I. R. S. A. C., Lwiro, C. B.; G. Rousseau, Conservateur, P. N. A., Rumangabo, C. B.; H. van Daele, Agronomist, I. N. E. A. C., Bikara, C. B.; Dr. L. van den Berghe, Dir., I. R. S. A. C., Lwiro, C. B.; Dr. V. Van Straelen, President, I. P. N. C. B., Brussels, Belgium; and J. Verscheuren, Biologist, P. N. A., Rutshuru, C. B.

Basic information on the region and on survey routes was provided by many of these men and also, among others, by H. Bjordahl, Mining Engineer, Kabale, U.; M. Kennedy, Missionary, Nkuba, C. B.; Dr. G. Marlier, Biologist, I. R. S. A. C., Uvira, C. B.; A. Meyer, Geologist, Goma, C. B.; A. Michelson, Forester, C. N. Ki, Bukavu, C. B.; M. Pierlot, Forest Ecologist, I. N. E. A. C., Mulungu, C. B., and by several dozen territorial administrators, assistant administrators, chefs de poste, district commissioners, native chiefs and village "capitas." Often these men went to considerable pains to round up knowledgeable field observers and put their experience at our disposal.

Valuable information on local conditions and on gorilla biology was given by many including G. Blondeau, Mining Engineer, Miya, C. B.; Prof. N. Bolwig, Makerere College, Kampala, U.; R. Christiaensen, Biologist, I. R. S. A. C., Lwiro, C. B.; Prof. K. Curry-Lindahl, Univ. Uppsala, Sweden; J. Engelbrecht, farmer, Lubero, C. B.,; Dr. M. Kawai, Japan Monkey Center, Inuyama, Japan; R. Kern, Engineer, M. G. L., Butembo, C. B.; Dr. H. Mitzuhara, Japan Monkey Center, Inuyama, Japan; A. Notte, Naturalist, Ihembe, C. B.; J. de Medina, G. C. E. O., Epulu, C. B.; W. Paar, Engineer, Nyakalonge, C. B.; Dr. A. Prigogine, Ornithologist, Kamituga, C. B.; Reuben Rwanzangire, gorilla guide, Kis-

oro, U.; R. Smeesters, Engineer, Kasese, C. B.; and Dr. Steensels, Physician, Hospital C. N. Ki, Kima, C. B. Space restrictions prohibit the listing of several dozen important contributors among missionaries, farmers, miners, engineers, forest rangers and others involved in various local enterprises.

Finally, our most valuable source of information on the actual details of distribution was the African native along the roads and in the villages. We interviewed more than 400 groups of these people and found them, with very few exceptions, to be extremely helpful and reliable.

LITERATURE CITED

COOLIDGE, H. J.
  1929. A revision of the genus Gorilla. Mem. Mus. Comp. Zool., 50: 291-381.

  1936. Zoological results of the George Vanderbilt African expedition of 1934. Part IV. Notes on four Gorillas from the Sanga River region. Proc. Acad. Nat. Sci. Phila., 88: 479-501.

KEAY, R. W. J.
  1959. Vegetation map of Africa south of the Tropic of Cancer. Oxford Univ. Press, London, 24 pp.

LEBRUN, J.
  1935. Les essences forestières des regions montagneuses du Congo oriental. Publ. Inst. Agron. Congo Belge, Ser. Sci. 1.

ROBYNS, W.
  1948. Les territoires phytogeographiques du Congo Belge et du Ruanda Urundi (Inst. Roy. Col. Belge, fasc. 1).

ROSS, R.
  1954. Ecological studies on the rain forest of Southern Nigeria. III. Secondary succession in the Shasha Forest Reserve. J. Ecol., 42: 259-282.

Author Citation Index

Subject Index